传感材料与传感技术丛书

# 化学传感器：传感材料基础

## 第1册
## 化学传感器基本原理及其材料

影印版

Ghenadii Korotcenkov　主编

哈尔滨工业大学出版社
HITP　HARBIN INSTITUTE OF TECHNOLOGY PRESS

黑版贸审字08-2013-062号

Ghenadii Korotcenkov

Chemical Sensors:Fundamentals of Sensing Materials,Vol 1:General Approaches

9781606501030

Copyright © 2010 by Momentum Press, LLC

All rights reserved.

Originally published by Momentum Press, LLC

English reprint rights arranged with Momentum Press, LLC through McGraw-Hill Education（Asia）

**This edition is authorized for sale in the People's Republic of China only, excluding Hong Kong, Macao SAR and Taiwan.**

本书封面贴有McGraw-Hill Education公司防伪标签，无标签者不得销售。
版权所有，侵权必究。

### 图书在版编目（CIP）数据

化学传感器基本原理及其材料＝Basic Principles and Materials of Chemical Sensors：英文／（摩尔）科瑞特森科韦（Korotcenkov,G.）主编． —影印本．—哈尔滨：哈尔滨工业大学出版社，2013.9

（传感材料与传感技术丛书；化学传感器：传感材料基础1）

ISBN 978-7-5603-4149-1

Ⅰ.①化… Ⅱ.①科… Ⅲ.①化学传感器－研究－英文 Ⅳ.①TP212.2

中国版本图书馆CIP数据核字（2013）第147633号

材料科学与工程
图书工作室

**责任编辑** 杨 桦 许雅莹 张秀华

**出版发行** 哈尔滨工业大学出版社

**社　　址** 哈尔滨市南岗区复华四道街10号 邮编 150006

**传　　真** 0451-86414749

**网　　址** http://hitpress.hit.edu.cn

**印　　刷** 哈尔滨市工大节能印刷厂

**开　　本** 787mm×960mm 1/16 印张 15.25

**版　　次** 2013年9月第1版 2013年9月第1次印刷

**书　　号** ISBN 978-7-5603-4149-1

**定　　价** 68.00元

（如因印刷质量问题影响阅读，我社负责调换）

# 影印版说明

1.《传感材料与传感技术丛书》为MOMENTUM PRESS的*SENSORS TECHNOLOGY SERIES*的影印版。考虑到使用方便以及内容统一,将原系列6卷分为10册影印。本册是

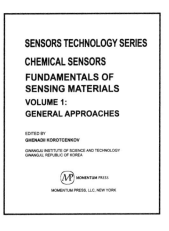

1~3章的内容。

2.原版各卷的文前介绍、索引、封底内容在其对应的影印版分册中均完整呈现。

3.各册均给出中文参考目录,方便读者快速浏览。

4.各册在页脚重新编排页码,该页码对应中文参考目录。保留了原版页眉及页码,其页码对应原书目录及索引。

5.各册的最后均给出《传感材料与传感技术丛书》的书目及各册的章目录。

**材料科学与工程图书工作室**

联系电话 0451-86412421
　　　　　0451-86414559
邮　　箱 yh_bj@aliyun.com
　　　　　xuyaying81823@gmail.com
　　　　　zhxh6414559@aliyun.com

# 目 录

## 1 化学传感器的基本工作原理 ... 1
1 引 言 ... 1
2 电化学传感器 ... 1
3 电容传感器 ... 8
4 功能传感器 ... 9
5 场效应晶体管型传感器 ... 12
6 化学场效应晶体管型传感器 ... 15
7 肖特基二极管型传感器 ... 16
8 催化传感器 ... 19
9 电导传感器 ... 20
10 声波传感器 ... 24
11 质量敏感型传感器 ... 31
12 光学传感器 ... 33
13 光声传感器 ... 45
14 热电传感器 ... 45
15 热电导传感器 ... 47
16 火焰离子化传感器 ... 49
17 LB膜传感器 ... 50
参考文献 ... 52

## 2 传感材料应具备的特性 ............................................. 63

    1 引 言 ............................................................. 63

    2 金属氧化物的共同特性 ............................................. 65

    3 传感材料的表面特性 ............................................... 77

    4 传感材料的稳定性参数 ............................................. 88

    5 传感材料的电物理特性 ............................................. 97

    6 传感材料的结构特性 ............................................... 112

    7 展 望 ............................................................. 144

    8 致 谢 ............................................................. 144

    参考文献 ............................................................. 145

## 3 传感材料的组合概念 ................................................. 159

    1 引 言 ............................................................. 159

    2 组合材料筛选的一般原则 ........................................... 160

    3 传感材料的机遇 ................................................... 162

    4 传感材料组合库的设计 ............................................. 163

    5 利用离散阵列探索和优化传感材料 ................................... 165

    6 利用梯度阵列优化传感材料 ......................................... 192

    7 用于传感材料组合筛选的新兴无线技术 ............................... 196

    8 总结和展望 ....................................................... 202

    9 致 谢 ............................................................. 203

    参考文献 ............................................................. 203

**索 引** ............................................................. **215**

**丛书书目**

# Contents

**1  Basic Principles of Chemical Sensor Operation**  1
*M. Z. Atashbar*
*S. Krishnamurthy*
*G. Korotcenkov*

| | | |
|---|---|---|
| 1 | Introduction | 1 |
| 2 | Electrochemical Sensors | 1 |
| | 2.1  Amperometric Sensors | 3 |
| | 2.2  Conductometric Sensors | 4 |
| | 2.3  Potentiometric Sensors | 5 |
| 3 | Capacitance Sensors | 8 |
| 4 | Work-Function Sensors | 9 |
| 5 | Field-Effect Transistor Sensors | 12 |
| 6 | chemFET-Based Sensors | 15 |
| 7 | Schottky Diode–Based Sensors | 16 |
| 8 | Catalytic Sensors | 19 |
| 9 | Conductometric Sensors | 20 |
| 10 | Acoustic Wave Sensors | 24 |
| | 10.1  Thickness Shear Mode Sensors | 25 |
| | 10.2  Surface Acoustic Wave Sensors | 27 |
| 11 | Mass-Sensitive Sensors | 31 |
| 12 | Optical Sensors | 33 |
| | 12.1  Fiber Optic Chemical Sensors | 36 |
| | 12.2  Fluorescence Fiber Optic Chemical Sensors | 38 |
| | 12.3  Absorption Fiber Optic Chemical Sensors | 39 |

|  |  | 12.4 Refractometric Fiber Optic Chemical Sensors | 39 |
|---|---|---|---|

        12.4 Refractometric Fiber Optic Chemical Sensors     39
        12.5 Absorption-Based Sensors     39
        12.6 Surface Plasmon Resonance Sensors     43

    13  Photoacoustic Sensors     45

    14  Thermoelectric Sensors     45

    15  Thermal Conductivity Sensors     47

    16  Flame Ionization Sensors     49

    17  Langmuir-Blodgett Film Sensors     50

    References     52

**2  DESIRED PROPERTIES FOR SENSING MATERIALS**     **63**
*G. Korotcenkov*

    1  Introduction     63

    2  Common Characteristics of Metal Oxides     65
        2.1 Crystal Structure of Metal Oxides     65
        2.2 Electronic Structure of Metal Oxides     68
        2.3 Role of the Electronic Structure of Metal Oxides in Surface Processes     70

    3  Surface Properties of Sensing Materials     77
        3.1 Electronic Properties of Metal Oxide Surfaces     77
        3.2 Role of Adsorption/Desorption Parameters in Gas-Sensing Effects     79
        3.3 Catalytic Activity of Sensing Materials     84

    4  Stability of Parameters in Sensing Materials     88
        4.1 Thermodynamic Stability     88
        4.2 Chemical Stability     92
        4.3 Long-Term Stability     94

    5  Electrophysical Properties of Sensing Materials     97
        5.1 Oxygen Diffusion in Metal Oxides     97
        5.2 Conductivity Type     101
        5.3 Band Gap     106
        5.4 Electroconductivity     107
        5.5 Other Important Parameters for Sensing Materials     110

    6  Structural Properties of Sensing Materials     112
        6.1 Grain Size     113
        6.2 Crystal Shape     120

|  |  | 6.3 | Surface Geometry | 123 |
|---|---|---|---|---|
|  |  | 6.4 | Film Texture | 127 |
|  |  | 6.5 | Surface Stoichiometry (Disordering) | 129 |
|  |  | 6.6 | Porosity and Active Surface Area | 131 |
|  |  | 6.7 | Agglomeration | 140 |
|  | 7 | Outlook | | 144 |
|  | 8 | Acknowledgments | | 144 |
|  |  | References | | 145 |

## 3   Combinatorial Concepts for Development of Sensing Materials    159
*R. A. Potyrailo*

| | 1 | Introduction | 159 |
|---|---|---|---|
| | 2 | General Principles of Combinatorial Materials Screening | 160 |
| | 3 | Opportunities for Sensing Materials | 162 |
| | 4 | Designs of Combinatorial Libraries of Sensing Materials | 163 |
| | 5 | Discovery and Optimization of Sensing Materials Using Discrete Arrays | 165 |
| | | 5.1  Radiant Energy Transduction Sensors | 165 |
| | | 5.2  Mechanical Energy Transduction Sensors | 177 |
| | | 5.3  Electrical Energy Transduction Sensors | 183 |
| | 6 | Optimization of Sensing Materials Using Gradient Arrays | 192 |
| | | 6.1  Variable Concentration of Reagents | 192 |
| | | 6.2  Variable Thickness of Sensing Films | 193 |
| | | 6.3  Variable 2-D Composition | 194 |
| | | 6.4  Variable Operation Temperature and Diffusion-Layer Thickness | 196 |
| | 7 | Emerging Wireless Technologies for Combinatorial Screening of Sensing Materials | 196 |
| | 8 | Summary and Outlook | 202 |
| | 9 | Acknowledgments | 203 |
| | | References | 203 |

**Index**

**Series Catalog**

# Preface to
# Chemical Sensors:
# Fundamentals of Sensing Materials

Sensing materials play a key role in the successful implementation of chemical and biological sensors. The multidimensional nature of the interactions between function and composition, preparation method, and end-use conditions of sensing materials often makes their rational design for real-world applications very challenging.

The world of sensing materials is very broad. Practically all well-known materials could be used for the elaboration of chemical sensors. Therefore, in this series we have tried to include the widest possible number of materials for these purposes and to evaluate their real advantages and shortcomings. Our main idea was to create a really useful "encyclopedia" or handbook of chemical sensing materials, which could combine in compact editions the basic principles of chemical sensing, the main properties of sensing materials, the particulars of their synthesis and deposition, and their present or potential applications in chemical sensors. Thus, most of the materials used in chemical sensors are considered in the various chapters of these volumes.

It is necessary to note that, notwithstanding the wide interest and use of chemical sensors, at the time the idea to develop these volumes was conceived, there was no recent comprehensive review or any general summing up of the fundamentals of sensing materials The majority of books published in the field of chemical sensors were dedicated mainly to analysis of particular types of devices. This three-volume review series is therefore timely.

This series, *Chemical Sensors: Fundamentals of Sensing Materials,* offers the most recent advances in all key aspects of development and applications of various materials for design of chemical sensors. Regarding the division of this series into three parts, our choice was to devote the first volume to the fundamentals of chemical sensing materials and processes and to devote the second and third volumes to properties and applications of individual types of sensing materials. This explains why, in *Volume 1: General Approaches,* we provide a brief description of chemical sensors, and then detailed discussion of desired properties for sensing materials, followed by chapters devoted to methods of synthesis, deposition, and modification of sensing materials. The first volume also provides general background information about processes that participate in chemical sensing. Thus the aim of this volume, although not exhaustive, is to provide basic knowledge about sensing materials, technologies used for their preparation, and then a general overview of their application in the development chemical sensors.

Considering the importance of nanostructured materials for further development of chemical sensors, we have selected and collected information about those materials in *Volume 2: Nanostructured Materials*. In this volume, materials such as one-dimension metal oxide nanostructures, carbon nanotubes, fullerenes, metal nanoparticles, and nanoclusters are considered. Nanocomposites, porous semiconductors, ordered mesoporous materials, and zeolites also are among materials of this type.

*Volume 3: Polymers and Other Materials,* is a compilation of review chapters detailing applications of chemical sensor materials such as polymers, calixarenes, biological and biomimetric systems, novel semiconductor materials, and ionic conductors. Chemical sensors based on these materials comprise a large part of the chemical sensors market.

Of course, not all materials are covered equally. In many cases, the level of detailed elaboration was determined by their significance and interest shown in that class of materials for chemical sensor design.

While the title of this series suggests that the work is aimed mainly at materials scientists, this is not so. Many of those who should find this book useful will be "chemists," "physicists," or "engineers" who are dealing with chemical sensors, analytical chemistry, metal oxides, polymers, and other materials and devices. In fact, some readers may have only a superficial background in chemistry and physics. These volumes are addressed to the rapidly growing number of active practitioners and those who are interested in starting research in the field of materials for chemical sensors and biosensors, directors of industrial and government research centers, laboratory supervisors and managers, students and lecturers.

We believe that this series will be of interest to readers because of its several innovative aspects. First, it provides a detailed description and analysis of strategies for setting up successful processes for screening sensing materials for chemical sensors. Second, it summarizes the advances and the remaining challenges, and then goes on to suggest opportunities for research on chemical sensors based on polymeric, inorganic, and biological sensing materials. Third, it provides insight into how to improve the efficiency of chemical sensing through optimization of sensing material parameters, including composition, structure, electrophysical, chemical, electronic, and catalytic properties.

We express our gratitude to the contributing authors for their efforts in preparing their chapters. We also express our gratitude to Momentum Press for giving us the opportunity to publish this series. We especially thank Joel Stein at Momentum Press for his patience during the development of this project and for encouraging us during the various stages of preparation.

Ghenadii Korotcenkov

# PREFACE TO
# VOLUME 1: GENERAL APPROACHES

This volume provides an introduction to the fundamentals of sensing materials. We have tried to to provide here the basic knowledge necessary for understanding chemical sensing through a brief description of the principles of chemical sensor operation and consideration of the processes that take place in chemical sensors and that are responsible for observed operating characteristics. In spite of the seeming extreme simplicity of chemical sensor operation and application, understanding the mechanisms involved in the process of chemical sensing is usually not so simple. Chemical sensing as a rule is a multistage and multichannel process, which requires a multidisciplinary approach. Therefore, in this volume we provide a description of the important electronic, electrophysical, and chemical properties, as well as diffusion, adsorption/desorption, and catalytic processes.

To our knowledge, this volume is the first attempt to analyze in detail the interrelationships between properties of sensing materials and operating parameters of chemical sensors. This volume describes the properties of sensing materials by emphasizing the specificities of these materials. We consider analyses that have been performed as bridging the gap between scientists studying properties of materials and researchers using these materials for actual chemical sensor design. We hope that the information included in this volume will help readers to approach soundly the selection of either sensing materials or technology for sensing material synthesis or deposition.

Detailed consideration of various materials properties with respect to their application in chemical sensors provides a clear idea of the complexity and ambiguity involved in selecting an optimal sensor material. Research has demonstrated that there is no universal sensing material, and selection of an optimal material is determined by the type of chemical sensor being designed and the requirements that device will have to meet. This volume also illustrates the complementary nature of functionality in sensing materials; for example, high sensitivity usually conflicts with stability. This richness and complexity in behavior cannot be ignored.

This volume is intended to provide readers with a good understanding of the techniques used for synthesis and deposition of sensing materials. Readers will find descriptions of different techniques such as various methods of film deposition, sol-gel technology, deposition from solutions, colloidal processing, the peculiarities of polymers synthesis, techniques used for depositing coatings on fibers, and so on. Description of various methods of synthesis and deposition, accompanied by detailed analysis of the advantages and shortcomings of those methods, provides the understanding necessary for considered selection of a technology for forming a sensitive layer.

Analysis of metal oxide modification methods highlights the opportunities for control of the properties of sensing materials, and demonstrates that a choice of methods should be based on consideration of all possible consequences of the technical decision that is made.

Combinatorial and high-throughput materials screening approaches analyzed in this volume will be also of interest to researchers working on materials design for chemical sensors.

We are confident that the present volume will be of interest of anyone who works or plans to start activity in the field of chemical sensor design, manufacturing, or application.

Ghenadii Korotcenkov

# About the Editor

***Ghenadii Korotcenkov*** received his Ph.D. in Physics and Technology of Semiconductor Materials and Devices in 1976, and his Habilitate Degree (Dr.Sci.) in Physics and Mathematics of Semiconductors and Dielectrics in 1990. For a long time he was a leader of the scientific Gas Sensor Group and manager of various national and international scientific and engineering projects carried out in the Laboratory of Micro- and Optoelectronics, Technical University of Moldova. Currently, he is a research professor at Gwangju Institute of Science and Technology, Gwangju, Republic of Korea.

Specialists from the former Soviet Union know G. Korotcenkov's research results in the study of Schottky barriers, MOS structures, native oxides, and photoreceivers based on Group III–V compounds very well. His current research interests include materials science and surface science, focused on metal oxides and solid-state gas sensor design. He is the author of five books and special publications, nine invited review papers, several book chapters, and more than 180 peer-reviewed articles. He holds 16 patents. He has presented more than 200 reports at national and international conferences. His articles are cited more than 150 times per year. His research activities have been honored by the Award of the Supreme Council of Science and Advanced Technology of the Republic of Moldova (2004), The Prize of the Presidents of Academies of Sciences of Ukraine, Belarus and Moldova (2003), the Senior Research Excellence Award of Technical University of Moldova (2001, 2003, 2005), a Fellowship from the International Research Exchange Board (1998), and the National Youth Prize of the Republic of Moldova (1980), among others.

# Contributors

**Beongki Cho** (Chapter 4)
Department of Material Science and Engineering
Gwangju Institute of Science and Technology
Gwangju, 500-712, Republic of Korea

**Ghenadii Korotcenkov** (Chapters 1, 2, 4, and 5)
Department of Material Science and Engineering
Gwangju Institute of Science and Technology
Gwangju, 500-712, Republic of Korea
*and*
Technical University of Moldova
Chisinau, 2004, Republic of Moldova

**Massood Zandi Atashbar** (Chapter 1)
Department of Electrical and Computer Engineering
Western Michigan University
Kalamazoo, Michigan 49008-5066, USA

Radislav A. Potyrailo (Chapter 3)
Chemical and Biological Sensing Laboratory
Chemistry Technologies and Material Characterization
General Electric Global Research
Niskayuna, New York 12309, USA

**Sridevi Krishnamurthy** (Chapter 1)
Department of Electrical and Computer Engineering
Western Michigan University
Kalamazoo, Michigan 49008-5066, USA

# CHAPTER 1

# BASIC PRINCIPLES OF CHEMICAL SENSOR OPERATION

M. Z. Atashbar
S. Krishnamurthy
G. Korotcenkov

## 1. INTRODUCTION

In recent years increased knowledge has led to significant development of chemical sensors for detection and quantification of chemical species. Chemical sensors have found a wide range of applications in clinical, industrial, environmental, agricultural, and military technologies.

Chemical sensors are characterized by parameters such as sensitivity, selectivity, response and recovery time, and saturation. Chemical sensors can be classified in a number of ways, depending on their principle of operation. In this chapter we describe the fundamental concepts of the following classes of chemical sensors: electrochemical, conductometric, capacitive, work function, chemFET, catalytic, Schottky diode, acoustic wave, mass-sensitive, optical, chemoluminesese, photoacoustic, thermal, surface plasmon resonance, thermoelectric, thermal conductivity–based, flame ionization, and Langmuir-Blodgett sensors.

## 2. ELECTROCHEMICAL SENSORS

Electrochemical sensors constitute the largest and most developed group of chemical sensors and have taken a leading position with respect to commercialization in the fields of clinical, industrial, environmental, and agricultural analysis (Korotcenkov et al. 2009). Electrochemical sensors are based on the detection of electroactive species involved in chemical recognition processes and make use of charge transfer from a solid or liquid sample to an electrode or vice versa. An electrochemical sensing method essentially requires a closed electrical circuit, which enables the flow of direct or alternating current to make measurements. Figure 1.1 illustrates a basic structure of an electrochemical sensor. For gas sensors, the top of the casing has a membrane which can be permeated by the gas sample.

For current to flow, at least two electrodes are needed. One electrode is an active electrode or working electrode (WE), where the chemical reaction takes place; the other is a return electrode, which is called a counter or auxiliary electrode (AE). A current is created as positive ions flow to the cathode and negative ions flow to the anode. To measure the electrochemical potentials produced by the electrodes and the electrolyte, a third electrode, called a reference electrode (RE), can be used. The reference electrode corrects the error introduced as a result of polarization of the working electrode. The working electrode is often made up of a layer of noble metal or catalytic metal such as platinum-, palladium-, or carbon-coated materials. This working electrode is covered with a hydrophobic membrane that acts as a transport barrier for the target chemical molecules (analyte), allowing the chemical species to diffuse to the working electrode and also preventing leakage of the electrolyte (Zao et al. 1992). To produce a measurable signal, the electrodes must have a large surface area in order to maximize the contact area with the analytes. Electrodes generally have a special coating to improve their rate of reaction and to extend their working life. An electrolytic medium is required to carry the ionic charges. Selective reactions can be accomplished by careful choice of the electrolyte and hence it is the first stage in enhancing selectivity to a particular analyte. The sensor formed by the combination of an electrolyte, electrodes, and an external circuit is called an electrochemical cell. The cell can be configured to extract electrical signals such as current, potential, conductance, or capacitance (Fradeen 2003).

A common application for potentiometric and amperometric sensors is for water analysis. The most common is the pH sensor system. A wide range of gaseous analytes such as oxygen, carbon oxides, nitrogen oxides, sulfur oxides, and combustible gases also can be detected and quantified using electrochemical sensors. Gases such as oxygen, nitrogen oxides, and chlorine, which are electrochemi-

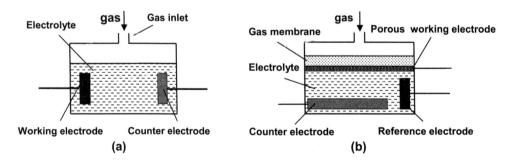

Figure 1.1. Schematic diagram of electrochemical gas sensors with (a) two- and (b) three-electrode configurations. (Reprinted with permission from Korotcenkov et al. 2009. Copyright 2009 American Chemical Society.)

cally reducible, are sensed at the cathode, while electrochemically oxidizable gases such as carbon monoxide, nitrogen dioxide, and hydrogen sulfide are sensed at the anode.

Remarkable detectability, simple construction, and low cost have been the important advantages of electrochemical sensors in comparison to optical, mass, and thermal sensing methods. Electrochemical sensors can be classified into three categories based on their operating principles: amperometric, conductometric, and potentiometric sensors (Janata 1989; Madou and Morrison 1989; Göpel et al. 1991; Gardner 1994).

## 2.1. AMPEROMETRIC SENSORS

Electrochemical amperometric sensing involves measuring the flow of current at a constant applied potential between working and auxiliary electrodes. The applied potential serves as the driving force for the electron-transfer reaction of the electroactive species. The resulting current is a direct measure of the rate of the electron-transfer reaction. It thus reflects the rate of the recognition event and is proportional to the concentration of the target analyte.

The earliest design example of this sensor type was the two-electrode Clark oxygen sensor (Clark et al. 1953). The counter electrode current arises by two oxygen–reduction steps:

$$\begin{aligned} O_2 + 2H_2O + 2e^- &\rightarrow H_2O_2 + 2OH^- \\ H_2O_2 + 2e^- &\rightarrow 2OH^- \end{aligned} \quad (1.1)$$

The basic structure of an amperometric sensor is similar to the one shown in Figure 1.1. A fixed potential is applied to the reference and working electrodes by means of a potentiostatic circuit. Current flow is recorded between the working and counter electrodes as a function of analyte concentration. Upon exposures to an electroactive analyte, the analyte diffuses into the electrochemical cell through the membrane and then to the working electrode. Depending on the type of analyte, the electrochemical reaction involves either oxidation or reduction, thus producing a current at the working electrode. For instance, carbon monoxide oxidizes to carbon dioxide, resulting in a flow of electrons from the working electrode to the counter electrode through the external circuit. On the other hand, gases such as oxygen reduce to water, resulting in a flow of electrons from the counter electrode to the working electrode. This current is related directly to the rate of reaction at the working electrode surface. If there are no other species other than the analyte, the current varies in proportion to the amount of analyte at the working electrode. Gold-based sensors respond to ozone and to nitrogen and sulfur compounds (e.g., $NO_2$, $NO$, $HNO_3$, $O_3$, $H_2S$, $SO_2$, $HCN$); platinum-based sensors respond to all of those gases plus carbon–oxygen compounds (CO, formaldehyde, alcohols, etc.). Some examples with the reactions at the counter and working electrodes are shown in Table 1.1.

Although amperometric sensors have long been in wide use, microfabrication of such sensors has recently started. To name a few, Somov and co-workers (2000) developed multielectrode zirconia electrolyte amperometric sensors, Benammar (2004) developed zirconia-based oxygen sensors, and Lee and co-workers (2004) developed solid-state amperometric $CO_2$ sensors. For example, Lee and co-workers (2004) used $Na_2CO_3$ as a porous coating, platinum paste as counter and working electrodes, and

**Table 1.1. Reactions at working and counter electrode in amperometric sensors, with their common electrocatalyst for the reaction mechanism**

| WORKING ELECTRODE (ANODE) | COUNTER ELECTRODE (CATHODE) | ELECTRODE ELECTROCATALYST |
|---|---|---|
| $CO + H_2O \rightarrow CO_2 + 2H^+ + 2e^-$ | $\frac{1}{2}O_2 + 2H^+ + 2e^- \rightarrow 2H_2O$ | Platinum |
| $SO_2 + H_2O \rightarrow H_2SO_4 + 2H^+ + 2e^-$ | $\frac{1}{2}O_2 + 2H^+ + 2e^- \rightarrow 2H_2O$ | Gold |
| $NO + 2H_2O \rightarrow HNO_3 + 3H^+ + 3e^-$ | $O_2 + 4H^+ + 4e^- \rightarrow 2H_2O$ | Gold |
| $NO_2 + 2H^+ + 2e^- \rightarrow NO + H_2O$ | $HO_2 \rightarrow \frac{1}{2}O_2 + 2H^+ + 2e^-$ | Gold |
| $Cl_2 + 2H^+ + 2e^- \rightarrow 2HCl$ | $HO_2 \rightarrow \frac{1}{2}O_2 + 2H^+ + 2e^-$ | Platinum |

NASICON ($Na_3Zr_2Si_2PO_{12}$, a sodium super-ionic conductor) as an electrolyte material with a constant potential of 0.1 V. Figure 1.2 shows the transient response of the sensor to various $CO_2$ concentrations. As can be seen from Figure 1.2, the current increases at higher concentrations of $CO_2$.

## 2.2. CONDUCTOMETRIC SENSORS

In an electrochemical conductometric sensor, the change in conductivity of the electrolyte in an electrochemical cell is measured at a series of frequencies. This type of sensor may have a capacitive component due to the polarization of the electrodes and the charge-transfer process. In a homogenous electrolytic solution, the conductance is (Fradeen 2003)

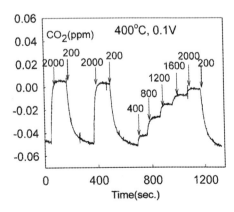

Figure 1.2. $CO_2$ amperometric transient response. (Reprinted with permission from Lee et al. 2004. Copyright 2004 Elsevier.)

$$G = \frac{\sigma A}{L} \tag{1.2}$$

where $\sigma$ is the specific conductivity of the electrolyte, which is related to the concentration of the charges and ionic species, $L$ is the segment of the solution along the electric potential; and $A$ is the cross-sectional area perpendicular to the electric field. For measurements, a Wheatstone bridge configuration is usually used, in which the electrochemical cell forms one of the arms of the bridge. Measurement of the conductivity of the liquid is complicated due to the charge-transfer process at the electrode surface and polarization of the electrodes at the operating voltage. Therefore the measurement needs to be performed at low voltage so that charge transfer does not take place. When an electrode is in an electrolyte solution, a potential is generated due to the unequal distribution of charges across the interface. Two oppositely charged layers, one on the electrode surface and one inside the electrolyte, form a "double layer," which behaves as a parallel-plate capacitor. This results in Warburg impedance, which depends on the frequency of the potential perturbation. At high frequencies the Warburg impedance is small, since diffusing reactants do not have to move very far. At low frequencies the reactants have to diffuse farther, thereby increasing the Warburg impedance. Employing a high-frequency, low-amplitude alternating signal can minimize both the charge-transfer and double-layer effects.

## 2.3. POTENTIOMETRIC SENSORS

In potentiometric sensors, the analytical information is obtained by converting the recognition process into a potential signal, which is proportional (in a logarithmic fashion) to the concentration (activity) of species generated or consumed in the recognition event. Such devices rely on the use of ion-selective electrodes to obtain the potential signal. A permselective ion-conductive membrane (placed at the tip of the electrode) is designed to yield a potential signal that is due primarily to the target ion. The response is measured under conditions of essentially zero current.

Upon equilibrium at the analye–electrolyte interface in the electrochemical cell, an electrical potential develops due to the "redox" reaction. The reaction occurs at one of the electrodes and is called a half-cell reaction. The reactions for $H_2$ sensors are shown in Figure 1.3.

Under quasi-thermodynamic equilibrium conditions, the Nernst equation is applicable (Janata 1989):

$$E = E_0 + \frac{RT}{nF} \ln\left(\frac{C_0^*}{C_R^*}\right) \tag{1.3}$$

where $R$, $T$, $n$, $F$, $C_0^*$, and $C_R^*$ are the gas constant (8.314 J mol$^{-1}$ K$^{-1}$), temperature (in K), number of electrons transferred, Faraday constant (96,485 C), concentrations of oxidant and reduced product, respectively, and $E_0$ is the electrode potential at the standard state.

In a potentiometric sensor, two half-cell reactions take place at each electrode; however, only the reaction at the working electrode needs to be involved in the measurements. The measurement of the cell potential should be made under zero-current or quasi-equilibrium condition. Therefore, a very-high-input impedance amplifier should be used.

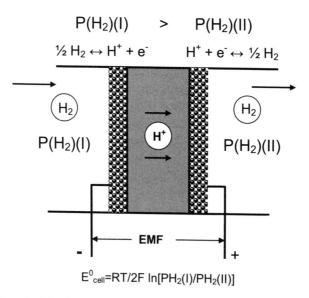

**Figure 1.3.** Working principle of hydrogen concentration cell using a proton-conducting membrane. (Adapted with permission from Sundmacher et al. 2005. Copyright 2005 Elsevier.)

Potentiometric sensors are very attractive for field operations because of their high selectivity, simplicity, and low cost. They are, however, less sensitive and often slower than their amperometric counterparts. In the past, potentiometric devices were more widely used, but the increasing amount of research on amperometric probes would gradually shift this balance. Detailed theoretical discussions of amperometric and potentiometric measurements are available in many textbooks and reference works (Korotcenkov et al. 2009).

## 2.3.1. Ion-Selective Electrodes

Ion-selective electrodes (ISEs) are potentiometric sensors that measure the electric potential of a specific ionic species in the solution. This potential is measured against a stable reference electrode with constant potential. These sensors are also known as specific-ion electrodes (SIEs). A schematic diagram illustrating the principle of operation of ISEs is shown in Figure 1.4.

The pH sensor based on a glass membrane is the oldest type of ISE. Credit for the first glass sensing pH electrode is given to Cremer, who first described it in a 1906 paper (Meyerhoff and Opdeycke 1986). In 1949, George Perley published an article on the relationship of glass composition to pH function (Frant 1994). Since then, many other types of ion-selective membranes have been developed and characterized, including polymeric membranes having various ionophores, which have extended the sensing capabilities of ISEs to a wide range of ions. Other ions that can be measured include ions such as metal fluorides (Apostolakis et al. 1991) and bromides (Shamsipur et al. 2000), cadmium (Hirata and Higashiyama 1971), copper (Gupta and D'Arc 2000), and gases in solution such as ammonia (Radomska et al. 2004) and carbon dioxide (Johan et al. 2003). Ion-selective electrodes offer direct and selective detection of ionic activities in water samples. Such potentiometric devices are simple, rapid, inexpensive, and compatible with online analysis. The inherent selectivity of these devices is attributed to highly selective interactions between the membrane material and the target ion. Depending on the nature of the membrane material used to impart the desired selectivity, ion-selective electrodes can be divided into three groups: glass, solid, and liquid electrodes.

The ISE ideally follows a Nernst equation (Taylor and Schultz 1996):

$$E = K + \left(2.303 \times \frac{RT}{ZF}\right) \log a_i \qquad (1.4)$$

where $E$ is the potential, $K$ is a constant representing the effect of various sources such as liquid junction potentials, $R$ is the universal gas constant (8.314 J mol$^{-1}$ K$^{-1}$), $T$ is the absolute temperature, $Z$ is ionic charge, and $a_i$ is the activity of ion $i$. The activity of an ion $i$ in solution is related to its concentration $c_i$:

$$a_i = \gamma_i c_i \qquad (1.5)$$

where $\gamma_i$ is the activity coefficient, which depends on the types of ions present and the total ionic strength of the solution.

At present, many ion-selective electrodes are commercially available and are used routinely in various fields. Electrochemical analysis systems based on ISEs may be applied to measure ionic concentrations in biomedical, food processing, water quality, and pollution monitoring applications.

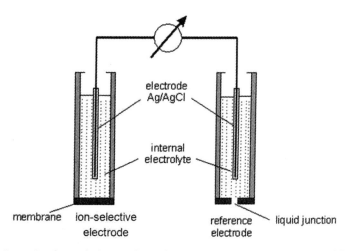

**Figure 1.4.** Schematic of a typical potentiometric cell assembly incorporating an ISE and reference electrode as the galvanic half-cells and a high-impedance voltmeter for measurement of the cell EMF.

## 3. CAPACITANCE SENSORS

A capacitor is an electrical circuit component which can store electrical energy in response to an applied electric field. It consists of a dielectric material which is sandwiched between two parallel metal electrodes. When a voltage is applied to these plates, equal and opposite charges accumulate on the electrodes. Capacitance can be expressed as

$$C = \frac{\varepsilon_0 \varepsilon_r A}{d} \qquad (1.6)$$

where $\varepsilon_0$ represents the permittivity in vacuum, $\varepsilon_r$ is the relative permittivity of the dielectric, $A$ is the area of the electrode, and $d$ is the distance between the two electrodes.

The main operating principle of capacitance sensors is that the analyte of interest can change $\varepsilon_r$, $A$, or $d$, and therefore, by measuring the change in the capacitance, the presence of the measurand can be detected (Ishihara et al. 1998). With reference to polymer coating, the mechanisms which can be accompanied by the change of these parameters are shown in Figure 1.5. Upon analyte absorption, the physical properties of the polymer layer change as a result of the incorporation of the analyte molecules into the polymer matrix. Changes in two physical properties influence the sensor capacitance: (1) volume (swelling) and (2) dielectric constant. The resulting capacitance change is detected by the read-out electronics (Kummer et al. 2004).

A change in area is not common in capacitor sensors, but changes in relative permittivity and dielectric thickness are commonly used (Seiyama et al. 1983; Matsuguch et al. 1998; Yamazoe et al. 1989; Harrey et al. 2002). Interdigitated electrode structures as shown in Figure 1.5 are normally used for capacitor sensors, where the capacitance of the sensor is measured rather than the conductance. In this case, the sensitive layer can be a polymer film or ceramic (Sheppard et al. 1982; Ishihara et al. 1998). The interdigitated capacitors used for this purpose have capacitance of the order of 1 pF and therefore the capacitance changes due to the sensing mechanism are in the range of attofarads ($10^{-18}$ F).

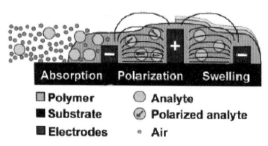

Figure 1.5. Schematic of sensing principle, showing analyte absorption and the two relevant effects that affect the sensor capacitance: a change in the dielectric constant and swelling. The interdigitated electrodes (+, −) on the substrate (black) are coated with a polymer layer (gray). Big and small globes represent analyte and air molecules, respectively. Analyte molecules are polarized (v) in the electric field (solid lines). Analyte-induced polymer swelling is indicated by dashed lines (right side). (Reprinted with permission from Kummer et al. 2004. Copyright 2004 American Chemical Society.)

Since it is difficult to transfer these small signals through wires and cables, on-chip circuitry is required (Hierlemann et al. 2003). For this purpose, two capacitors, one a sensing capacitor and the other a reference capacitor, are used in a switched capacitor configuration along with a sigma-delta modulator for analog-to-digital conversion (Hagleitner et al. 1999). The differential capacitance changes are converted to frequency changes by the on-chip circuitry.

Capacitive microsensors are promising devices for chemical sensing, because no micromechanical parts, such as membranes or cantilevers, are needed to render the chip more rugged, and there are fewer postprocessing steps, and hence, lower costs. Moreover, the low power consumption favors their use in hand-held devices.

One of the most well known capacitive gas sensors based on changes in relative permittivity is the humidity sensor. Polymer films such as polyimides or cellulose acetates are used as humidity-sensitive dielectrics because the difference in their relative permittivities at room temperature is very large. Pure water has a dielectric constant of 78.5, whereas those of the polyimides are in the range of 3–6. Therefore, the capacitance in these sensors is highly sensitive to the adsorption of water (Ishihara et al. 1998; Hierlemann et al. 2003).

Capacitive sensors based on changes in thickness rely on the formation of depletion layers in the dielectric of the capacitor. Depletion layers formed between $p$- and $n$-type semiconductors, metal oxide semiconductors, or metal semiconductors have been investigated for this purpose. The electronic interaction of the analyte and the interface changes the carrier concentration in the semiconductor, resulting in changes in the depletion-layer thickness (Schoeneberg et al. 1990; Ishihara et al. 1998; Nagai et al. 1998). Chemical capacitance sensors have also been investigated for detection of other gases such as CO (Balkus et al. 1997), ammonia and methane (Domansky et al. 2001), and $NO_x$ (Ishihara et al. 1995, 1996).

## 4. WORK-FUNCTION SENSORS

In the solid phase, the work function ($\Phi$) of a material is defined as the minimum energy required by an electron to escape from the Fermi level of the bulk material to the vacuum level (Streetman 1990).

$$\Phi = -q\theta - E_F \qquad (1.7)$$

$$E_F = -\Phi - q\theta \qquad (1.8)$$

The Fermi energy ($E_F$) level corresponds to the electrochemical potential of an electron. At a given temperature, the Fermi level is the highest occupied energy level in the band gap. The energy difference between $E_F$ and the vacuum (Volta) energy level ($-q\theta$) is the work function, as illustrated in Figure 1.6 (Janata and Josowicz 2002; Anh et al. 2004).

When an electroactive material is brought into electrical contact with a metallic electrode, charge transfer occurs until the Fermi energies of both materials reach thermal equilibrium and the Fermi levels of the two materials become equal. However, their Volta energy levels are different. The difference in their work functions is equal to their difference in Volta energy. This can be represented as (Anh et al. 2004)

**Figure 1.6.** Illustration of energy levels in an energy-band diagram.

$$-\phi_{Met} - q\theta_{Met} = -\phi_{Mat} - q\theta_{Mat} \qquad (1.9)$$

$$\phi_M^{Met} = \phi_{Met} - \phi_{Mat} = -q(\theta_{Met} - \theta_{Mat}) \qquad (1.10)$$

where $\phi_{Met}$ and $\phi_{Mat}$ are the work functions of electrons in the metal and the electroactive material, respectively; $-q\theta_{Met}$ and $-q\theta_{Mat}$ are the Volta energies of the metal and the electroactive material, respectively; and $\phi_{Mat}^{Met}$ is the difference between the work function of the electrons in the metal and that of the material (Anh et al. 2004). The redox properties of materials can be studied by measuring their work function using a Kelvin probe, which consists of a vibrating capacitor (Janata and Josowicz 1997). Using this method, the change in the work function due to the redox reaction of the material can be measured. When the surface of the metal and the electroactive material are held parallel to each other and separated by a dielectric material, there is a potential difference across the surface due to transfer of charges (Bergstrom et al. 1997). This potential difference is given by the difference between the work functions of the two materials and is given by a bulk term and a surface dipole term. Gases interact with the surface dipole of one or both materials that constitute the Kelvin probe. The extent to which the gases can interact with the surface dipole determines the amount of work function shift (Bergstrom et al. 1997).

In measuring the contact potential differences between two surfaces using the Kelvin probe method (see Figure 1.7), one of the two surfaces oscillates periodically with respect to the other. This oscillation induces a charge across the surface. This structure can be modeled as a capacitor with an induced charge $Q(t)$ that depends on the time-varying displacement of the two plates. This induced charge can be expressed in terms of the capacitance $C(t)$ and the work function difference $V_{12}$ between the sensing plate and the reference plate as (Bergstrom et al. 1997)

$$Q(t) = C(t)(V_{12} + V_b) = \frac{\varepsilon_0 A}{d(t)}(V_{12} + V_b) \qquad (1.11)$$

where $V_b$ is the backing potential which is used to nullify the charge when $V_{12} = -V_b$, $A$ is the area of the plates, and $d(t)$ is the time-varying distance between the plates and is given by

$$d(t) = d_0 + d_1 \sin \omega t \qquad (1.12)$$

# BASIC PRINCIPLES OF OPERATION • 11

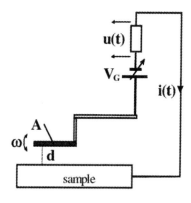

**Figure 1.7.** Setup of a Kelvin probe. A metallic grid oscillates over a sample. The counter voltage $V_G$ is adjusted so that the current *i* vanishes. (Reprinted with permission from Barsan and Weimar 2003. Copyright 2003 IOP.)

where $d_0$ is the plate spacing, $d_1$ is the displacement, and $\omega$ is the frequency of oscillation. The oscillating plate capacitively generates a sinusoidal current in the sensing film, and this current is proportional to the work function difference $V_{12}$. Therefore, the resolution of the current measurement determines the overall resolution of the system (Bergstrom et al. 1997). For a given work function difference, resolution improves with the magnitude of the current. An example of $NO_2$ sensing using the work function principle is presented by Karthigeyan and co-workers (2001), who used nanoparticulate $SnO_2$ films. The $SnO_2$ films used in the sensor contained particles with an average size of 15 nm. The work function measurements were performed with a Kelvin probe. Figure 1.8 shows the response of the $SnO_2$ work function–based sensor to 100 ppm $NO_2$ at a temperature of 130°C. It can be seen from this figure that sensors have almost a negligible baseline drift after repeated exposure to $NO_2$ (Karthigeyan et al. 2001).

Work function–type gas sensors for detection of $CO_2$ (Ostrick et al. 1999), ammonia (Ostrick et al. 2000), oxygen (Bergstrom et al. 1997), and ozone (Doll et al. 1996; Zimmer et al. 2001) have been explored. Work-function sensors based on field-effect transistors (FETs) have also been studied in

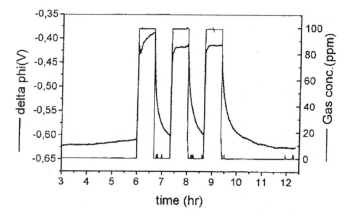

**Figure 1.8.** Response of $SnO_2$ sensor to 100-ppm $NO_2$ at a working temperature of 130°. (Reprinted with permission from Karthigeyan et al. 2001. Copyright 2001 Elsevier.)

detail for gas sensing. These are called adsorption-based FETs (ADFETs) (Middlehoek and Audet 1989; Janata 1989; Madou and Morrison 1989), ion-sensitive FETs (ISFETs) (Harame et al. 1987; Bergveld 1991), or chemically modified FETs (chemFETs) (Eddowes 1987). ChemFETs are discussed in detail in the following section.

## 5. FIELD-EFFECT TRANSISTOR SENSORS

The basic metal-oxide semiconductor field-effect transistor (MOSFET) consists of four terminals, a source, drain, gate, and substrate, as shown in Figure 1.9. The source and drain are highly conducting semiconductor regions, whereas the gate is a metal region separated from the source and drain by a gate oxide. The most important parameter of a MOSFET is the threshold voltage ($V_T$), which is defined as the gate voltage at which there is an onset of inversion layer at the surface of the semiconductor. The carrier concentration of the inversion layer has opposite polarity to that of the carrier concentration in the bulk of the semiconductor. The formation of the inversion layer is controlled by the applied gate voltage. The net result is that the current between the drain and source is controlled by the voltage that is applied to the gate and the drain-to-source voltage.

The drain current of a MOSFET is given by the following equation (Razavi 2000):

$$I_D = \beta \left( V_{GS} - V_T - \frac{1}{2} V_{DS} \right) V_{DS} \tag{1.13}$$

where $V_{GS}$ and $V_{DS}$ are the gate-to-source and drain-to-source voltages, respectively; $V_T$ is the threshold voltage of the MOSFET; and $\beta$ is the sensitivity parameter determined by the dimensions of the gate and given by

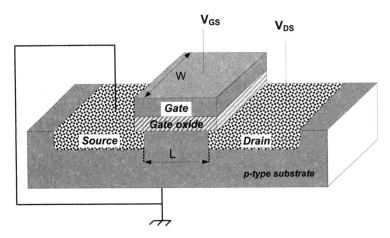

**Figure 1.9.** Schematic of a four-terminal MOSFET.

## BASIC PRINCIPLES OF OPERATION • 13

$$\beta = \mu C_{ox} \frac{W}{L} \quad (1.14)$$

where $\mu$ is the mobility of electrons in the channel, $C_{ox}$ is the gate oxide capacitance per unit area, $W$ is the channel's width, and $L$ is the length of the channel. The threshold voltage ($V_T$) of the MOSFET is given by (Razavi 2000)

$$V_T = \phi_{MS} + 2\phi_F + \frac{Q_{dep}}{C_{ox}} \quad (1.15)$$

where $\phi_{MS}$ is a function of the difference between the work function of the gate and the silicon substrate, $Q_{dep}$ is the charge in the depletion region, and

$$\phi_F = \frac{kT}{q} \ln\left(\frac{N_{sub}}{n_i}\right) \quad (1.16)$$

where $N_{sub}$ is the doping concentration of the substrate and $n_i$ is the intrinsic carrier concentration.

A theory for the operation of these FET-based devices and metal insulator semiconductor (MIS) capacitors as gas sensors was developed by Lundström and co-workers. (Lundström 1991; Ekedahl et al. 1998). The basic theory regarding a gas FET or MIS capacitor with a Pd electrode (or similar catalytic metal, such as Pt) sensitivity to hydrogen is produced by a modulation in the flat band potential. If we consider a gas FET with a thick Pd-gate film ($\approx$ 100 nm), hydrogen gas molecules are absorbed at the surface of the palladium by disassociating into atomic hydrogen. These atoms diffuse through the metal and absorb on the inner surface of the gate insulator junction, where they become polarized as shown in Figure 1.10. The resulting dipole layer is in equilibrium with the outer layer of chemisorbed hydrogen

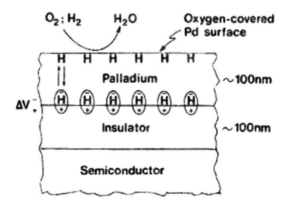

**Figure 1.10.** Mechanism of MOSFET operation in a hydrogen atmosphere. (Reprinted with permission from Lundstrom et al. 2007. Copyright 2007 Elsevier.)

and hence in phase with the gas. The dipole layer produces an abrupt rise in the surface potential at the metal oxide interface.

When the hydrogen is removed from the ambient atmosphere, the atomic hydrogen within the MOSFET recombines into hydrogen molecules, which desorb from the surface. If the atmosphere contains oxygen, then this recombination is dominated by the production of water. Once the hydrogen has been removed from the metal–insulator interface, the potential between the Pd and the insulator returns to its initial level. It has been shown that the reaction is totally reversible, though the recovery rate can be very slow and may take a number of hours. The reaction described above occurs at room temperature, though to increase the speed of the response and to remove water molecules rapidly from the metal surface, the sensors are normally operated at 150–200°C.

The gas FET response is measured as a shift in either gate-source voltage ($V_{GS}$) or drain-source current ($I_{DS}$) (see Figure 1.11), and it has been shown that this response is related to a shift in the threshold voltage. This shift in threshold voltage is usually defined as

$$V_T = V_{T0} - \Delta V \qquad (1.17)$$

where $V_{T0}$ is the initial threshold voltage and $\Delta V$ is the potential produced by the formation of the dipole layer.

Experiments have shown that this sensitivity to hydrogen can be measured over a dynamic pressure range of 14 orders of magnitude without the sensor saturating.

Earlier gas FET sensors were fabricated mainly based on silicon. However, in recent years, great success has been achieved in development of such devices on the basis of silicon carbide. The use of SiC as the semiconductor substrate in field-effect devices led to the development of gas sensors

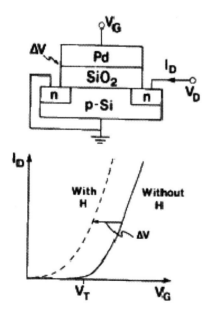

**Figure 1.11.** "Classical" schematic drawing illustrating the hydrogen-sensitive field-effect devices, in which hydrogen atoms adsorbed at the metal–oxide interface cause a shift of the electrical characteristics along the voltage axis in devices having catalytic metal (Pd) gates. (Reprinted with permission from (Lundstrom et al. 2007. Copyright 2007 Elsevier.)

# BASIC PRINCIPLES OF OPERATION • 15

operating at high temperatures. It was demonstrated that SiC-based sensors respond to a change in ambient between oxygen and propane even at 1000°C (Lundstrom et al. 2007).

## 6. chemFET-BASED SENSORS

The schematic of a chemFET is similar to that of a MOSFET except that the gate structure is slightly modified. Based on the gate structure, the family of chemFETs can be divided into three main types, ion-sensitive field-effect transistors (ISFETs), enzymatically selective field-effect transistors (ENFETs), and basic or work function chemFETs (Wilson et al. 2001).

The ion-sensitive field-effect transistor is a chemFET structure without a conductive gate. The ion-selective layer is placed on top of the insulator layer of the FET structure. The enzymatically selective field-effect transistor is also a chemFET structure without a conductive gate. The chemically sensitive enzyme layer forms a part of or the entire insulator layer of the FET. The basic or work function chemFET is a chemFET structure with a conductive gate. The conductive gate is the chemically selective layer (Wilson et al. 2001). All of these chemFET structures work on the basic principle that the surface charge changes at the interface between the insulator and the overlying layer (the overlying layer is the ion-selective layer in an ISFET, the enzyme-selective layer in the ENFET, and the gate layer in the basic or work function chemFET) (Yotter and Wilson 2004). This results in a change in the work function, and therefore, the threshold voltage changes in accordance with Eq. (1.15).

Extensive research has been conducted on chemFETs for monitoring ammonium (Brzozka et al. 1997; Senillou et al. 1998), cadmium and lead (Reinhoudt 1995; Ali et al. 2000), $Cu^{2+}$ (Taillades et al. 1999), hydrogen (Domansky et al. 1998), and pH (Bausells et al. 1999; Cho and Chiang 2000).

Ion-sensitive field-effect transistors were first developed in the early 1970s (Bergveld 1970). Figure 1.12 is a schematic diagram of an ISFET. As can be seen, its structure is the same as that of a MOSFET. The main component of an ISFET is an ordinary metal-oxide silicon field-effect transistor (MOSFET) with the gate electrode replaced by a chemically sensitive membrane, solution, and a reference electrode

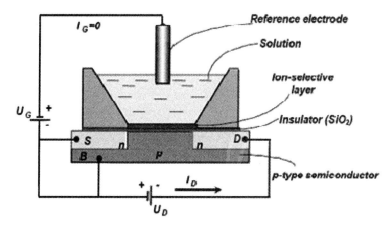

**Figure 1.12.** Schematic diagram of an ISFET.

(RE). As in a MOSFET, the channel resistance in an ISFET depends on the electric field perpendicular to the direction of the current. Charges from solution accumulate on top of this insulating membrane and do not pass through the ion-sensitive membrane. Anisotropic ions accumulate at the contact interface between an electrochemically active surface and a liquid electrolyte. Because of their different sizes and charges, the ions form a well-confined electric double layer close to the surface, and a diffuse layer of outer charges exists between the Helmholtz planes and the neutral bulk of the solution (Yuqing et al. 2003). Of course, to build a selective sensor, it is necessary to select a membrane that allows only one chemical reaction to occur.

Because of the similarity between the ISFET and the MOSFET, most papers dealing with the ISFET operational mechanism start with the theoretical description of a MOSFET. Therefore, a general expression for the drain current of the ISFET and MOSFET in nonsaturated mode is as described before [see Eq. (1.13)].

The fabrication processes for MOSFETs and ISFETs are the same, which results in identical threshold voltages as shown in Eq. (1.18). However, in the case of ISFETs, two additional parameters are involved: the constant potential of the reference electrode, $E_{ref}$, and the interfacial potential at the solution/oxide interface, which is the chemical input parameter, shown to be a function of the solution pH, and is the surface dipole potential of the solvent and thus has a constant value. Hence the expression for the ISFET threshold voltage becomes (Bergveld 2003)

$$V_T = E_{ref} - \Psi + \chi^{sol} + 2\phi_F + \frac{Q_{dep}}{C_{ox}} \qquad (1.18)$$

The fabrication of chemical microsensors based on ISFETs and selective ionic membranes has been extensively studied (Bergveld 2003). Among the various aspects to consider for a good implementation of selective membranes on such devices, the most critical are the adhesion of the film to the sensor surface and the reproducibility of the deposition process, which in most cases is carried out manually. The use of Langmuir-Blodgett (LB) nanostructures is a good technique, since the deposition method provides films with well-ordered molecular orientation and good thickness control. Thin selective monolayers can be obtained using a phospholipid matrix containing ionophores as specific recognition molecules (Jimenez et al. 2004).

Because of their robustness and the need for minimal maintenance, ISFET sensors can be used for online process measurements, eliminating the need for sampling and later analysis in the laboratory. In addition, ISFETs have very fast response, high sensitivity, batch processing capability, microsize, and the potential for on-chip circuit integration. What is most attractive is that ISFETs are a preferred transducing element for biosensors because the $SiO_2$ surface contains reactive SiOH groups, which can be used for covalent attachment of organic molecules and polymers. Also, ISFETs can be used to develop immunobiosensors (Yuqing et al. 2003).

## 7. SCHOTTKY DIODE–BASED SENSORS

The Schottky diode is a semiconductor-based diode and consists of a metal–semiconductor junction instead of a semiconductor–semiconductor junction as in conventional diodes. Schottky diodes use the metal–semiconductor junction as the Schottky barrier. Since the barrier height is lower in metal–semiconductor junctions than in conventional *p–n* junctions, Schottky diodes have lower forward voltage drop. An *n*-type semiconductor is normally used in Schottky diodes. Due to the absence of the *p*-type semiconductor region in these diodes, faster switching times are possible since the mobility of electrons is three times that of holes. These diodes can be used as sensors for various chemicals if the adsorbed analyte changes the electrical characteristics of the Schottky diode.

Based on the thermionic field emission conduction mechanism of the Schottky diode, the *I–V* characteristics of the diode for forward bias voltage greater than 3 kT is given by (Chen and Chou 2003)

$$I = I_{SAT} \exp\left(\frac{qV}{nKT}\right) \quad (1.19)$$

where *K* is the Boltzmann constant, *T* is the temperature in kelvin, *n* is the ideality factor, and $I_{SAT}$ is the saturation current defined as (Chen and Chou 2003)

$$I_{SAT} = SA^{**}T^2 \exp\left(\frac{-q\phi_b}{KT}\right) \quad (1.20)$$

where *S* is the area of the junction, $A^{**}$ is the effective Richardson constant, and $\phi_b$ is the barrier height. The ideality factor can be extrapolated from Eq. (1.19), and the barrier height can be calculated from Eq. (1.20).

A cross section of a typical chemical sensor based on a Schottky barrier is shown in Figure 1.13.

Each Schottky diode comprises two contacts, or junction areas, by definition: (1) between metal I and the semiconductor, forming the Schottky barrier junction, which is the origin of the sensor signal; and (2) between metal II and the semiconductor, forming the ohmic contact, which has to be inactive and hence should not contribute to the sensor response. The type of contact is defined by the work function of the materials used for fabrication.

The changes in the *I–V* characteristics of the Schottky diode in the presence of the analyte as a rule are conditioned by the change of the Schottky barrier height. The response can be due either to *adsorption* of the species of interest at the metal surface affecting interfacial polarization by formation of a dipole layer or to *absorption* of gases or vapors of interest by the semiconductor and their interaction

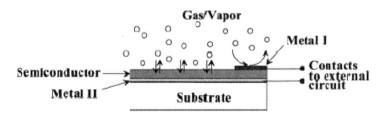

**Figure 1.13.** Cross-sectional view of a typical geometric arrangement of a Schottky barrier diode used for chemical sensing applications. (Reprinted with permission from Potje-Kamloth 2008. Copyright 2008 American Chemical Society.)

with the semiconductor, which changes its work function and hence the contact potential or built-in voltage of the diode (Trinchi et al. 2004; Potje-Kamloth, 2008)

It is necessary to note that Schottky diodes with catalytic contact metals have been widely studied mainly for their application in the detection of hydrogen and hydrogen-producing compounds (see Figure 1.14) (Trinchi et al. 2003; Tsai et al. 2003; Lin et al. 2004). Schottky diodes are advantageous for gas-sensing applications due to the simple electrical circuitry required to operate them. In hydrogen sensors based on Schottky diodes, the key role is played by the Schottky metallization, mostly belonging to the platinum group metals (Pd, Pt), which ensures the catalytic dissociation of molecular hydrogen into hydrogen atoms. It should be noted that the semiconducting substrate is only used to provide sufficiently high Schottky barrier heights, necessary for good sensing performance, but is not the gas-sensitive part of the sensor.

According to the dipole model, originally developed by Lundstrom, the hydrogen sensitivity of Schottky diode gas sensors is based on the dissociation of hydrogen molecules on the catalytic metal surface and the diffusion through the metal film to form a polarized layer at the metal/insulator interface. When atomic hydrogen has been formed on the outer metal surface, which is exposed to the ambient (see Figure 1.14), an equilibrium between the hydrogen concentration at this metal surface and that at the metal/semiconductor interface is reached. Based on this model, the change in the $I$–$V$ characteristics and the decrease in the Schottky barrier height are strongly related to the hydrogen concentration.

Schottky-based sensors also respond to hydrogen-containing molecules such as hydrocarbons, provided that the molecules are also dissociated on the catalytic metal surface (Trinchi et al. 2003). Oxygen atoms and $NO_x$ (Tuyen et al. 2002) are also dissociated on the catalytic metal surface. Water formation with oxygen atoms from oxygen-containing molecules consumes hydrogen and therefore decreases the sensor response. In other words, catalytic metal gates have a direct response to hydrogen

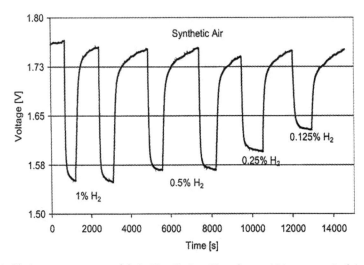

**Figure 1.14.** Hydrogen response of Schottky diode with a forward bias current of 1 mA at 310°C. (Reprinted with permission from Trinchi et al. 2004. Copyright 2004 Elsevier.)

and hydrocarbons as well as an indirect response to oxygen molecules, the effect of which is to decrease the direct response.

Besides Si, compound semiconductors such as GaAs, InP, GaN, $Ga_2O_3$, and SiC have been alternatively employed as substrate materials for Schottky diode–type hydrogen sensors (Chen and Chou 2003; Trinchi et al., 2004). GaN, $Ga_2O_3$, and SiC have a wide bandgap, and hence operating temperatures up to 900°C are achievable, as compared to silicon substrates, for which operating temperatures are, limited up to 250°C (Trinchi et al. 2003).

## 8. CATALYTIC SENSORS

The catalytic sensors widely known as "catalytic bead" or "pellistors" were among the first chemical sensors. Such detectors have been in widespread use for more the 50 years in portable, transportable, and fixed multipoint gas alarms. State-of-the-art catalytic sensors are stable, reliable, accurate, rugged, and have a long operating life. The output is linear because the platinum wire has a good linear coefficient of thermal resistance. Although many design improvements have been made to detectors of this type over the years, in essence, the basic concept has not changed (Symons 1992; Miller 2001; Korotcenkov 2007).

In operation, the pellet and consequently the catalyst layer is heated by passing a current through the underlying coil. In the presence of a flammable gas or vapor, the hot catalyst allows oxidation to occur in a chemical reaction similar to combustion. Just as in combustion, the catalytic reaction releases heat, which causes the temperature of the catalyst together with that of its underlying pellet and coil to rise. This rise in temperature results in a change in the electrical resistance of the coil, and it is this change in electrical resistance which constitutes the signal from the sensor.

The sensor temperature rise can be detected via an increase in the Pt coil resistance, typically by incorporating the sensing element in a Wheatstone bridge circuit. A measurement voltage is applied across both arms of the bridge, and the resistance in each arm is matched so that the potential difference measured across the center of the bridge is zero. Any change in the resistance of the platinum wire will now result in a change in this measured voltage.

The detection elements may take various forms according to the origin of their design. Possible versions of detection elements in simplified form are presented in Figure 1.15. The simplest form is

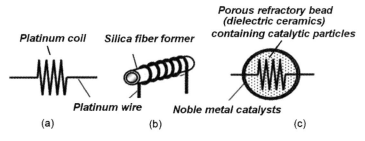

**Figure 1.15**. Schematic views of various pellistors. (Reprinted with permission from Korotcenkov 2007b. Copyright 2007 Elsevier.)

the platinum wire filament (Figure 1.15a), still used in some gas detection instruments. However, the "bead" construction is the most effective and promising. The "bead" (see Figure 1.15c) takes the form of a ceramic pellet supported on a platinum wire coil. Usually, pellistors consist of small-diameter Pt spirals (approximately 20 turns) surrounded by a porous refractory bead, typically alumina impregnated with a precious metal catalyst such as palladium, thorium, or another catalyst. A noble metal acts as a catalyst to promote exothermic oxidation of flammable gases. The presence of catalyst coating reduces the temperature needed to achieve a stable signal for hydrocarbons. In this case the sensitive element is normally electrically self-heated to between 400 and 800°C for methane and other gases. This helps in extending the life of the filament.

We need to note that pellistors are sensitive only to flammable gases and vapors. Hot-wire sensors (pellistors) are traditionally used to detect the percentage of the lower explosive limit (% LEL) levels of flammable gases and vapors. The LEL is usually expressed as percent by volume of combustible gas in air. Of course, pellistors have been used for detection of parts-per-million concentrations of some gases and vapors, but in this case they require active sampling and other instrument design considerations. Moreover, the pellistors are nonselective sensors (Korotcenkov 2007).

## 9. CONDUCTOMETRIC SENSORS

Conductometric sensors have a simple structure and their operating principle is based on the fact that their electrical conductivity (or electrical conductance) can be modulated by the presence or absence of some chemical species that comes in contact with the device (Wohltjen et al. 1985). Figure 1.16 shows a typical structure of a conductometric sensor. It consists of two elements, a sensitive conducting layer and contact electrodes. These electrodes are often interdigitated and embedded in the sensitive layer. To make the measurement, a DC voltage is applied to the device and the current flowing through the electrodes is monitored as the response. The main advantages of these sensors are easy fabrication, simple operation, and low production cost, which means that well-engineered metal-oxide conductometric sensors can be mass produced at reasonable cost. Moreover, these sensors are compact and durable. As a result, they are amenable to being placed *in situ* in monitoring wells.

The basis of the operation of conductometric sensors is the change in resistance under the effect of reactions (adsorption, chemical reactions, diffusion, catalysis) taking place on the surface of

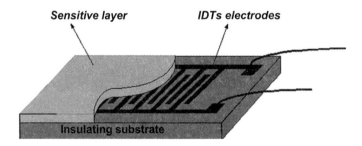

**Figure 1.16.** Structure of a typical conductometric sensor.

the sensing layer. The chemical species interact with the sensitive layer and thus modulate its electrical conductivity. This can be measured as a change in the current, which is correlated to the concentration of the chemical species.

The two main categories of conductometric gas sensors are thin- and thick-film sensors. Their preparation involves different techniques. The different structural properties due to the fabrication and film thickness lead to different sensing properties.

In thick-film sensors the film thickness is typically in the range of 2–300 μm. Thick-film technology, based mainly on screen printing, is the most common fabrication technique for such sensors. The layer itself is a porous body, so the inner surface also becomes a working surface. Therefore, the gases diffuse into it, leading to good sensitivity. The microstructure of the sensing layer is a function of temperature parameters of grains sintering. The conductance usually is controlled by the contact resistance between the grains.

In thin-film sensors the film thickness is typically 5–500 nm, in special cases even 1 μm. The most common preparation process is thin-film technology based on sputtering, evaporation, chemical vapor deposition, spray pyrolysis, chemical deposition, or laser ablation methods of forming the sensing layer. Sensing layers in thin-film sensors have more dense structure in comparison with thick-film sensors.

At present, for conductometric sensors, mainly oxide semiconductors—electron or mixed conductors—are used as base materials (Korotcenkov 2007a). Polymers, thin-film metals, and the carbon nanotubes, which exhibit changes in their conductivity when exposed to an analyte, also can be used for fabricating such sensors. Extensive literature is available for a variety of polymers (Harris et al. 1997; Wang et al. 1997; Su and Tsai 2004), metal oxides (Patel et al. 1997; Atashbar et al. 1998; Tan et al. 2003; Yu et al. 2005), and mixed oxides (Moon et al. 2002). Metals are normally substituted into polymers, and the choice of metal governs the sensitivity of the conductometric sensors. Metal-substituted phthalocyanines are commonly used due to their stability toward oxidation at high temperatures (Wohltjen et al. 1985). Middlehoek and Audet (1989) used Cu-substituted phthalocyanine conductive polymer as a $CCl_4$ sensor. Previous studies used sublimation techniques to deposit thin films for chemical vapor sensing. These devices, however, had problems due to poor control of film thickness, size, and morphology (Van Oirschot et al. 1972; Bott and Jones 1984). To overcome these problems, Wohltjen and co-workers (1985) deposited monolayers of Cu-substituted phthalocyanine on interdigitated Au electrodes using Langmuir-Blodgett methods to sense ammonia and $NO_2$. Ultrathin metallic films have been used successfully for detecting variuous species. Platinum thin films were employed by Majoo and Patel for hydrogen gas sensing (Majoo et al. 1996; Patel et al. 1999). A variety of metallic nanowires has also been reported for gas sensing, such as palladium nanowires for hydrogen sensing (Yang et al. 2009; Atashbar and Singamaneni 2005) and silver nanowires for amine vapor (Murry et al. 2005).

The most accepted mechanism, explaining sensitivity of $n$-type metal oxide–based sensors, includes consideration of the role of the chemisorbed oxygen (Barsan et al., 1999; Gurlo, 2006). Oxygen chemisorption means the formation of $O_2^-$, $O^-$, $O^{2-}$ species on the surface. Among these, $O^-$ proved to be more reactive than $O_2^-$, while $O^{2-}$ is not stable. So the dominant species is the $O^-$ species. The oxygen chemisorption results in a modification of the space charge region toward depletion. The resistance corresponding to this state is considered the base resistance. The appearance of a reducing gas leads to partial consumption of the adsorbed oxygen, resulting in a decrease in resistance, while the appearance of oxygen increases the surface oxygen coverage, and hence the resistance. The above mechanisms suggest the existence of a grain boundary (connected to the modification of the space charge region (see Figure 1.17). The relationship between the amount of change in resistance to the concentration of a combustible gas can be expressed by a power-law equation (Barsan et al. 1999):

$$G = \frac{1}{R} = kC^n \qquad (1.21)$$

where $C$ is the concentration of analyte and $k$ and $n$ are individual constants, which depend on the mechanism of sensitivity and must be determined empirically by calibration.

To activate reactions of oxygen chemisorptions and surface catalysis, high temperature (>200°C) is required. For these purposes, metal-oxide gas sensors have incorporated heaters, which are electrically isolated from the sensing layer. One possible variant of conductometric sensor configuration is shown in Figure 1.18.

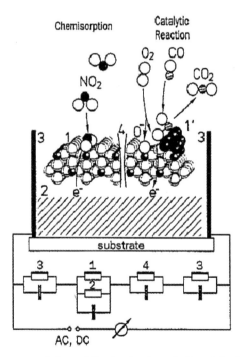

**Figure 1.17.** Schematic representation of the basic steps in the detection of gas molecules using conductometric metal-oxide gas sensors. Surface and bulk reactions lead to changes of the overall DC or AC conductance, which may include frequency-dependent contributions from (1) an undoped or doped surface, (2) bulk, (3) three-phase boundary or contacts, and (4) grain boundaries. Equivalent circuits with different resistance/capacitance (RC) units describe the frequency behavior formally. Each unit corresponds to a characteristic charge carrier transport. (Reprinted with permission from Gopel and Schierbaum 1995. Copyright 1995 Elsevier.)

# BASIC PRINCIPLES OF OPERATION • 23

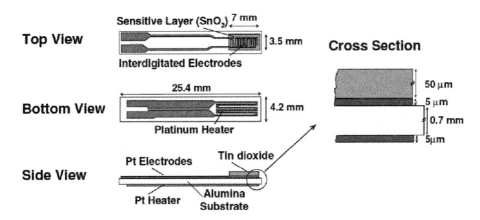

**Figure 1.18.** Layout of planar alumina substrate with Pt electrodes and Pt heater. The SnO$_2$ layer is printed on top of the interdigitated electrodes. The heater on the back keeps the sensor at the operational temperature. (Reprinted with permission from Barsan and Weimar 2003. Copyright 2003 IOP.)

The sensitivity of conductometric metal-oxide sensors can be improved using different approaches based on the tools of materials engineering: bulk doping, surface modification, structure optimization, etc. (Korotcenkov 2005). The selectivity can be improved by selecting or preparing proper materials (Korotcenkov 2007a). The incorporation of a catalyst results in improvement of surface reaction selectivity. An applied filter selectively removes the undesired component(s). However, the filter saturates and requires regeneration. To improve the stability, a materials science approach, i.e., the use of chemically stable materials, seems to be the most efficient.

The main considerations in the fabrication of a stable conductometric sensor are careful selection of the sensitive layer, film-coating technique, bias potential, and measurement scheme. The sensitive layer should have affinity to the target measurand, and the film coating technique needs to be compatible with the device fabrication process. In addition, its chemical composition, thickness, and morphology should be reproducible. A thorough review paper has been published by Korotcenkov (2008) on the influence of morphology and crystallographic structure of metal oxides and characteristics such as thickness, grain size, agglomeration, porosity, faceting, grain network, surface geometry, and film texture on conductometric gas sensors. With regard to the measurement scheme, Harris and co-workers (1997) presented signal-to-noise ratio (SNR) analysis for conductivity measurements of thin films deposited between microelectrodes. They compared the AC and DC excitation of the sensor, and their noise measurements revealed that passage of DC current generates a high level of $1/f$ noise characteristic. Remarkable improvement in SNR can be obtained by applying low-frequency (a few kilohertz) excitation. They also found that the conduction mechanism is nonohmic at frequencies below 1 MHz, but considerably more ohmic at frequencies of tens of MHz. In the case of an AC excitation, however, the distinction between the conductivity and capacitive measurements becomes challenging, since the sensitive layers have both resistive and capacitive elements (Kovacs 1998). Wohltjen and co-workers (1985) also noted that a differential DC conductivity measurement can be made to reduce temperature-dependent errors by keeping the reference sensor isolated from the analyte.

The major advantages of metal oxides are reversibility, rapid response, longevity, low cost, and robustness. The materials and manufacturing techniques easily lend themselves to batch fabrication. Unfortunately, several disadvantages limit more widespread sensor utilization and continue to challenge sensor development. The metal-oxide sensor generally requires a high temperature for measurable gas response, often leading to high power consumption. Metal-oxide sensors are not highly selective, and much effort has been involved in devising materials and methods of operation to improve specificity. These sensors are generally not appropriate for applications in vacuum or inert atmospheres. Baseline sensor drift may require periodic calibration for certain applications.

Unlike the metal-oxide gas sensors, devices based on polymers can operate at room temperature. A more versatile sensing system that may work at room temperature is based on a carbon black–polymer blend, in which the carbon particles provide electrical conductivity and the polymer (any polymer) provides the sensor function. The sensor response is a result of the swelling of the polymer, which causes the conductivity of the sensor to change. The amount of swelling corresponds to the concentration of the chemical vapor in contact with the absorbent. Weak van der Waals forces between the polymer and the target gas molecules are responsible for the swelling of the sensing layer; therefore, the change is purely physical and is reversible, which makes the sensor reusable. Some hysteresis can occur when the sensor is exposed to high concentrations.

Polymer-based conductometric gas sensors have better selectivity. So far, however, problems such as low stability, high sensitivity to humidity, and degradation under UV radiation and during interaction with ozone have not been resolved for polymers.

## 10. ACOUSTIC WAVE SENSORS

Acoustic wave–based sensors offer a simple, direct, and sensitive method for probing the chemical and physical properties of materials. Acoustic wave sensors use piezoelectric material as substrates to generate acoustic waves. The type of acoustic wave generated in a piezoelectric material depends mainly on the substrate material properties, the crystal cut, and the structure of the electrodes utilized to transform the electrical energy into mechanical energy.

Acoustic wave sensors have been broadly used for many applications in detecting chemical and biological components in either gas or liquid media. To monitor a specific gas or vapor, a sensitive layer is generally employed. The utilization of acoustic wave devices for chemical sensing applications relies on their sensitivity toward small changes (perturbations) occurring at the "active" surface. The interface selectively adsorbs target molecules in the analyte to the surface of the sensing area. This adsorption results in changes in the electrical and mechanical properties of the device (acoustic wave–based gas and vapor sensors measure the change in the mass at the surface or in the sheet conductivity of the sensitive layer), which in turn causes changes in the acoustic wave characteristics, namely, amplitude, velocity, and the resonance frequency of the device. In the presence of an analyte species, the waves' properties become perturbed in a measurable way that can be correlated to the analyte concentration (Ippolito et al. 2009).

It has been established that acoustic-wave-based sensors have high sensitivity toward surface perturbations and exhibit good linearity with low hysteresis. They are also typically small, relatively inexpensive, and inherently capable of measuring a wide variety of input quantities. They offer a simple means of analyte monitoring, with several advantages over other solid-state sensors (Ippolito et al. 2009). These include well-established fabrication processes, chemical inertness of the substrate materials, and high structural rigidity. Additionally, the microelectronics industry has developed numerous piezoelectric transducer platforms, which in turn has widely encouraged the development and improvement of various acoustic wave–based chemical sensors.

However, it is necessary to note that, similar to other solid-state chemical sensors, with acoustic wave–based sensors it is inherently difficult to satisfy the requirements for all sensing applications with a single sensor design. Therefore, in most cases, the selection of materials and structural parameters results in design trade-offs among device sensitivity, long-term stability, selectivity, and operating temperature.

Acoustic wave devices come in a number of configurations, each with its own distinct acoustic and electrical characteristics. Major classes of acoustic devices for sensor applications are thickness shear mode, Rayleigh and shear horizontal surface acoustic wave, acoustic plate mode, Love wave mode, and flexural plate wave devices. Several different groups of acoustic wave devices that are commonly employed for gas sensing will be discussed in the following.

## 10.1. THICKNESS SHEAR MODE SENSORS

The thickness shear mode (TSM) resonators were first studied and used by Sauerbrey (1959) to determine the thickness of thin layers adhering to the surface of the device. Over time, the use of TSM resonators has been extended for detection of various chemical and biological species by coating the surface of the device with a suitable sensitive layer. These devices are also referred to as quartz crystal microbalances (QCMs) and consist of an AT-cut quartz crystal with thin-film metal electrodes patterned on both sides forming a single electrical port (typical electrode materials are gold and aluminum). These devices fall into the category of bulk wave acoustic sensors, as application of a voltage produces an electric field throughout the bulk of the device. The piezoelectric nature of quartz in turn causes this electric field to produce a shearing motion transverse to the normal of the crystal. The resonant frequency of the device is inversely proportional to the thickness of the crystal (which is half a wavelength) and is typically in the range of 5–15 MHz. QCMs have been used extensively for thickness monitoring in metal evaporation (McCallum 1989); protein sensing (Barnes et al. 1992; Imai et al. 1994; Atashbar et al. 2005), and gravimetric immunoassays (Muramatsu et al. 1987; Caruso et al. 1996; Anzai et al. 1996; Sakti et al. 2001). One of the advantages the QCM is the small temperature dependence of the operating frequency. The electronic circuitry and the measurements associated with QCMs are relatively simple and straightforward. Further, these devices have a very high quality factor. The most serious limitation of these devices is their low sensitivity, which is limited by the thickness of the crystal. Although the device's sensitivity can be increased by decreasing the crystal thickness, it makes the crystal much more fragile. Figure 1.19 shows top and side views of a typical QCM crystal. Useful reviews of TSM devices have been provided by Lu and co-workers (1984) and by Benes and co-workers (1995).

If we assume that the adsorbed mass is of infinitely small thickness with negligible phase change across the film thickness and there is no acoustic energy dissipation, then it is easy to derive the Sauerbrey equation (Cheeke and Wang 1999):

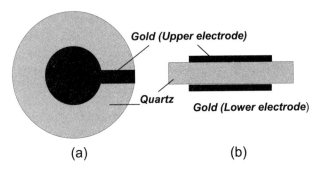

**Figure 1.19.** (a) Top view of a QCM crystal. (b) Side view of a QCM crystal.

$$\frac{\Delta f}{f_0} = \frac{-2 f_0 \Delta m_S}{\sqrt{\mu_q \rho_q}} = \frac{-2 \Delta m_S}{\rho_q \lambda_0} \tag{1.22}$$

where $f_0$ is the resonance frequency, $\mu_q$ is the substrate shear stiffness, $\rho_q$ is the substrate mass density, $\lambda_0$ is the acoustic wavelength of the resonator, and $m_S$ is the area mass loading on the surface of the resonator. Hence, a measurement of the resonant frequency will give a measure of $m_S$. In practice, the device must be calibrated.

More sensors designed to respond to different gas components are made by coating the metallizing layers with an additional acceptor layer, which can absorb the sample component. This acceptor layer should interact with the desired component as selectively as possible, without being changed by other parameters. If an uncoated QCM with a fundamental frequency of $f_a$ is coated with a acceptor layer, the fundamental frequency of the QCM will drop to $f_b$. When this coated QCM is then exposed to a vapor, some of the vapor will be absorbed into the acceptor layer, changing the mass of the crystal and hence its fundamental frequency.

Ideal coatings for QCM applications should be nonvolatile so that the coating stays on the crystal surface and allows rapid and easy diffusion of vapors into and out of the material. The material should be stable over prolonged use and not undergo any hysteresis effects. Selectivities of QCM sensors differ according to their acceptor layer. For instance, palladium is highly selective for hydrogen, organic amines are only weakly selective for sulfur dioxide, and carbowaxes (commonly used as the stationary phase in chromatography) interact with numerous nonpolar gases, hence they are nonselective sensors for hydrocarbons. We need to note that polymers are the most widely usable coatings for QCM-based sensors.

QCM sensors are fast, cheap, reliable sensors that can be used for a variety of applications ranging from solvent detection to food quality analysis. They are inexpensive because they are mass produced for oscillator circuits, and their lifetime is approximately 1 year. They have good linearity and can be incorporated into an array and used to build up a library of characteristic responses for individual analytes, allowing for rapid and reliable identification of unknown compounds. They can be used for early-warning detection of possible contaminants in many industrial processes, such as beer and tobacco production.

Application of piezoelectric sensors in liquids is much more difficult than working with gases, since the surrounding liquid phase acts strongly, damping the oscillating crystal. Although the problems have been solved, real analytical applications are rarely encountered even today. The best-investigated applications are antibody layers immobilized on quartz crystals, which are used in immunoassays. Reaction with the corresponding antigen is highly selective, and the resulting mass change is measurable. Piezoelectric sensors in liquid phase do not obey the Sauerbrey equation.

## *10.2. SURFACE ACOUSTIC WAVE SENSORS*

Surface acoustic wave (SAW) or Rayleigh sensors consist of a piezoelectric substrate with electrodes patterned on the surface in the form of interdigital transducers (IDTs), as shown in Figure 1.20.

This is the most commonly utilized structure for gas-sensing applications, with the sensitive layer normally deposited in between the two IDT ports. However, if the layer is of low conductivity, it may be deposited over the entire device. This configuration generates the Rayleigh-mode SAW, which predominantly has two particle displacement components in the sagittal plane. Surface particles move in elliptical paths with a surface-normal and a surface-parallel component. The surface-parallel component is parallel to the direction of propagation. The electromagnetic field associated with the acoustic wave travels in the same direction. The wave velocity is determined by the substrate material and the cut of the crystal. Most of the energy of the SAW is confined to a few wavelengths depth of the substrate (Wohltjen 1984). The acoustic wavelength is determined by the IDT periodicity. These sensors are normally operated in a "delay line" configuration, with one set of IDTs acting as a transmitter and the other set acting as a receiver. These devices work at higher frequencies (a typical frequency of operation being 30–500 MHz) and in comparison to TSM resonators have higher sensitivity, because the wave propagates not through the bulk of the device but near the surface. The use of Rayleigh SAW sensors is applicable only to gas media, as the Rayleigh wave is severely attenuated in liquid media (Calabrese et al. 1987).

The basic principle in a SAW-based gas sensor is the detection of small deviations in acoustic wave propagation characteristics, which are caused by perturbations on the active surface of the device. The most common technique is to measure the deviation in propagation velocity of the surface wave as a change in resonant frequency of the device. However, perturbations that affect acoustic phase

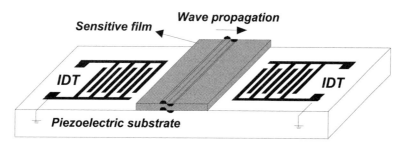

**Figure 1.20.** Schematic view of a SAW sensor.

velocity can be caused by many factors, each of which represents a potential sensor response (Ricco and Martin 1991): $T$ (temperature), $\varepsilon$ (permittivity), $E$ (electric field), $\sigma$ (electrical conductivity), $c$ (stiffness), $\eta$ (viscosity), $\rho$ (density), $p$ (pressure), $\mu$ (shear elastic modulus), $m$ (mass). These interactions change the boundary conditions, producing a measurable shift in the propagating SAW mode's phase velocity. Equation (1.23) illustrates the change in acoustic phase velocity as a result of possible external perturbations, under the assumption that the perturbations are small and linearly combined:

$$\frac{\Delta v}{v_0} \cong \frac{1}{V_P}\left(\frac{\delta v}{\delta T}\Delta T + \frac{\delta v}{\delta \varepsilon}\Delta \varepsilon + \frac{\delta v}{\delta E}\Delta E + \frac{\delta v}{\delta \sigma}\Delta \sigma + \frac{\delta v}{\delta m}\Delta m + \frac{\delta v}{\delta p}\Delta p + \ldots\right) \quad (1.23)$$

Since acoustic wave devices use piezoelectric materials for the excitation and detection of acoustic waves, the nature of almost all the parameters involved in sensor applications concerns either mechanical or electrical perturbations. Furthermore, the temperature dependence of each parameter and the overall temperature coeffcient of the structure must also be considered. Therefore, the sensor response may be due to a combination of the above parameters. However, it is necessary to note that during chemical sensing the change of mass and conductivity of functionalizing (sensing) layer has the strongest influence.

In the simplest case, mass loading (ML) of adsorbed species reduces the phase velocity in the substrate, leading to a reduction in frequency so that

$$\frac{\Delta v}{v_0} = -c_m f_0 \cdot \Delta m_S \quad (1.24)$$

where $c_m$ is a mass sensitivity factor.

Some excellent articles describing the application of SAW sensors for the measurement of parameters such as temperature, pressure, electrical conductivity, mass, and viscoelastic changes of thin films have been reported in a number of reviews (D'Amico and Verona 1989; McCallum 1989; Ballantine and Wohltjen 1989; Khlebarov et al. 1992; Grate et al. 1993; Caliendo et al. 1997; Ballantine et al. 1999). SAW sensors have found extensive use in gas sensing (Atashbar and Wlodarski 1997; Varghese et al. 2003; Fontecha et al. 2004; Shen et al. 2004).

SAW devices are well established in the electronics industry and have been employed for gas-sensing applications for approximately 30 years. They can be designed to offer high sensitivity with a reasonably large dynamic range, good linearity, and low hysteresis. This makes them extremely suitable for sensing applications at parts-per-million (ppm) and parts-per-billion (ppb) concentrations (Ippolito et al. 2009). When a good coating is available, it is usually possible to detect species at concentration levels of 10–100 ppb. The mass detection limit is in the range of 0.05 pg/mm$^2$.

Many SAW sensor applications use a dual-delay-line configuration; one delay line can be coated with the sorptive or reactive film, while the other remains inert or protected from environmental effects. Typically, the frequency difference is measured, which is of the order of kilohertz and can be easily sampled. The uncoated or protected resonator acts as a reference to compensate undesired effects, for instance, frequency fluctuations caused by temperature or pressure changes.

### 10.2.1. Shear-Horizontal Surface Acoustic Wave Sensors

Shear-horizontal surface acoustic wave (SH-SAW) sensors belong to the category of SAW sensors in which the application of a potential to the IDTs produces an acoustic wave having particle displacement perpendicular to the direction of wave motion and in the plane of the crystal (shear displacement). As in SAW devices, the acoustic wavelength is determined by the transducer periodicity. An example of a piezoelectric substrate that supports SH-SAW is lithium tantalate (LiTaO$_3$), in which the dominant acoustic wave propagating on 36° rotated Ycut, X propagating LiTaO$_3$ is a shear horizontal mode. In the SH-SAW, the acoustic wave propagation is not severely attenuated when the surface is loaded with a liquid, as is the case with the Rayleigh-mode SAW-based devices. SH-SAW sensors have been applied to a number of enzyme-detecting (Kondoh et al. 1994) and liquid-sensing applications (Nomura et al. 2003; Martin et al. 2004).

### 10.2.2. Shear-Horizontal Acoustic Plate Mode Sensors

Shear-horizontal acoustic plate mode (SH-APM) sensors are essentially Rayleigh sensors with the difference that their substrate thickness is of the order of a few acoustic wavelengths (Déjous et al. 1995; Esteban et al. 2000). Under such conditions, IDTs in addition to the Rayleigh waves also generate the shear horizontal waves. These waves are not confined to the surface only but travel through the bulk, being reflected between the top and the bottom surfaces of the substrate which now acts as an acoustic waveguide (see Figure 1.21). The particle displacement for SH-APM devices is parallel to the surface (in plane) and transverse to the direction of wave propagation. The Rayleigh waves typically propagate at much lower velocities than those of the SH-APM modes. The frequency of operation is determined by the material properties, the thickness-to-wavelength ratio of the substrate, and the IDT finger spacing.

SH-APM devices are used mainly for liquid sensing and offer the advantage of using the back surface of the plate as the active sensing area. Consequently, the IDTs can be isolated from the liquid media, thereby preventing direct chemical attack on the IDTs. The interfaces where the wave interacts with liquid and air should be polished and smooth, since any irregularities will produce additional noise signals (Gardner et al. 2001). Considerable research has been devoted to developing various sensing applications. Applications such as vapor and viscosity measurements (Ricco et al. 1988; Esteban et al.

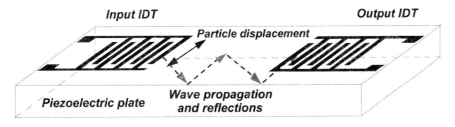

**Figure 1.21.** Schematic view of a SH-APM device structure.

2000), identification via the electrical surface perturbation (Kelkar et al. 1991; Dahint et al. 1993), and biosensing (Andle et al. 1995; Déjous et al. 1995) have been reported.

### 10.2.3. Love-Mode Acoustic Wave Sensors

Love wave devices are characterized by acoustic waves that propagate in a layered structure consisting of a piezoelectric substrate and a guiding layer (Bender et al. 2000; Kalantar-zadeh et al. 2001) (Figure 1.22). They have a pure shear polarization, with the particle displacements perpendicular to the normal of the surface plane. An essential condition for existence of Love waves is that the shear acoustic velocity in the guiding layer is less than the shear acoustic velocity in the substrate (Haueis et al. 1994; White et al. 1997). Under such conditions, the elastic waves generated in the substrate are coupled to this surface guiding layer, which traps the acoustic energy near the surface of the device and thereby increases the sensitivity of the device to mass loadings.

The mass sensitivity of a Love-mode sensor is determined mainly by the thickness of the guiding layer and its operating frequency. Hence the guiding layer should be one with low density and low shear velocity. The guiding layer also serves to passivate the IDTs from the contacting liquid (Josse et al. 2001). The most commonly used substrates for SH-SAW sensors and Love-mode devices are quartz, lithium niobate ($LiNbO_3$), and lithium tantalate ($LiTaO_3$). The guiding layer may be $SiO_2$ (Harding 2001), certain polymers such as poly(methyl methacrylate), also known as PMMA (Gizelli 2000), or ZnO (Kalantar-zadeh et al. 2001). Top and side views of a Love-wave sensor are shown in Figure 1.22.

Love mode sensors are advantageous for gravimetric sensing applications such as organic vapors absorbed by a thin polymer film (Jakoby et al. 1999).

### 10.2.4. Lamb Wave Sensors

Lamb wave sensors are essentially Rayleigh wave sensors in which the waves propagate in a thin plate (membrane), which has a thickness less than an acoustic wavelength. Particle displacement is transverse

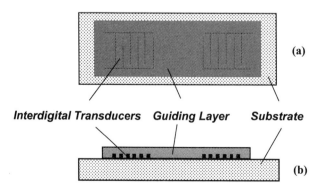

**Figure 1.22.** (a) Top view of Love wave sensor. (b) Side view of a Love wave sensor.

# BASIC PRINCIPLES OF OPERATION • 31

to the direction of wave propagation and is parallel and normal to the plane of the surface. Application of a potential across the electrodes gives rise to both symmetric and antisymmetric plate modes. The lowest-order antisymmetric mode ($A_0$) of the Lamb wave is known as the flexural plate wave (FPW), and it is this mode that is used in liquid sensing applications (Wang et al. 1998). The IDTs form a delay line for launching the flexural plate waves that set the whole membrane into motion. The wave velocities and hence the frequency of operation of these devices therefore depends on the plate material and its thickness. Figure 1.23 shows top and side views of a flexural plate wave sensor. These FPW devices have been investigated for applications such as chemical gas sensors (Wenzel and White 1989; Grate et al. 1991), pressure sensors (Tirole et al. 1993, Choujaa et al. 1995), and for the determination of liquid properties (Vellekoop et al. 1994). Development of a FPW-based biosensor has also been investigated (Andle and Vetelino 1995; Wang et al. 1997).

## 11. MASS-SENSITIVE SENSORS

Mass-sensitive chemical sensors are the simplest form of gravimetric sensors that respond to the mass of species attached to the sensing layer (Ballantine et al. 1999). In principle, any species that can be immobilized on the sensor can be sensed. The mass changes can be monitored by either deflecting a micromechanical structure due to stress changes or mass loading, or by assessing the frequency characteristics of a resonating structure or a traveling acoustic wave upon mass loading. As we know, the change of the resonant frequency $f$ is given by (Cimalla et al. 2007)

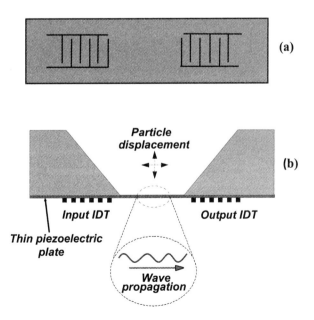

**Figure 1.23.** (a) Top view of a FPW sensor. (b) Side view of a FPW sensor.

$$\Delta f = -\frac{1}{2} f \frac{\Delta m}{m} \qquad (1.25)$$

Among the different geometric shapes of mass-sensitive components, clamped-free beams, also called cantilevers, represent the simplest microelectromechanical system (MEMS). Although cantilevers play a role as a basic building block for complex MEMS devices, they are also widely used as force-sensor probes in atomic force microscopy (AFM). Recent developments in AFM have enabled mass fabrication of micrometer-sized cantilevers with various geometric shapes and made of various materials (silicon, silicon nitride, etc.). The free-standing cantilever beam end is coated with a sensitive layer (see Figure 1.24). The detection principle of such devices is based on measurement of the change in cantilever deflection or the change in resonance frequency of the cantilever (Lavrik et al. 2004). These changes are induced by the adsorption of chemical species on the functionalized surface of a microcantilever (Chen 1995). Micromachined cantilever arrays have already proven their suitability as chemical and biological sensors for gas- (Battiston et al. 2001; Kim et al. 2001; Lange et al. 2001; Raiteri et al. 2001; Moulin et al. 2003; Then et al. 2006) and liquid-phase interactions (Raiteri et al. 2002).

The sensitivity of such sensors to additional mass loadings can be increased by reducing the inherent active mass of the resonator, which is the reason for the ongoing research toward nano-electromechanical systems (NEMS). The detection of monolayers of molecules, single viruses, and carbon particles are outstanding results that have been demonstrated recently (Cimalla et al. 2007).

At present, the measurement of microcantilever deflection with subangstrom resolution can be organized into two approaches: optical and electrical (Raiteri et al. 2001). The most widely used method in AFM is the so-called beam bounce or optical lever technique (see Figure 1.25). The visible light from a low-power laser diode is focused on the free apex of the cantilever, which acts as a mirror. The reflec-

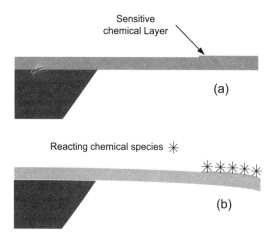

**Figure 1.24.** Schematic view of the working principle of a microcantilever-based chemical sensor: The mass deposition that follows the chemical reaction causes a stress that leads to a detectable bending of the cantilever.

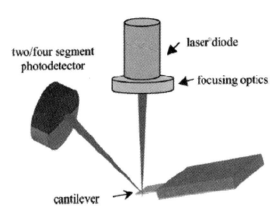

**Figure 1.25.** Optical lever deflection detection method. (Reprinted with permission from Raiteri et al. 2001. Copyright 2001 Elsevier.)

tion beam hits a position-sensitive photodetector or a split photodetector, as illustrated schematically in Figure 1.25. When the cantilever bends, the reflected laser light moves along the photodetector surface. The distance traveled by the laser beam is proportional to the cantilever deflection and linearity magnified by the cantilever–photodetector distance as the arm of a lever.

The interference of a reference laser beam with the one reflected by the cantilever is another optical deflection method that has been implemented. Intrrerferometry is highly sensitive and provides a direct and absolute measurement of the displacement. The capacitance method (electrical method) measures displacement as a change in the capacitance of a plane capacitor. This technique is highly sensitive and can provide absolute displacement.

## 12. OPTICAL SENSORS

Optical sensors can be used for detection and determination of chemical or physical parameters by measuring changes in optical properties. Techniques for using optical chemical sensors have been a growing research area over the last three decades. Two basic operation principles are utilized for optically sensing chemical species (Baldini et al. 2006):

- An intrinsic optical property of the analyte is utilized for its detection.
- Indicator (or label) sensing is used when the analyte has no intrinsic optical property. For example, pH is measured optically by immobilizing a pH indicator on a solid support and observing changes in the absorption or fluorescence of the indicator as the pH of the sample varies with time.

The first principle is used mainly in infrared and ultraviolet spectroscopes. The most widely used techniques employed in optical chemical sensors are optical absorption, fluorescence, and chemiluminescence. In this case, optical chemical sensors detect the intensity of photon radiation that arrives at a sensor. Sensors based on other optical parameter spectroscopies, such as refractive index and reflectivity, have also been well developed.

Absorption is based on the transmitted light remaining after incident light is passed through a medium. The medium can be a gas, a liquid, or a solid. There are no perfectly transparent materials that would allow light or electromagnetic energy to pass through them without any change. Some energy must be absorbed, depending on the physical properties of the transmitting medium. Variations in light absorption can be due to the vibrational and rotational movements for different chemical bonds in the molecules in the light path which absorb energy at different wavelengths. For example, water is fairly transparent to visible and ultraviolet light, but it strongly absorbs infrared radiation and begins to absorb in the far-ultraviolet range. The optical properties of a liquid medium can also be affected by the concentration ($C$, moles per liter) of other chemical solutes that are dissolved in it. The Lambert-Beer law is often used to characterize the intensity ($I$) of light transmitted through a uniform, homogenous medium as a function of the incident light ($I_0$). The intensity is given by

$$I = I_0 \cdot \exp(-k' C \cdot \Delta x) \qquad (1.26)$$

where $k'$ is an extinction coefficient and $\Delta x$ is the thickness of the medium. Light scattering in the sample is not considered.

Fluorescent chemical sensors occupy a prominent place among the optical devices because of their superb sensitivity, combined with the required selectivity that photo- or chemiluminescence imparts to the electronic excitation. This is due to the fact that the excitation and emission wavelengths can be selected from those of the absorption and luminescence bands of the luminophore molecule (the emitted photon has a fluorescence peak at longer wavelength than the absorption peak); in addition, the emission kinetics and anisotropy features of the latter add specificity to luminescent measurements (Baldini et al. 2006). One can simply measure the intrinsic fluorescence of the target analyte or design sensors based on the variation of the fluorescence of an indicator dye with the determinand concentration. In the latter case, the probe molecule can be immobilized onto a (thin) polymer support (sometimes even the waveguide surface itself) and placed at the distal end or in the evanescent field of an optical fiber or integrated optics sensor.

Provided it is optically diluted, the relationship between a luminophore ("luminescence bearer") concentration and the intensity of its emission is a linear one:

$$I_L = \Phi_L I_0 \kappa \varepsilon_\lambda \cdot lc \qquad (1.27)$$

where $\Phi_L$ is the luminescence quantum yield (i.e., the ratio of emitted to absorbed photons per unit of time, an intrinsic feature of the luminophore), $I_0$ is the intensity of the excitation light, $\kappa$ is an instrumental parameter (related to the geometry of the luminescent sample, the source, and the detector, i.e., the emission collection efficiency), $\varepsilon_\lambda$ is the luminophore absorption coefficient at the excitation wavelength (in $dm^3 \cdot mol^{-1} \cdot cm^{-1}$), $l$ is the optical path length (in cm), and $c$ is the luminophore concentration (in $mol \cdot dm^{-3}$). Such relationship helps us to understand potential problems that may arise when designing and operating optical sensors based on polymer-supported luminescent probes (Lakowicz 1994).

Both absorption and fluorescence signals need an excitation light source, so they are not as compatible with other miniaturized solid-state chemical sensors. Chemiluminescence does not need an excitation light source. Chemiluminescence occurs in the course of some chemical reactions when an electronically excited state is generated. Chemiluminescence measurements consist of monitoring the rate of production of photons, and thus, the light intensity depends on the rate of the luminescent reaction. Consequently, light intensity is directly proportional to the concentration of a limiting reactant involved in a luminescence reaction. With modern instrumentation, light can be measured at a very low level, and this has allowed the development of very sensitive analytical methods based on these light-emitting reactions (Baldini et al. 2006). The first chemiluminescence sensor for hydrogen peroxide analysis was reported by Freeman and Seitz in 1978.

In many cases, optical sensors are based on optical fiber technology and planar waveguide configurations (Figure 1.26). Generally, an optical fiber consists of a fiber as the core surrounded by a cladding layer and a light-impermeable jacket. The refractive index of the core ($n_1$) is always higher than that of the cladding ($n_2$) layer. As a result, light beams are reflected toward the inside of the guide by internal total reflection. A planar waveguide sensor can consist of thin planar layers instead of optical fibers. The principle of measurement is shown in Figure 1.26b. The evanescent field is the basis of such sensors. There are different methods for inputting light into the active layer. Commonly, light is coupled to the device by means of prisms (prism coupler sensors). Alternatively, the light can be coupled by miniature diffraction gratings at the sensor ends (grating coupler sensors, GCS). With grating couplers, preferably, laser beams are used which are coupled to the planar medium by diffraction at the grating with a precisely defined angle of incidence. Light is totally reflected numerous times at the interfaces as a result of the small thickness of the active layer. This means an extraordinarily intensive interaction with the interface, i.e., with the sample medium. The reason is the energy quantum which is transferred at each reflection from the extension of the evanescent field across the interface (Gründler 2006).

Increasing flexibility in measurement modes, such as evanescent wave, surface reflectance, and surface plasmon reflectance, has resulted in their rapid development. Recent developments in the field have been driven by factors such as the availability of low-cost miniature optoelectronic components (light sources, waveguides, and detectors). Additional driving factors are the need for multianalyte array-based sensors, particularly in the area of biosensing, as well as advances in microfluidics and imaging technology. Portable UV/visible, Raman, and fluorescence instruments are now available for field measurements.

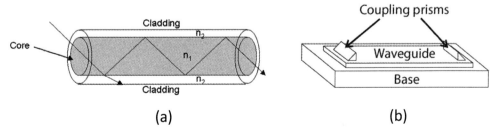

**Figure 1.26.** Basic setup of (a) cylindrical light guides, e.g., fiber optic guides. (b) Planar waveguide configuration.

## 12.1. FIBER OPTIC CHEMICAL SENSORS

Fiber optic sensors are a class of sensors that use optical fibers to detect chemical contaminants. Light is generated by a light source and is sent through an optical fiber. The light then returns through the optical fiber and is captured by a photodetector. According to Rogers and Poziomek (1996), fiber optic sensor configurations fall into three main groups: end-of-fiber, side-of-fiber, and porous or interrupted fiber configurations (see Figure 1.27). For end-of-fiber sensors, the optical fiber acts as a conduit to carry light to and from the sample. This method involves sending a light source directly through the optical fiber and analyzing the light that is reflected or emitted by the contaminant. The refractive index of the material at the tip of the optical fiber can be used to determine what phases (vapor, water, or any solution) are present. The modulation of intensity for a given range of wavelengths is dependent on the absorbance or fluorescence of the analyte, indicator, or analyte–indicator complex. The indicator compound can be trapped behind a membrane, in a polymer, or covalently immobilized to the end of the fiber. The side-of-fiber configuration typically relies on the use of the evanescent wave. This effect occurs when light is propagated down an optical waveguide. An electromagnetic wave is generated at the fiber surface and decays exponentially into the medium surrounding the fiber. This evanescent zone,

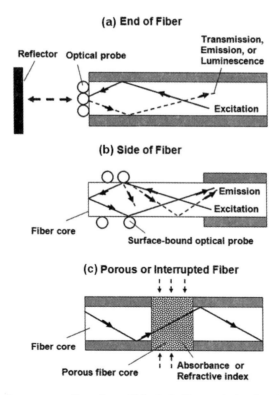

**Figure 1.27.** Fiber optic sensor configurations. (Adapted with permission from Rogers and Poziomek 1996. Copyright 1996 Elsevier.)

which is usually limited to less than 100 nm, can be used to detect the presence of optical indicators or changes in refractive index at the surface of the unclad fiber. In the case of fluorescent indicators, the method is very sensitive.

Cladding-based chemical fiber-optic sensors are examples of the side-of-fiber configuration. These sensors can be made using microporous or other types of sensitive claddings (Lieberman 1993). Evanescent wave refractometric cladding–based sensors detect the absorption of the species in the polymeric cladding, which leads to a variation of its refractive index and thus to a variation of the overall transmission efficiency of the fiber. Porous polymer cladding sensors can be used, for instance, for humidity measurements (Ogawa et al. 1988). The quantity of moisture absorbed by the microporous cladding varies with humidity. The optical power level of the transmitted light in the core varies according to the moisture absorption because of the change in the refractive index of the cladding. Another example is the application of organopolysiloxanes as cladding materials, which can change refractive index when they adsorb and/or absorb molecules from the surrounding medium (Harsanyi 2000).

For porous or interrupted fiber configurations, indicator chemistry is typically incorporated directly into the structure. For several reasons, this configuration has the potential to be extremely versatile. For example, because of the large surface area provided by the porous fiber core, this method is particularly well suited for absorbance measurements. Further, because the porous regions are intrinsically coupled with the fiber (i.e., they are part of the fiber), measurements can be made at multiple locations along a single fiber. Porous fibers may exhibit very high gas permeability and liquid impermeability, so they can be used for the detection of gases in liquids. Vapors permeating into the porous zone can produce a spectral change in transmission. For better sensitivity and selectivity, colorimetric reagents can be trapped in the pores.

However, we need to note that, in general, optical fiber sensors are usually classified as intrinsic and extrinsic types. In the extrinsic type, the optical fiber is used only as a means of light transport to an external sensing system, i.e., the fiber structure is not modified in any way for the sensing function.

Regarding intrinsic fiber optic chemical sensors, one can note that there are four general sensor designs: fiber refractometers, evanescent spectroscopy, active coating, and active core. Peculiarities of the design of these sensors are discussed in detail by Baldini et al. (2006). Coating-based sensors are the largest class of intrinsic fiber optic chemical sensors; in this design, a small section of the optical fiber passive cladding is replaced by an active coating. The analyte reacts with the coating to change the optical properties of the coating—refractive index, absorbance, fluorescence, etc.—and is then coupled to the core to change the transmission through the optical fiber.

The basic fiber optic chemical sensor (optode) design consists of a source fiber and a receiver fiber connected to a third optical fiber by a special connector as shown in Figure 1.28. The tip of the third fiber is coated with a sensitive material, usually by a dip coating procedure. The chemical to be sensed may interact with the sensitive tip by changing the absorption, reflection, scattering properties, luminescence intensity, refractive index, or polarization behavior, hence changing the reflected light properties. The fiber in this case acts as a light pipe transporting light to and from the sensing region.

Commonly, the following advantages of optodes are cited (Wolfbeis 1990):

- Optodes do not require a "reference electrode." There is no electric circuit to close, hence it is not necessary to work with two probes.

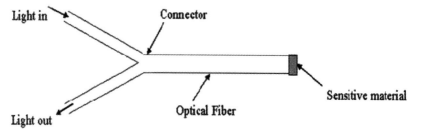

**Figure 1.28.** Fiber optic optode for chemical sensing.

- Optodes can be miniaturized easily.
- Optical fibers with high transparency allow signal transmission over long distances (up to 1000 m) without loss of quality. This characteristic is valuable in particular for long-distance measurements when the instrument cannot be placed near the sensor.
- The sensor signal is primarily optical in nature and consequently insufficient to block electrical interference from the environment. Such interference may provoke serious problems for electrochemical sensors.
- Materials for optical sensors, such as fused silica, are highly inert and may remain in contact with aggressive media for a long time.

The main problem associated with this design is that the area of interaction between the chemical and the material is very small (Cimalla et al. 2007), i.e., 8–10 μm in diameter in the case of a single-mode fiber and 50–200 μm in diameter in the case of a multimode fiber, which directly affects the sensitivity achieved using this type of sensor. In addition, ambient light may distort the measurement, and some of the reagents immobilized at the sensor surface, mainly dyes, are unstable and can be bleached by UV radiation or washed out by solvents.

Fiber optic chemical sensors can also be categorized based on the signal detection mechanism employed. These include fluorescence, absorption (colorimetric and spectroscopic), and refractometry.

## 12.2. FLUORESCENCE FIBER OPTIC CHEMICAL SENSORS

In order to transform a standard optical fiber into an intrinsically fluorescence-based optical chemical sensor, it is necessary to impart analyte-sensitive fluorescence to the fiber in some manner. This may be achieved by replacing the cladding of the fiber over a portion of its length with a solid matrix that contains a fluorescent compound. This process involves removing a portion of the original cladding of the fiber and coating the decladded region with a sensing material, which is subsequently cured to form a solid fluorescent cladding. A variation of the fluorescent cladding configuration was employed by Ahmad et al. (2005), who described a system based on two "positive" fibers attached to either end of a "negative" fiber. Park et al. (2005) described development of fiber optic sensors for detection of inter- and intracellular dissolved oxygen. Preejith et al. (2006) realized a tapered fluorescent fiber optic evanescent wave—based sensor for serum protein.

## 12.3. ABSORPTION FIBER OPTIC CHEMICAL SENSORS

Absorption-based optical sensors can be colorimetric or spectroscopic in nature. Colorimetric sensors, as the name suggests, are based on detection of an analyte-induced color change in the sensor material, while spectroscopic absorption–based sensors rely on detection of the analyte by probing its intrinsic molecular absorption. The sensors that utilize these techniques have been modified to facilitate the interaction of the light guided within the fiber with either the color-changing indicator or the analyte itself, depending on the nature of the sensor. In fact, the modified cladding configuration described above has been used extensively in a number of modes for development of absorption-based sensors. Examples include sensors for ammonia (Moreno et al. 2005), pH (Gupta and Sharma 2002; James and Tatam, 2006), and ethanol (King et al. 2004).

## 12.4. REFRACTOMETRIC FIBER OPTIC CHEMICAL SENSORS

A range of refractometric optical sensors has been developed in the past decade, which involves addition of refractive index–sensitive optical structures to the optical fiber. Such structures include fiber Bragg gratings. Fiber optic chemical sensors based on fiber Bragg gratings detect changes in effective refractive index and require decladding of the fiber section bearing the fiber Bragg gratings in order to impart sensitivity toward the refractive index of the environment in which the fiber is placed. Changes in the intensity of reflected light may accurately represent physical as well as chemical events that occur. It is possible to have an evanescent wave that penetrates only a very short distance (<100 nm) into the medium, allowing study of localized surface phenomena. In other applications, it may be desirable to have the light penetrate as deeply as possible into the medium.

Examples of refractometric-based sensors include sugar and propylene detection in glycol solutions (Sang et al. 2006), detection of analytes such as sodium chloride (Falciai et al. 2001), ethylene glycol, antibodies (DeLisa et al. 2000), dissolved ammonia (Pisco et al. 2006), organic solvents (Liu et al. 2006), and a humidity sensor for breath monitoring (Kang et al. 2006).

## 12.5. ABSORPTION-BASED SENSORS

### 12.5.1. Infrared and Near-Infrared Absorption

For many applications, infrared (IR) and near-infrared (NIR) spectroscopic detection has been a reliable method of detecting chemical species. The growing interest in IR and NIR spectroscopy is probably a direct result of its major advantages over other analytical techniques, namely, easy sample preparation without any pretreatments, the possibility of separating the sample measurement position and the spectrometer by use of fiber optic probes, and the prediction of chemical and physical sample parameters from a single spectrum (Reich 2005), which means that a single spectrum allows several analytes to be determined simultaneously. Moreover, gas absorption spectroscopy enables identification and quantification of chemicals in liquid and gas mixtures with little interference from other gases.

IR and NIR measure the vibrational transitions of molecular bonds. In the mid-IR region of the spectrum, the incident radiation excites the fundamental transitions between the ground state of a vibrational mode and its first excited state. A broad range of molecules, including most organic compounds, absorb in the IR and NIR range, and therefore IR and NIR spectroscopy is a valuable analytical tool in the quantitative and qualitative analysis of organic matter. Typical infrared spectra are shown in Figure 1.29.

NIR absorption bands are typically broad, overlapping, and 10–100 times weaker than their corresponding fundamental mid-IR absorption bands. These characteristics severely restrict sensitivity in the classical spectroscopic sense and call for chemometric data processing to relate spectral information to sample properties. The low absorption coefficient, however, permits high penetration depth and, thus, an adjustment of sample thickness. This aspect is actually an analytical advantage, since it allows direct analysis of strongly absorbing and even highly scattering samples, such as turbid liquids or solids in either transmittance or reflectance mode without further pretreatments.

The most prominent absorption bands occurring in the NIR region are related to overtones and combinations of fundamental vibrations of –CH, –NH, –OH (and –SH) functional groups, characteristic of organic matter (Reich 2005). Therefore, IR and NIR spectroscopy have been widely used for detection of gases, for example, $CO_2$ (Zhang et al. 2000), CO (Zhang and Wu 2004), $NO_2$ (Ashizawa et al. 2003), and $CH_4$ (Chévrier et al. 1995).

In its simplest form, the technique involves confining a sample of target material (gas) in an optical absorption cell and measuring its absorption at specific IR wavelengths, which are characteristic of the vibrational modes of the molecule. The system components usually include an IR source, a monochromator (or optical filters to select specific absorption wavelengths), a sample holder or a sample

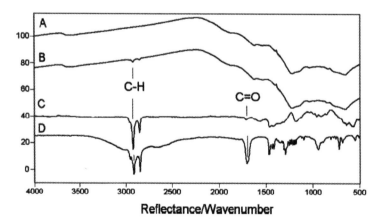

**Figure 1.29.** Infrared spectra of stearic acid in a basalt matrix: A, basalt powder with 50 ppm stearic acid; B, basalt powder with 50 ppm stearic acid and reference basalt spectrum subtracted; C, reference spectrum of stearic acid; D, samples acquired at four-wavenumber resolution with a Biorad FTS 6000 FTS and a Pike diffuse reflectance attachment, with KBr as background and offset and scaled for clarity. (Reprinted with permission from Anderson et al. 2005. Copyright 2005 American Institute of Physics.)

BASIC PRINCIPLES OF OPERATION • 41

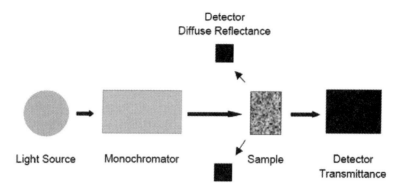

**Figure 1.30.** Basic near-infrared spectrometer configuration. (Reprinted with permission from Reich 2005. Copyright 2005 Elsevier.)

presentation interface, and a detector, allowing for transmittance or reflectance measurements at the wavelength of interest (see Figure 1.30). The light source is usually a tungsten halogen lamp, since it is small and rugged (Reich 2005).

The appropriate NIR measuring mode will be dictated by the optical properties of the sample. Transparent materials are usually measured in transmittance. When, for example, gas in the path absorbs energy from the source, the detector receives less radiation than without the gas present, and the detector can quantify the difference. Turbid liquids or semisolids and solids may be measured in diffuse transmittance, diffuse reflectance, or transflectance, depending on their absorption and scattering characteristics (Reich 2005). Qualitative analysis involves comparison of the sample spectrum with spectral libraries set up previously for known samples of known products. Sophisticated software assists in spectral matching, utilizing preset threshold values which provide stringent identification and discrimination criteria (Rinnan and Rinnan 2007).

As we indicated earlier, NIR spectroscopy requires no sample preparation and no reagents. This has helped the development of quartz fiber optic–based devices that allow the spectra of samples with different characteristics to be recorded simply by selecting the most suitable mode for each (usually, reflectance for solids, transmittance for liquids, and transflectance for emulsions and turbid liquids).

The ability to perform field measurements instead of having to collect samples for subsequent analysis in the laboratory is another advantage of NIR spectroscopy. Some NIR spectrophotometers can make measurements on site. The miniaturization of optical components has boosted development of portable NIR spectrophotometers. Currently available models (www.asdi.com) using such optical devices include hand-held instruments and equipment that can be carried in a backpack or mounted on a vehicle. Moreover, because water shows relatively weak absorption in the NIR region and NIR radiation is relatively low-energy, NIR spectroscopy enables noninvasive analysis. Therefore, NIR spectroscopy has become a widely used analytical method in the agricultural, pharmaceutical, chemical, medical, and petrochemical industries (Siesler et al. 2002; Ciurczak and Drennen, 2002).

However, it is necessary to know that IR sensors can only monitor specific gases that have nonlinear molecules. In addition, they can be affected by humidity and water; they can be expensive; and dust and dirt can coat the optics and impair response, which is a concern in in-situ environments.

## 12.5.2. UV Absorption

Many molecules absorb ultraviolet or visible light of different wavelengths. Therefore, these molecules can be detected using UV spectroscopy. When a molecule absorbs a photon from the UV/visible region, the corresponding energy is captured by one (or several) of its outermost electrons. Ultraviolet-visible spectroscopy (UV = 200–400 nm, visible = 400–800 nm) corresponds to electronic excitations between the energy levels that correspond to the molecular orbitals of the systems. In particular, transitions involving π orbitals and ion pairs ($n$ = nonbonding) are important, and so UV/visible (UV/Vis) spectroscopy is used mostly for identifying conjugated systems which tend to have stronger absorptions (Baldini et al. 2006). Absorbance is directly proportional to the path length and the concentration of the absorbing species. This means that an absorption spectrum can show a number of absorption bands corresponding to structural groups within the molecule (Rouessac and Rouessac 2007).

Modern absorption instruments can usually display the data as transmittance, percent transmittance, or absorbance. The transmittance ($T$) is a measure of the attenuation of a beam of monochromatic light based on comparison between the intensities of the transmitted light ($I$) and the incident light ($I_0$) according to whether the sample is placed, or not, in the optical pathway between the source and the detector. $T$ is expressed as a fraction or a percentage:

$$T = \frac{I}{I_0} T \qquad (1.28)$$

Absorbance (formerly called optical density) is defined by

$$A = -\log T \qquad (1.29)$$

For compounds of simple atomic composition, provided the spectrometer possesses high enough resolution, the fundamental transitions appear as if isolated. In extreme situations the positions of the absorptions are recorded in cm$^{-1}$ (wavenumber), a unit better adapted for precise pointing than nm.

An unknown concentration of an analyte can be determined by measuring the amount of light that a sample absorbs and applying Beer's law (Baldini et al. 2006), which is presented here in its current form:

$$A = \varepsilon_\lambda l C \qquad (1.30)$$

where $A$ is the absorbance, an optical parameter without dimension that is accessible with a spectrophotometer, $l$ is the thickness (in cm) of the solution through which the incident light is passed, $C$ is the molar concentration, and $\varepsilon_\lambda$ is the molar absorption coefficient (L mol$^{-1}$ cm$^{-1}$) at wavelength λ, at which the measurement is made. If the absorptivity coefficient is not known, the unknown concentration can be determined using a working curve of absorbance versus concentration derived from standards. The light source is usually a hydrogen or deuterium lamp for UV measurements and a tungsten lamp for visible measurements. The wavelengths of these continuous light sources are selected with a wavelength separator such as a prism or grating monochromator.

At present, direct UV absorption sensing has been used mainly in environmental applications to monitor pollutants such as heavy metals, hydrocarbons, and volatile organic compounds (VOCs) in air and water. Examples include a sensor to detect Cr in water (Tau and Sarma 2005), ozone and $NO_2$ in the atmosphere (Wu et al. 2006), and VOCs and aromatic hydrocarbons in ambient air in the vicinity of a refinery (Lin et al. 2004).

## 12.6. SURFACE PLASMON RESONANCE SENSORS

Over the past two decades, the surface plasmon resonance (SPR) technique has been developed for detection of chemical and biological species. In principle, SPR sensors are thin-film refractometers that measure changes in the refractive index occurring at the surface of a metal film supporting a surface plasmon. Surface plasmon resonance is a charge-density oscillation that may exist at the interface of two media with dielectric constants of opposite signs, for instance, a metal and a dielectric. The resonance condition depends on the wavelength, the angle at which the light strikes the substrate–metal interface, and the dielectric constants of all the materials involved (Karlsen et al. 1995; Bardin et al. 2002). It was established that at optical wavelengths, this condition is fulfilled by several metals, of which gold and silver are the most commonly used. The charge-density wave is associated with an electromagnetic wave, the field vectors of which reach their maxima at the interface and decay evanescently into both media. This surface plasma wave (SPW) is a TM-polarized wave (the magnetic vector is perpendicular to the direction of propagation of the SPW and parallel to the plane of the interface). The propagation constant of the surface plasma wave propagating at the interface between a semi-infinite dielectric and a metal is given by the following expression (Homola et al. 1999):

$$\beta = k\sqrt{\frac{\varepsilon_m n_s^2}{\varepsilon_m + n_s^2}} \tag{1.31}$$

where $k$ denotes the free-space wave number, $\varepsilon_m$ is the dielectric constant of the metal, and $n_s$ is the refractive index of the dielectric. A surface plasmon excited by a light wave propagates along the metal film, and its evanescent field probes the medium (sample) in contact with the metal film. A change in the refractive index of the dielectric gives rise to a change in the propagation constant of the surface plasmon, which, through the coupling conditions, alters the characteristics of the light wave coupled to the surface plasmon (e.g., coupling angle, coupling wavelength, intensity, phase).

So, the operational principle of SPR is based on an interface between two transparent media with different refractive indices. The light coming from the side of higher refractive index is partly reflected and partly refracted. Above a certain critical angle of incidence, no light is refracted across the interface, and total internal reflection is observed. While incident light is totally reflected, the electromagnetic field component penetrates a short distance (tens of nanometers) into the medium of a lower refractive index, creating an exponentially evanescent wave (Homola et al. 1999). If the interface between the media is coated with a thin layer of metal (e.g., gold), and the light is monochromatic and $p$-polarized (the electric field vector component is parallel to the plane of incidence), then the energy carried by photons of the light can be coupled to the electrons of the metal film. Energy transfer takes place when there is a match (resonance) between the incident light energy photons and the electrons at the metal film surface. The coupling occurs at a particular wavelength which is dependent on the metal and the environment of the illuminated metal surface. The coupling phenomenon results in creation of plasmons, which are a group of highly excited electrons that behave like a single electrical entity. All the light except that at the resonant wavelength is absorbed. This beam loses most of its energy to the surface plasmon wave, and this is called surface plasmon resonance. This wavelength modulation depends on the magnitude of chemical changes. The material adsorbed onto the thin metal film influences the resonance conditions (Pockrand et al. 1979; Jorgenson and Yee 1993). Generally, a linear relationship is found between resonance energy and mass concentration of (bio)chemically relevant molecules such as proteins, sugars, and DNA. The SPR signal, which is expressed in resonance units, is therefore a measure of mass concentration at the sensor chip surface. This means that the analyte and ligand association and dissociation can be observed, and ultimately, rate constants as well as equilibrium constants can be extracted. These SPR biosensors have a wide variety of applications; the most popular commercial instruments for SPR biosensing are Biocore instruments.

Generally, an SPR optical sensor comprises an optical system, a transducing medium which interrelates the optical and (bio)chemical domains, and an electronic system that supports the optoelectronic components of the sensor and allows data processing. The transducing medium transforms changes in the quantity of interest into changes in the refractive index, which may be determined by optically interrogating the SPR. The optical part of the SPR sensor contains a source of optical radiation and an optical structure in which SPW is excited and interrogated. In the process of interrogating the SPR, an electronic signal is generated and processed by the electronic system. Major properties of an SPR sensor are determined by properties of the sensor's subsystems. The sensor sensitivity, stability, and resolution depend on properties of both the optical system and the transducing medium. The selectivity and the response time of the sensor are determined primarily by the properties of the transducing medium (Homola et al. 1999).

This enhancement and subsequent coupling between light and a surface plasmon are performed in a coupling device (coupler). The most common couplers used in SPR sensors include a prism coupler, a waveguide coupler, and a grating coupler (Figure 1.31). Prism couplers represent the most frequently used means of optical excitation of surface plasmons.

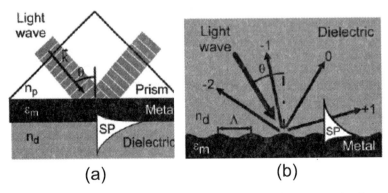

**Figure 1.31.** Most widely used configurations for SPR sensors: (a) prism coupler–based SPR system (ATR method); (b) grating coupler–based SPR system. (Reprinted with permission from Homola 2008. Copyright 2008 American Chemical Society.)

Considerable literature is available on the principle of SPR sensing and various configurations (Homola et al. 1999). Applications such as hydrogen sensing (Benson et al. 1999), methylene sensing (Ideta and Arakawa 1993), sensing of hydrocarbons such as ethane and methane (Urashi and Arakawa 2001), and nitrogen dioxide sensing (Wright et al. 1995), to name a few, have been reported.

## 13. PHOTOACOUSTIC SENSORS

The generation of acoustic waves by modulated optical radiation is called the *photoacoustic effect*. This phenomenon was discovered in 1880 by A. G. Bell. Bell found that sound could be generated by exposing optically absorbing materials to chopped sunlight. Today's photoacoustic gas sensors utilize absorption of light by a certain species. The absorbed light leads to an increase in the temperature of the sample and the surrounding gas and results in thermal expansion of the latter. If the light source is modulated with a certain frequency, then the thermal expansion will also be modulated and form a sound wave proportional to the absorbed energy. To improve the detection limit, in most cases the acoustical signal is further amplified by an acoustic resonator (Keller et al. 2005).

There are excellent review articles describing the principle and application of photoacoustic sensors (Miklós et al. 2001; Elia et al. 2006) for the measurement of gases. Photoacoustic sensors have been used for sensing a wide range of species, including $H_2O$, $NH_3$ (Paldus 1999), $NH_3$, $CO_2$, $CH_3OH$ (Hofstetter et al. 2001), $O_3$ (Da Silva et al. 2004), NO (Elia et al. 2005), HMDS (Elia et al. 2006), and smoke (Keller et al. 2005).

## 14. THERMOELECTRIC SENSORS

The thermoelectric heat power transducer for calorimetric chemical sensors is gaining more attention as an alternative to conductometric gas sensors. The *thermoelectric effect*, also known as the *Seebeck effect*, is the conversion of temperature differences directly into electricity. For small temperature differences, the relation can be approximated as follows (Göpel et al. 1994):

$$U_S = (\alpha_A - \alpha_B)(T_1 - T_0) = \alpha_{A/B}(T_1 - T_0) \tag{1.32}$$

where $\alpha_A$ and $\alpha_B$ denote the Seebeck coefficients (thermopower or thermoelectric power) of materials A and B. They express specific transport properties determined by the band structure and the carrier transport mechanisms of the materials. $U_s$ is always created in an electrically conducting material when a temperature gradient is maintained along the sample, but it cannot be observed with two legs of the same material, for reasons of symmetry.

The use of the Seebeck effect to detect thermal responses from calorimetric gas sensors was first reported by McAleer and co-workers (1985). In their work, the thermoelectric sensors consisted of monolithic structures in which a catalyst layer was coated on one side of a tin dioxide pellet. Seebeck voltage was measured when hydrogen gas reacted with oxygen and generated heat. To achieve selectivity and reversibility, these devices required temperatures in excess of 240°C. More recently, a number of researchers have reported planar devices, in both thick tin oxide (Širok 1993; Ionescu

1996) and thin indium oxide film forms (Papadopoulos et al. 1996; Vlachos et al. 1997). Although these devices are undoubtedly better than monolithic devices in terems of sensor size, sensitivity, and in some cases selectivity, they all require temperatures in excess of 240°C for the reversible detection of trace combustibles.

Thermoelectric metal oxide sensors fabricated using thick-film technology, as well as Si or SiGe thin-film thermopile sensors prepared by conventional CMOS processes, provide, for example, for hydrogen, detection limits in the low parts-per-million range. Because of the small thermal inertia of silicon thermopile sensors and the inherent reference temperature compensation effects, they can be operated in a high-speed temperature scanning regime up to 600°C and with cycle times of a few milliseconds (Schreiter et al. 2006). With respect to differences in the activation energies of the involved gas components, the shape of the signals is a characteristic of the investigated mixture; i.e., the application of pattern-recognition methods should lead to enhanced selectivity of the sensor.

A thermoelectric sensor that operated at room temperature was also reported by Markinkowska et al. (1991). Their devices were coated with hydrophobic oxidation catalyst granules, with the other material surrounded by a bed of noncatalytic hydrophobic catalyst support material.

*Pyroelectricity* is related to *piezoelectricity*. Sometimes, both effects can be encountered in the same material. One could imagine that a volume increase of a piezo crystal at a higher temperature should result in effects similar to those caused by mechanical compression or expansion. Spontaneous polarization will be a strong function of temperature, since the atomic dipole moments vary as the crystal expands or contracts. Heating the crystal tends to desorb the surface neutralizing ions as well as change the polarization, so that a surface charge may then be detected. Thus, the crystal appears to have been charged by heating. This is called the *pyroelectric effect*.

The electric field developed across a pyroelectric crystal can be remarkably large when it is subjected to a small change in temperature. We define a pyroelectric coefficient, $p$, as the change in flux density ($D$) in the crystal due to a change in temperature ($T$) (Anderson et al. 1990):

$$p = \frac{\delta D}{\delta T} \tag{1.33}$$

Using a capacitor, the pyroelectric voltage signal $\Delta U$ is

$$\Delta U = pd \frac{\Delta T}{\varepsilon_r \varepsilon_0} \tag{1.34}$$

where $\varepsilon_r \varepsilon_0$ is the permittivity and $d$ is the thickness of the pyroelectric film.

The performance characteristics of the sensors depend on the sensing material, the preparation and geometry of the electrodes, the use of absorbing coatings, the thermal design of the structure, and the nature of the electronic interface as well. Since the output is dependent on the rate of change, the radiant flux incident on the sensor must be chopped, pulsed, or otherwise modulated. Important parameters of the materials are the heat capacity per volume, the heat conductivity, and the penetration depth of the temperature wave in the pyroelectric material at a given modulation frequency (Harsanyi, 2000).

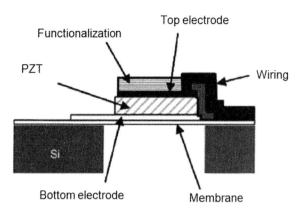

**Figure 1.32.** Schematic of pyroelectric sensor layout. (Reprinted with permission from Schreiter et al. 2006. Copyright 2006 Elsevier.)

One example of a pyroelectric sensor is shown in Figure 1.32. The array is formed by a thin-film capacitor deposited on a supporting $SiO_2/Si_3N_4$ membrane. The dielectric is made of a thin film of ferroelectric lead zirconate titanate (PZT). To provide sufficiently high thermal isolation, the underlying silicon is removed by bulk micromachining. To enable elevated working temperatures, a local low-power thin-film heater arranged on bulk-micromachined membranes and integrable with the pyroelectric sensor structure was developed and tested. The functionalization of active elements of the pyroelectric detector array was accomplished by coating them with a chemically sensitive layer of polydimethylsiloxan (PDMS) for the detection of VOCs or with Pt-cluster arrays grown on bacterial surface layers (S-layers) serving as catalyst for hydrogen oxidation. Another commonly used pyroelectric material is lithium tantalate ($LiTaO_3$). For use in sensors, the material is doped with lanthanum traces. Polyvinylidene fluoride (PVDF) is a synthetic organic pyroelectric polymer.

## 15. THERMAL CONDUCTIVITY SENSORS

Thermal conduction is one form of thermal energy transfer from one object to another. Physical contact between two bodies is required for heat conduction. Kinetic energy is transferred to a cooler body from a warmer body by thermally agitating its particles. As a result, the cooler body gains heat while the warmer body loses heat. For instance, heat passage through a rod is governed by a law similar to Ohm's law; the heat flow rate is proportional to the thermal gradient across the material $(dT/dx)$ and its cross-sectional area $(A)$, or

$$H = \frac{dQ}{dT} = -kA\frac{dT}{dx} \qquad (1.35)$$

where $k$ is called the thermal conductivity. The minus sign indicates that heat flows in the direction of temperature decrease. A good thermal conductor has a high value of thermal conductivity, whereas thermal insulators have low values of $k$.

The general principle of thermal conductivity sensors is as follows. A known temperature difference is maintained between a "cold" element and a "hot" element (Devoret et al. 1980). Heat is transferred from the "hot" element to the "cold" element via thermal conduction through the investigated gas. The power needed to heat the "hot" element is therefore a direct measure of the thermal conductivity. Heat loss due to radiation, convection, and heat conduction through the suspensions of the "hot" element need to be minimized by sensor design. The precise mechanism is quite complex, because the thermal conductivity of gases varies with temperature and convection as well as conductivity. Most gases produce a linear output signal, but not all.

In the simplest case, power loss of a single filament thermistor by heat conduction via the ambient gas can be expressed as

$$P = k_{TC} \lambda \cdot \Delta T \quad (1.36)$$

where $P$ is the power dissipation of the heater by thermal conduction of the gas, $\lambda$ is the thermal conductivity of the given gas–gas mixture, $k_{TC}$ is a constant that is characteristic for a given geometry, and $\Delta T$ is the temperature difference between the heater and the ambient gas.

In the sensor, a heated thermistor or platinum filament is mounted so that it is exposed to the sample. Another element, which acts as a reference, is enclosed in a sealed compartment. If the sample gas or vapor has a thermal conductivity higher than the reference, heat is lost from the exposed element and its temperature decreases, whereas if its thermal conductivity is lower than that of the reference, the temperature of the exposed element increases. These temperature changes alter the electrical resistance, which is measured using a bridge circuit.

Figure 1.33 shows a schematic diagram of a thermistor-based sensor which can be used to measure humidity, using the thermal conductivity of gas (Miura 1985). Because of the amount of power required to heat these sensors, they have to be mounted in a flameproof enclosure, in the same way as pellistors are. Two tiny thermistors ($T_1$ and $T_2$) are supported by thin wires to minimize thermal conductivity loss to the housing. Small venting holes are used to expose thermistor $T_1$ to the outside gas, and thermistor $T_2$ is sealed in dry air. Both thermistors are connected into a bridge circuit ($R_1$ and $R_2$), which is powered by voltage $+E$. Due to the passage of electric current, the thermistors develop self-heating. Their temperatures rise up to 170°C over the ambient temperature. To establish a zero reference point initially, the bridge is balanced in dry air. As absolute humidity rises from zero, the output of the sensor gradually increases.

It is necessary to know that the presence of gases that have thermal conductivities relative to air of >1 leads to cooling of the exposed thermistor or filament. These gases are often measured by the TC technique—the higher their thermal conductivity, the lower the concentration which can be measured. Gases with thermal conductivities of <1 are more difficult to measure, partly because water vapor may cause interference. Gases with thermal conductivities close to 1 cannot be measured by this technique. These include ammonia, carbon monoxide, nitric oxide, oxygen, and nitrogen (Barsony et al. 2009).

Thermometric and calorimetric sensors are discussed here only for their use for gases. In the liquid phase, the conditions are much less advantageous, since the thermal conductivities of liquids are orders of magnitude higher than those of gases. Calorimetric measurements must be performed in closed

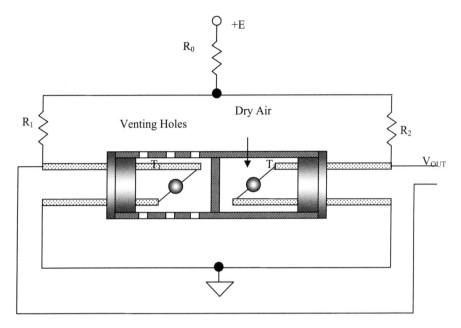

**Figure 1.33.** Schematic diagram of a thermistor-based humidity sensor with self-heating thermistors.

vessels with thermal isolation, or at least in a flowing stream. In some applications, thermometric sensors are used as detectors for flow injection analysis.

## 16. FLAME IONIZATION SENSORS

The presence of ions in flames has long been known, and a large volume of literature exists on the subject of flame ionization. The literature includes descriptions of the mechanisms of ion formation and the electrical properties of the flame. There are thorough review articles (Calcote 1957; Fialkov 1997) on this subject, and it is widely accepted that the key mechanism enabling the flow of electrical current through hydrocarbon flames results from the chemi-ionization of the formyl radical, CHO*

$$CH + O \rightarrow CHO^* \rightarrow CHO^+ + e^- \qquad (1.37)$$

Flame ionization–based sensors are used in gas chromatography (Morgan 1961). The first flame ionization detector was developed more than five decades ago by Commonwealth Scientific and Industrial Research Organization (CSIRO) scientists in Melbourne, Australia. A typical flame ionization sensing technique consists of at least two electrodes arranged so that a voltage potential can be applied across the flame, or a part of the flame. The electrical properties of the flame facilitate a measurable current that is related to a parameter of interest. For example, a typical flame ionization detector (FID) used in a hydrocarbon analyzer has electrodes arranged so that a voltage potential is applied across the entire flame, and the amount of electrical current flow through the flame is linearly proportional to the hydrocarbon concentration (Cheng et al. 1998).

Most commonly, flame ionization–based sensors are attached to a gas chromatography system. Figure 1.34 shows a general schematic diagram of a of a flame ionization sensor. The gas enters the detector's oven through the gas chromotography column, where it is mixed with the hydrogen fuel and then with the oxidant as it moves up. The effluent/fuel/oxidant mixture continues to travel up to the nozzle head, where a positive bias voltage exists. This positive bias helps to repel the reduced carbon ions created by the flame, thus pyrolyzing the eluent. The ions are then repelled up toward the collector plates, which are connected to a very sensitive ammeter, which detects the ions hitting the plates and then feeds that signal to an amplifier, integrator, and display system. The products of the flame are finally vented out of the detector through the exhaust port.

## 17. LANGMUIR-BLODGETT FILM SENSORS

Langmuir-Blodgett sensors are named for Irving Langmuir and Katherine Blodgett, who discovered unique properties of thin films in the early 1900s. Langmuir's original work involved the transfer of monolayers from liquid to solid substrates. Blodgett expanded on Langmuir's research to include the deposition of multilayer films on solid substrates (Blodgett 1935). The structure of the film can be controlled at the molecular level by transferring monolayers of organic material from a liquid to a solid substrate. These *Langmuir-Blodgett films* exhibit various electrochemical and photochemical properties.

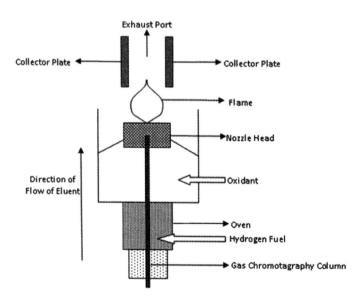

**Figure 1.34.** Schematic diagram of a flame ionization sensor.

Langmuir-Blodgett (LB) films basically consist of molecules which are amphiphilic in nature, i.e., they have a hydrophilic (polar) head group that can be bonded to various solid surfaces and a long hydrophobic (nonpolar) tail that extends outward (Grandke and Ko 1996). Amphiphilic substances are generally saturated fatty acids such as palmitic acid, magaric acid, stearic acid, arachidic acid, etc. The head group is a carboxylic acid and the tail is a straight-chain hydrocarbon. When these amphiphilic molecules come in contact with water, they accumulate at the air/water interface, causing a decrease in the surface tension of the water and exhibiting a tendency to orient with the head group into the water and the tail out of the water (see Figure 1.35b). The hydrophilic forces of the head group and the hydrophobic forces of the tail group allow the material to form oriented monolayers. The material dissolves in water if the head group is too strongly attracted to the water, and there is no layer formation if the tail group is too hydrophobic. LB films can be deposited on a substrate by dipping it up and down through the monolayer while simultaneously keeping the surface pressure constant. Multilayers can be formed by successive deposition of these monolayers on the same substrate. Figure 1.35c shows a schematic representation of the deposition of the floating monolayer on a solid substrate.

Comprehensive information about the properties and methods of preparation of LB films can be found in the literature; for example, see Roberts (1990). LB films have been widely studied, because the end of the tail group can be functionalized to form precisely arranged molecular arrays for various applications ranging from simple ultrathin insulators and lubricants to complex chemical and biological sensors. LB films have been deposited on conductometric, capacitive, optoelectric, and FET-based

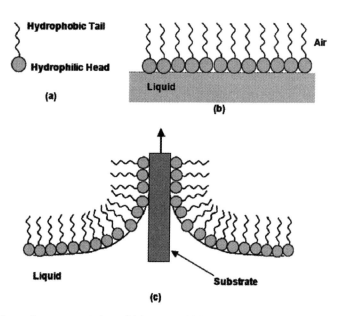

**Figure 1.35.** Schematic representation of (a) an amphiphilic component, (b) the orientation of amphiphilic molecules, with their head groups on the water surface and their tails sticking up, and (c) deposition of the floating monolayer on a solid substrate.

devices for sensing applications. Review papers describe the fundamental scheme of LB films for chemical sensors, fields of major interest in chemical sensors, and future trends in LB sensor studies (Roberts 1983; Bubeck 1988; Moriizumi 1988; Barraud 1990; Izumi et al. 1998).

## REFERENCES

Ahn S., Kulis D.M., Erdner D.L., Anderson D.M., and Walt D.R. (2006). Fiber-optic microarray for simultaneous detection of multiple harmful algal bloom species. *Appl. Environ. Microbiol.* **72**, 5742–5749.

Ahmad M., Chang K.P., King T.A., and Hench L.L. (2005) A compact fibre-based fluorescence sensor. *Sens. Actuators A* **119**, 84–89.

Ali M.B., Kalfat R., Sfihi H., Chovelon J.M., Ouada H.B., and Jaffrezic-Renault N. (2000) Sensitive cyclodextrin-polysiloxane gel membrane on EIS structure and ISFET for heavy metal ion detection. *Sens. Actuators B* **62**, 233–237.

Anderson J.C., Leaver K.D., Rawlings R.D., and Alexander J.M. (1990) *Materials Science.* Chapman & Hall, London.

Anderson M.S., Andringa J.M., Carlson R.W., Conrad P., Hartford W., Shafer M., Soto A., Tsapin A.I., Dybwad J.P., Wadsworth W., and Hand K. (2005) Fourier transform infrared spectroscopy for Mars science. *Rev. Sci. Instrum.* **76**, 034101.

Andle J.C. and Vetelino J.F. (1995) Acoustic wave biosensors. *Sens. Actuators A* **44**, 167–176.

Andle J.C., Weaver J.T., Vetelino J.F., and McAllister D.J. (1995) Selective acoustic plate mode DNA sensor. *Sens. Actuators B* **24–25**, 129–133.

Anh D.T.V., Olthuis W., and Bergveld P. (2004) Work function characterization of electroactive materials using an EMOSFET. *IEEE Sensors J.* **4**, 284–287.

Anzai J., Guo B., and Osa T. (1996) Quartz-crystal microbalance and cyclic voltammetric studies of the adsorption behaviour of serum albumin on self-assembled thiol monolayers possessing different hydrophobicity and polarity. *Bioelectrochem. Bioenerg.* **40**, 35–40.

Ashizawa H., Yamaguchi S., Endo M., Ohara S., Takahashi M, Nanri K., and Fujioka T. (2003) development of a nitrogen dioxide gas sensor based on mid-infrared absorption spectroscopy. *Rev. Laser Eng.* **31**, 151–155.

Atashbar M.Z., Bejcek B., Vijh A., and Singamaneni S. (2005) QCM biosensor with ultra thin polymer film. *Sens. Actuators B* **107**, 945–951.

Atashbar M.Z., Sun H.T., Gong B., Wlodarski W., and Lamb R. (1998) XPS study of Nb-doped oxygen sensing $TiO_2$ thin films prepared by sol-gel method. *Thin Solid Films* **326**, 238–244.

Atashbar M.Z. and Wlodarski W. (1997) Design, simulation and fabrication of doped $TiO_2$-coated surface acoustic wave oxygen sensor. *J. Intell. Mater. Syst. Struct.* **8**, 953–959.

Atashbar M.Z., Sadek A.Z., Wlodarski W., Sriram S., Bhaskaran M., Cheng C.J., Kaner R.B., and Kalantar-zadeh K. (2009) Layered SAW gas sensor based on CSA synthesized polyaniline nanofiber on AlN on 64 degrees YX $LiNbO_3$ for $H_2$ sensing. *Sens. Actuators B* **138**, 85–89.

Apostolakis J.C., Georgiou C.A., and Koupparis M.A. (1991) Use of ion-selective electrodes in kinetic flow injection: determination of phenolic and hydrazino drugs with 1-Fluoro-2,4-dinitrobenzene using a fluoride-selective electrode. *Analyst* **116**, 233–237.

Baldini F., Chester A.N., Homola J., and Martellucci S. (eds.) (2006) *Optical Chemical Sensors*. Springer-Verlag, Dordrecht, The Netherlands.

Balkus K.J., Ball L.J., Gnade B.E., and Anthony J.M. (1997) A capacitive type chemical sensor based on $AlPO_4$-5 molecular sieves. *Chem. Mater.* **9**, 380–386.

Ballantine D.S., White R.M., Martin S.J., Ricco A.J., Frye G.C., Zellers E.T., and Wohltjen H. (1997) *Acoustic Wave Sensors: Theory, Design, and Physico-Chemical Applications*. Academic Press, San Diego, CA.

Ballantine D.S. Jr. and Wohltjen H. (1989) Surface acoustic wave devices for chemical analysis. *Anal. Chem.* **61**, 704–712.

Bardin F., Kašik I., Trouillet A., Matějec A., Gagnaire H., and Chomát M. (2002) Surface plasmon resonance sensor using an optical fiber with an inverted graded-index profile. *Appl. Opt.* **41**, 2514–2520.

Barnes C., D'Silva C., Jones J.P., and Lewis T.J. (1992) Lectin coated piezoelectric crystal biosensors. *Sens. Actuators B* **7**, 347–350.

Barraud A. (1990) Chemical sensors based on LB films. *Vacuum* **41**, 1624–1628.

Barsan N., Schweizer-Berberich M., and Göpel W. (1999) Fundamental and practical aspects in the design of nanoscaled $SnO_2$ gas sensors: a status report. *Fresenius J. Anal. Chem.* **365**, 287–304.

Barsan N. and Weimar U. (2003) Understanding the fundamental principles of metal oxide based gas sensors; the example of CO sensing with $SnO_2$ sensors in the presence of humidity. *J. Phys.: Condens. Matter* **15**, R813–R829.

Bausells J., Carrabina J., Errachid A., and Merlos A. (1999) Ion-sensitive field-effect transistors fabricated in a commercial CMOS technology. *Sens. Actuators B* **57**, 56–62.

Benammar M. (2004) Design and assembly of miniature zirconia oxygen sensors. *IEEE Sensors J.* **4**(1), 3–8.

Battiston F.M., Ramseyer J.-P., Lang H.P., Baller M.K., Gerber C., Gimzewski J.K., Meyer E., and Güntherodt H.-J. (2001) A chemical sensor based on a microfabricated cantilever array with simultaneous resonance-frequency and bending readout. *Sens. Actuators B* **77**, 122–131.

Barsony I., Ducso C., and Furjes P. (2009) Thermometric gas sensing. In: Comini E., Faglia G., and Sberveglieri G. (eds.), *Solid State Sensing*. Science-Business Media, chap. 7.

Bell A.G. (1880) On the production and reproduction of sound by light: the photophone. *Am. J. Sci.* **3**(20), 305–324.

Bender F., Cernosek R.W., and Josse F. (2000) Love wave biosensors using cross-linked polymer waveguide on $LiTaO_3$ substrates. *Electron. Lett.* **36**(19), 1–2.

Benes E., Gröschl M., Burger W., and Schmid M. (1995) Sensors based on piezoelectric resonators. *Sens. Actuators A* **48**, 1–21.

Benson D.K., Tracy C.E., Hishmeh G.A., Ciszek P.A., Lee S.-H., Pitts R., and Haberman D.P. (1999) Low cost fiber-optic chemochromic hydrogen gas detector. In: *Proceedings of the 1999 US DOE Hydrogen Program Review*, NREL/CP-570-26938.

Bergstrom P.L., Patel S.V., Schwank J.W., and Wise K.D. (1997) A micromachined surface work-function gas sensor for low-pressure oxygen detection. *Sens. Actuators B* **42**, 195–204.

Bergveld P. (1970) Development of an ion-sensitive solid-state device for neurophysiological measurements. *IEEE Trans. Biomed. Eng.* **17**, 70–71.

Bergveld P. (1991) Future applications of ISFETs. *Sens. Actuators B* **4**, 125–133.

Bergveld P. (2003) Thirty years of ISFETOLOGY What happened in the past 30 years and what may happen in the next 30 years. *Sens. Actuators B* **88**, 1–20.

Blodgett K.B. (1935) Films built by depositing successive monomolecular layers on a solid surface. *J. Am. Chem. Soc.*, **57**, 1007–1022.

Bott B. and Jones T.A. (1984) A highly sensitive $NO_2$ sensor based on electrical conductivity changes in phthalocyanine films. *Sens. Actuators* **5**, 43–52.

Bowden M., Song L.N., and Walt D.R. (2005) Development of a microfluidic platform with an optical imaging microarray capable of attomolar target DNA detection. *Anal. Chem.* **77**, 5583–5588.

Brzozka Z., Dawgul M., Pijanowska D., and Trobicz W. (1997) Durable $NH_4^+$ sensitive chemFET. *Sens. Actuators B* **44**, 527–531.

Bubeck C. (1988) Reactions in monolayers and Langmuir-Blodgett films. *Thin Solid Films* **160**(1–2), 1–14.

Calabrese G., Wohltjen H., and Roy M.K. (1987) Surface acoustic wave devices as chemical sensors in liquids. Evidence disputing the importance of Rayleigh wave attenuation. *Anal. Chem.* **59**, 833–837.

Calcote, H.F. (1957) Mechanism for the formation of ions in flames. *Combust. Flame* **1**, 385–403.

Caliendo C., Verardi P., Verona E., D'Amico A., Di Natale C., Saggio G., Serafini M., Paolesse R., and Huq S.E. (1997) Advances in SAW-based gas sensors. *Smart Mater. Struct.* **6**, 689–699.

Caruso F., Rodda E., and Furlong D.N., (1996) Orientational aspects of antibody immobilization and immunological activity on quartz crystal microbalance electrodes. *J. Colloid Interface Sci.* **178**, 104–115.

Cimalla V., Niebelschütz F., Tonisch K., Foerster Ch., Brueckner K., Cimalla I., Friedrich T., Pezoldt J., Stephan R., Hein M, and Ambacher O. (2007) Nanoelectromechanical devices for sensing applications. *Sens. Actuators B* **126**, 24–34.

Cheeke J.D.N. and Wang Z. (1999) Acoustic wave gas sensors. *Sens. Actuators B* **59**, 146–153.

Chen H.I. and Chou Y.I. (2003) A comparative study of hydrogen sensing performances between electroless plated and thermal evaporated Pd/InP Schottky diodes. *Semicond. Sci. Technol.* **18**, 104–110.

Chen G.Y., Thundat T., Wachter E.A., and Warmack R.J. (1995). Adsorption-induced surface stress and its effects on resonance frequency of microcantilevers. *J. Appl. Phys.* **77**(8), 3618–3622.

Cheng W.K., Summers T., and Collings N. (1998) The fast-response flame ionization detector. *Prog. Energy Combust. Sci.* **24**, 89–124.

Chévrier J.-B., Baert K., Slater T., and Verbist A. (1995) Micromachined infrared pneumatic detector for gas sensor. *Microsyst. Technol.* **1**, 71–74.

Cho J.-C. and Chiang J.-L. (2000) Ion sensitive field effect transistor with amorphous tungsten trioxide gate for pH sensing. *Sens. Actuators B* **2**, 81–87.

Choujaa A., Tirole N., Bonjour C., Martin G., Hauden D., Blind P., Cachard A., and Pommier C. (1995) AlN/Silicon Lamb-wave microsensors for pressure and gravimetric measurements. *Sens. Actuators A* **46**, 179–182.

Ciurczak E.W. and Drennen J.K. (2002) *Pharmaceutical and Medical Applications of Near-Infrared Applications (Practical Spectroscopy)*. Marcel Dekker, New York.

Clark L.C., Wolf R., Granger D., and Taylar Z. (1953) Continuous recording of blood oxygen tensions by polarography. *J. Appl. Physiol.* **6**, 189–193.

D'Amico A. and Natale C.D. (2001) A contribution on some basic definitions of sensors properties. *IEEE Sensors J.* **1**, 183–190.

D'Amico A. and Verona E. (1989) SAW sensors. *Sens. Actuators* **17**, 55–66.

Dahint R., Shana Z.A., Josse F., Riedel S.A., and Grunze M. (1993) Identification of metal ion solutions using acoustic plate mode devices and pattern recognition. *IEEE Trans. Ultrason. Ferroel. Freq. Contr.* **40**(2), 114–120.

Davide F., Di Natale C., and D'Amico A. (1993) Sensor arrays figure of merits: definitions and properties. *Sens. Actuators B* **13**, 327–332.

DeLisa M.P., Zhang Z., Shiloach M., Pilevar S., Davis C.C., Sirkis J.S., and Bentley W.E. (2000) Evanescent wave long period fiber Bragg grating as an immobilized antibody biosensor. *Anal. Chem.* **72**, 2895–2900.

Déjous C., Savart M., Rebière D., and Pistré J. (1995) A shear-horizontal acoustic plate mode (SH-APM) sensor for biological media. *Sens. Actuators B* **26–27**, 452–456.

Devoret M., Sullivan N.S., Esteve D., and Deschamps P. (1980) Simple thermal conductivity cell using a miniature thin film printed circuit for analysis of binary gas mixtures. *Rev. Sci. Instrum.* **51**(9), 1220–1224.

Doll T., Lechner J., Eisele I., Schierbaum K.-D., and Göpel W. (1996) Ozone detection in the ppb range with work function sensors operating at room temperature. *Sens. Actuators B* **34**, 506–510.

Domansky K., Baldwin D.L., Grate J.W., Hall T.B., Li J., Josowicz M., and Janata J. (1998) Development and calibration of field-effect transistor-based sensor array for measurement of hydrogen and ammonia gas mixtures in humid air. *Anal. Chem.* **70**, 473–481.

Domansky K., Liu J., Wang L.-Q., Engelhard M.H., and Baskaran S. (2001) Chemical sensors based on dielectric response of functionalized mesoporous silica films. *J. Mater. Res.* **16**, 2810–2816.

Eddowes M.J. (1987) Response of an enzyme-modified pH sensitive ion-selective device; analytical solution for the response in the presence of pH buffer. *Sens. Actuators* **11**, 265–274.

Elia A., Lugarà P.M., and Giancaspro C. (2005) Photoacoustic detection of nitric oxide by use of a quantum cascade laser. *Opt. Lett.* **30**(9), 988–990.

Elia A., Rizzi F., Di Franco C., Lugarà P.M., and Scamarcio G. (2006) Quantum cascade laser-based photoacoustic spectroscopy of volatile chemicals: application to examethyldisilazane. *Spectrochim. Acta A* **64**, 426–429.

Ekedahl L.-G., Eriksson M., and Lundström I. (1998) Hydrogen sensing mechanisms of metalinsulator interfaces. *Accounts Chem. Res.* **31**, 249–256.

Esteban I., Déjous C., Rebière D., Pistré J., Planade R., and Lipskier J.F. (2000) Mass sensitivity of SH-APM sensors: potentialities for organophosphorous vapors detection. *Sens. Actuators B* **68**, 244–248.

Falciai R., Mignani A.G., and Vannini A. (2001) Long period gratings as solution concentration sensors. *Sens. Actuators B* **74**, 74–77.

Fialkov A.B. (1997) Investigations on ions in flames. *Prog. Energy Combust. Sci.* **23**, 399–528.

Fontecha J., Fernández M.J., Sayago I., Santos J.P., Gutiérrez J., Horrillo M.C., Gràcia I., Cané C., and Figueras E. (2004) Fine-tuning of the resonant frequency using a hybrid coupler and fixed components in SAW oscillators for gas detection. *Sens. Actuators B* **103**, 139–144.

Freeman T.M. and Seitz W.R. (1978) Chemiluminescence fiber optical probe for hydrogen peroxide based on the luminol reaction. *Anal. Chem.* **50**, 1242–1246.

Fradeen J. (2003) *Handbook of Modern Sensors*, 3rd ed. Springer-Verlag, New York.

Frant M.S. (1994) History of the early commercialization of ion-selective electrodes. *Analyst* **199**, 2293–2301.

Sauerbrey G. (1959) Verwendung von Schwingquarzen zur Wägung dünner Schichten und zur Mikrowägung. *Z. Physik* **155**, 206–222.

Gardner J.W. (1994) *Microsensors*. Wiley, Chichester, UK.

Gardner J.W., Varadan V.K., and Awadelkarim O.O. (2001) *Micro Sensors, MEMS and Smart Devices*. Wiley, Chichester, UK.

Gizelli E. (2000) Study of the sensitivity of the acoustic waveguide sensor. *Anal. Chem.* **72**, 5967–5972.

Göpel W., Hesse J., and Zemel J.N. (eds.) (1991) *Sensors: A Comprehensive Survey. Chemical and Biological Sensors*. VCH, Weinhiem.

Gopel W. and Schierbaum K.D. (1995) $SnO_2$ sensors: current status and future prospects. *Sens. Actuators B* **26–27**, 1–12.

Göpel W., Hesse J., and Zemel J.N. (eds.) (1994) *Sensors: A Comprehensive Survey,* Vol. **1**. VCH, Weinheim.

Grate J.W., Martin S.J., and White R.M. (1993) Acoustic wave microsensors—Part II. *Anal. Chem.* **65**(22), 987–996.

Grate J.W., Wenzel S.W., and White R.M. (1991) Flexural plate wave devices for chemical analysis. *Anal. Chem.* **63**, 1552–1561.

Gründler P. (2006) *Chemical Sensors: An Introduction for Scientists and Engineers*. Springer-Verlag, Berlin.

Gupta K.C. and D'Arc M.J. (2000) Performance evaluation of copper ion selective electrode based on cyanocopolymers. *Sens. Actuators B* **62**, 171–176.

Gupta B.D. and Sharma N.K. (2002) Fabrication and characterization of U-shaped fiber-optic pH probes. *Sens. Actuators B* **82**, 89–93.

Gurlo A. (2006) Interplay between $O_2$ and $SnO_2$: Oxygen ionosorption and spectroscopic evidence for adsorbed oxygen. *ChemPhysChem.* **7**, 2041–2052.

Hagleitner C., Koll A., Vogt R., Brand O., and Baltes H. (1999) CMOS capacitive chemical microsystem with active temperature control for discrimination of organic vapors. *Tech. Dig. Transducers* **2**, 1012–1015.

Harame D.L., Bousse L.J., Shott J.D., and Meindl J.D. (1987) Ion-sensing devices with silicon nitride and borosilicate glass insulators. *IEEE Trans. Elect. Dev.* **34**, 1700–1707.

Harding G.L. (2001) Mass sensitivity of Love-mode acoustic sensors incorporating silicon dioxide and silicon-oxy-fluoride guiding layers. *Sens. Actuators B* **88**, 20–28.

Harrey P.M., Ramsey B.J., Evans P.S.A., and Harrison D.J. (2002) Capacitive-type humidity sensors fabricated using the offset lithographic printing process. *Sens. Actuators B* **87**, 226–232.

Harris P.D., Arnold W.M., Andrews M.K., and Partridge A.C. (1997) Resistance characteristics of conducting polymer films used in gas sensors. *Sens. Actuators B* **42**, 177–184.

Harsanyi G. (2000) *Sensors in Biomedical Applications: Fundamentals, Technology and Applications*. CRC Press, Boca Raton, FL.

Haueis R., Vellekoop M.J., Kovacs G., Lubking G.W., and Venema A. (1994) A Love-wave based oscillator for sensing in liquids. In: *Proceedings of the Fifth International Meeting of Chemical Sensors*, **1**, Lake Tahoe, CA, pp. 126–129.

Hierlemann A., Brand O., Hagleitner C., and Baltes H. (2003) Microfabrication techniques for chemical/biosensors. *Proc IEEE* **91**, 839–863.

Hirata H. and Higashiyama K. (1971) Analytical study of a cadmium ion-selective ceramic membrane electrode. *J. Anal. Chem.* **257**, 104–107.

Hofstetter D., Beck M., Faist J., Nagele M., and Sigrist M.W. (2001) Photoacoustic spectroscopy with quantum cascade distributed-feedback lasers. *Opt. Lett.*, **26**, 887–889.

Homola J., Yee S.S., and Gaugiltz G. (1999) Surface plasmon sensors: review. *Sens. Actuators B* **54**, 3–15.

Homola J. (2008) Surface plasmon resonance sensors for detection of chemical and biological species. *Chem. Rev.* **108**(2), 462–493.

Ideta K. and Arakawa T. (1993) Surface plasmon resonance study for the detection of some chemical species. *Sens. Actuators B* **13–14**, 384–386.

Imai S.H., Mizuno H., Suzuki M., Takeuchi T., Tamiya E., Mashige F., Ohkubo A., and Karube I. (1994) Total urinary protein sensor based on a piezoelectric quartz crystal. *Anal. Chim. Acta* **292**, 65–70.

Ionescu R. (1998) Combined Seebeck and resistive $SnO_2$ gas sensors, a new selective device, *Sens. Actuators B* **48**, 392–394.

Ippolito S.J., Trinchi A., Powell D.A., and Wlodarski W. (2009) Acoustic wave gas and vapor sensors. In: Comini E., Faglia G., and Sberveglieri G. (eds.), *Solid State Gas Sensing*. Science-Business Media, New York, pp. 261–304.

Ishihara T. and Matsubara S. (1998) Capacitive type gas sensors. *J. Electrocer.* **2**, 215–228.

Ishihara T., Sato S., Fukushima T., and Takita Y. (1996) Capacitive gas sensor of mixed oxide $CoO$-$In_2O_3$ to selectively detect nitrogen monoxide. *J. Electrochem. Soc.* **143**, 1908–1914.

Ishihara T., Sato S., and Takita Y. (1996) Capacitive-type sensors for the selective detection of nitrogen oxides. *Sens. Actuators B* **24–25**, 392–395.

Izumi M., Ohnuki H., Kato R., Imakubo T., Nagata M., Noda T., and Kojima K. (1998) Recent progress in metallic Langmuir-Blodgett films based on TTF derivatives. *Thin Solid Films* **327–329**, 14–18.

Jakoby B., Ismail G.M., Byfield M.P., and Vellekoop M.J. (1999) A novel molecularly imprinted thin film applied to a love wave gas sensor. *Sens. Actuators A* **76**, 93–97.

James S.W. and Tatam R.P.J. (2006) Fibre optic sensors with nano-structured coatings. *Opt. A: Pure Appl. Opt.* **8**, S430–S445.

Janata J. (1989) *Principles of Chemical Sensors*. Plenum Press, New York.

Janata J. and Josowicz M. (1997) A fresh look at some old principles: the Kelvin probe and the Nernst equation. *Anal. Chem.* **69**, 293A–296A.

Janata J. and Josowicz M. (2002) Conducting polymers in electronic chemical sensors. *Nature Mater.* **2**, 19–24.

Jimenez C., Rochefeuille S., Berjoan R., Seta P., Desfours J.P., and Dominguez C. (2004) Nanostructures for chemical recognition using ISFET sensors. *Microelectron. J.* **35**, 69–71.

Johan B., Magdalena M.Z., and Andrzej L. (2003) Carbonate ion-selective electrode with reduced interference from salicylat. *Biosens. Bioelectron.* **18**, 245–253.

Jorgenson R.C. and Yee S.S. (1993) A fiber optic chemical sensor based on surface plasmon resonance. *Sens. Actuators B* **12**, 213.

Josse F., Bender F., and Cernosek R.W. (2001) Guided shear horizontal surface acoustic wave sensors for chemical and biochemical detection in liquids. *Anal. Chem.* **73**, 5937–5944.

Kalantar-zadeh K., Trinchi A., Wlodarski W., Holland A., and Atashbar M.Z. (2001) A novel SAW Love mode device with nanocrystalline ZnO film for gas sensing applications. In: *Proceedings of the 1st IEEE Conference on Nanotechnology, IEEE-NANO*, Hawaii, pp. 556–561.

Kang Y., Ruan H., Wang Y., Arregui F.J., Matias I.R., and Claus R.O. (2006) Nanostructured optical fibre sensors for breathing airflow monitoring. *Meas. Sci. Technol.* **17**, 1207–1210.

Karlsen S.R., Johnston K.S., Jorgenson R.C., and Yee S.S. (1995) Simultaneous determination of refractive indices and absorbance spectra of chemical samples using surface plasmon resonance. *Sens. Actuators B* **24–25**, 747–749.

Karthigeyan A., Gupta R.P., Scharnagl K., Burgmair M., Zimmer M., Sharma S.K., and Eisele I. (2001) Low temperature $NO_2$ sensitivity of nano-particulate $SnO_2$ film for work function sensors. *Sens. Actuators B* **78**, 69–72.

Keller A, Ruegg M, Forster Loepfeb M., Pleischb M., Nebikerb P., and Burtschera H. (2005) Open photoacoustic sensor as smoke detector. *Sens. Actuators B* **104**, 1–7.

Kelkar U.R., Josse F., Haworth D.T., and Shana Z.A. (1991) Acoustic plate waves for measurements of electrical properties of liquids. *Microchem. J.* **43**, 155–164.

Khlebarov Z.P., Stoyanova A.I., and Topalova D.I. (1992) Surface acoustic wave sensors. *Sens. Actuators B* **8**, 33–40.

Kim B.H., Prins F.E., Kern D.P., Raible S., and Weimar U. (2001) Multicomponent analysis and prediction with a cantilever array based gas sensor. *Sens. Actuators B* **78**, 12–18.

King D., Lyons W.B., Flanagan C., and Lewis E. (2004) Interpreting complex data from a three-sensor multipoint optical fibre ethanol concentration sensor system using artificial neural network pattern recognition. *Meas. Sci. Technol.* **15**, 1560–1567.

Kondoh J., Matsui Y., Shiokawa S., and Wlodarski W.B. (1994) Enzyme-immobilized SH-SAW biosensor. *Sens. Actuators B* **20**, 199–203.

Kovacs G.T.A. (1998) *Micromachined Transducers Source Book*. McGraw-Hill, New York.

Kang Y., Ruan H., Wang Y., Arregui F.J., Matias I.R., and Claus R.O. (2006) Nanostructured optical fibre sensors for breathing airflow monitoring. *Meas. Sci. Technol.* **17**, 1207–1210.

Korotcenkov G., Han S.-D., and Stetter J.R. (2009) Review of electrochemical hydrogen sensors. *Chem. Rev.* 109(3), 1402–1433.

Korotcenkov G. (2008) The role of morphology and crystallographic structure of metal oxides in response of conductometric-type gas sensors. *Mater. Sci. Eng. R* **61**, 1–39.

Korotcenkov G. (2007a) Metal oxides for solid state gas sensors. What determines our choice? *Mater. Sci. Eng. B* **139**, 1–23.

Korotcenkov G. (2007b) Practical aspects in design of one-electrode semiconductor gas sensors: status report. *Sens. Actuators B* **121**, 664–678.

Korotcenkov G. (2005) Gas response control through structural and chemical modification of metal oxides: state of the art and approaches. *Sens. Actuators B* **107**, 209–232.

Kummer A.M., A. Hierlemann A., and Baltes H. (2004) Tuning sensitivity and selectivity of complementary metal oxide semiconductor-based capacitive chemical microsensors. *Anal. Chem.* **76**, 2470–2477.

Lakowicz J.R. (ed.) (1994) *Topics in Fluorescence Spectroscopy, Vol. 4: Probe Design and Chemical Sensing*. Plenum Press, New York.

Lange D., Hagleitner C., Brand O., and Baltes H. (2001) CMOS resonant beam gas sensing system with on-chip self excitation. In: *Proceedings of IEEE Micro Electro Mechanical Systems 14th International Conference,* Interlaken, Switzerland, pp. 547–552.

Lavrik N., Sepaniak M., and Datskos P. (2004) Cantilever transducers as a platform for chemical and biological sensors. *Rev. Sci. Instrum.* **75**, 2229–2253.

Lee J.S., Lee J.H., and Hong S.H. (2004) Solid state amperometric $CO_2$ sensor using a sodium ion conductor. *J. Eur. Cer. Soc.* **24**, 1431–1434.

Lieberman R.A. (1993) Recent progress in intrinsic fiber-optic chemical sensing II. *Sens. Actuators B* **11**, 43–55.

Lin K.-W, Chen H.-I., Chuang H.-M., Chen C.-Y., Lu C.-T., Cheng C.-C., and Liu W.-C. (2004) Characteristics of Pd/InGaP Schottky diodes hydrogen sensors. *IEEE Sensors J.* **4**, 72–79.

Lin T., Sree R., Tseng S., Chiu K., Wu C., and Lo J. (2004) Volatile organic compound concentrations in ambient air of Kaohsiung petroleum refinery in Taiwan. *Atmos. Environ.* **38**, 4111.

Liu N., Hui J., Sun C.Q., Dong J.H., Zhang L.Z., and Xiao H. (2006) Nanoporous zeolite thin film-based fiber intrinsic Fabry-Perot interferometric sensor for detection of dissolved organics in water. *Sensors* **6**, 835–847.

Lu C., Czanderna W., and Townshend A. (1984) Applications of piezoelectric quartz crystal microbalances: book review. *Anal. Chim. Acta* **199**, 279.

Madou M.J. and Morrison S.R. (1989) *Chemical Sensing with Solid Devices.* Academic Press, Boston.

Lundström I. (1991) Field effect chemical sensors. In: Gopel W., Hesse J., and Zemel J.N. (eds.), *Sensors: A Comprehensive Survey.* VCH, Weinheim, Vol. 1, pp. 467–529.

Lundstrom I., Sundgren H., Winquist F., Eriksson M., Krantz-Rulcker C., and Lloyd-Spetz A. (2007) Twenty-five years of field effect gas sensor research in Linkoping. *Sens. Actuators B* **121**, 247–262.

Majoo S, Gland J.L., Wise K.D., and Schwank J.W. (1996) A silicon micromachined conductometric gas sensor with a maskless Pt sensing film deposited by selected-area CVD. *Sens. Actuators B* **36**, 312–319.

Markinkowska K., McGauley M.P., and Symons E.A. (1991) A new carbon monoxide sensor based on a hydrophobic oxidation catalyst. *Sens. Actuators B* **5**, 91–96.

Martin F., Newton M. I., McHale G., Melzak K.A., and Gizeli E. (2004) Pulse mode shear horizontal-surface acoustic wave (SH-SAW) system for liquid based sensing applications. *Biosens. Bioelectron.* **19**, 627–632.

Matsuguch M., Kuroiwa T., Miyagishi T., Suzuki S., Ogura T., and Sakai Y. (1998) Stability and reliability of capacitive-type relative humidity sensors using crosslinked polyimide films. *Sens. Actuators B* **52**, 53–57.

McCallum J.J. (1989) Piezoelectric devices for mass and chemical measurements: an update. *Analyst* **114**, 1173–1189.

Meyerhoff M.E. and Opdycke W.N. (1986) Ion selective electrodes. *Adv. Clin. Chem.* **25**, 1–47.

Middlehoek S. and Audet S.A. (1989) *Silicon Sensors for Chemical Signals.* Academic Press, Boston.

Miller J.B. (2001) Catalytic sensors for monitoring explosive atmospheres. *IEEE Sensors J.* **1**(1), 88–93.

Miura T. (1985) Thermistor humidity sensor for absolute humidity measurements and their applications. In: Chaddock, J.B. (ed.), *Proceedings of International Symposium on Moisture and Humidity,* ISA, Washington, DC, pp. 555–573.

Moon W.J., Yu J.H., and Choi G.M. (2002) The CO and $H_2$ gas selectivity of CuO-doped $SnO_2$–ZnO composite gas sensor. *Sens. Actuators B* **87**, 464–470.

Moreno J., Arregui F.J., and Matias I.R. (2005) Fiber optic ammonia sensing employing novel thermoplastic polyurethane membranes. *Sens. Actuators B* **105**, 419–424.

Morgan D.J. (1961) Construction and operation of a simple flame-ionization detector for gas chromatography. *J. Sci. Instrum.* **38**, 501–503.

Moriizumi T. (1988) Langmuir-Blodgett films as chemical sensors. *Thin Solid Films* **160**, 413–429.

Moulin A.M., O'Shea S.J., and Welland M.E. (2000). Microcantilever-based biosensors. *Ultramicroscopy* **82**, 23–31.

Muramatsu H.M., Tamiya E., and Karube I. (1987) Determination of microbes and immunoglobulins using a piezoelectric biosensor. *J. Membr. Sci.* **41**, 281–290.

Nagai M. and Nishio T. (1988) A new type of $CO_2$ gas sensor comprising porous hydroxyapatite ceramics. *Sens. Actuators* **15**, 145–151.

Nomura T., Saitoh A., and Miyazaki T. (2003) Liquid sensor probe using reflecting SH-SAW delay line. *Sens. Actuators B* **91**(1–3), 298–302.

Ogawa K., Tsuchiya S., Kawakami H., and Tsutsui T. (1988) Humidity-sensing effects of optical fibers with microporous $SiO_2$ cladding. *Electron. Lett.* **24**(1), 42–43.

Ostrick B., Muhlsteff J., Fleischer M., Meixner H., Doll T., and Kohl C.-D. (1999) Adsorbed water as key to room temperature gas-sensitive reactions in work function type sensors: the carbonate-carbon dioxide system. *Sens. Actuators B* **57**, 115–119.

Ostrick B., Pohle R., Fleischer M., and Meixner H. (2000) TiN in work function type sensors: a stable ammonia sensitive material for room temperature operation with low humidity cross sensitivity. *Sens. Actuators B* **68**, 234–239.

Paldus B.A., Spence T.G., Zare R.N., Oomens J., Harren F.J.M., Parker D.H., Gmachl C., Capasso F., Sivco D.L., Baillargeon J.N., Hutchinson, A.L., and Cho A.Y. (1999) Photoacoustic spectroscopy using quantum cascade lasers. *Opt. Lett.* **24**, 178–180.

Papadopoulos C.A, Vlachos D.S., and Avaritsiotis J.N. (1996) A new planar device based on Seebeck effect for gas sensing applications. *Sens. Actuators B* **34**, 524–527.

Patel S.V., Wise K.D., Gland J.L., Zanini-Fisher M., and Schwank J.W. (1997) Characteristics of silicon-micromachined gas sensors based on $Pt/TiO_x$ thin films. *Sens. Actuators B* **42**, 205–215.

Park E.J., Reid K.R., Tang W., Kennedy R.T., and Kopelman R. (2005) Ratiometric fiber optic sensors for the detection of inter- and intra-cellular dissolved oxygen. *J. Mater. Chem.* **15**, 2913–2919.

Patel S.V., Gland J.L., and Schwank J.W. (1999) Film structure and conductometric hydrogen-gas-sensing characteristics of ultrathin platinum films. *Sens. Actuators B* **15**, 3307–3311.

Pisco, M., Consales, M., Campopiano, S., Viter, R., Smyntyna, V., Giordano, M., and Cusano, A. J. (2006) A novel optochemical sensor based on $SnO_2$ sensitive thin film for ppm ammonia detection in liquid environment. *Lightwave Technol.* **24**, 5000–5007.

Pockrand I., Swalen J.D., Gordon JG. II, and Philpott M.R. (1979) Exciton-surface plasmon interactions. *J. Chem. Phys.* **70**, 3401.

Potje-Kamloth K. (2008) Semiconductor junction gas sensors. *Chem. Rev.* **108**, 367–399.

Preejith P.V., Lim C.S., and Chia T.F. (2006) Serum protein measurement using a tapered fluorescent fibre-optic evanescent wave-based biosensor. *Meas. Sci. Technol.* **17**, 3255–3260.

Radomska A., Bodenszac E., Glab S., and Koncki R. (2004) Creatinine biosensor based on ammonium ion selective electrode and its application in flow-injection analysis. *Talanta* **64**(3), 603–608.

Raiteri R., Grattarola M., Butt H.-J., and Skládal P. (2001). Micromechanical cantilever-based biosensors. *Sens. Actuators B* **79**, 115–126.

Raiteri R., Grattarola M., and Berger R. (2002) Micromechanics senses biomolecules. *Mater. Today* **5**(1), 22–29.

Razavi B. (2000) *Design of Analog CMOS Integrated Circuits*. McGraw-Hill, New York.

Reich G. (2005) Near-infrared spectroscopy and imaging: basic principles and pharmaceutical applications. *Adv. Drug Deliv. Rev.* **57**, 1109–1143.

Reinhoudt D.N. (1995) Durable chemical sensors based on field-effect transistors. *Sens. Actuators B* **24–25**, 197–200.

Ricco A.J., Martin S.J., Frye G.C., and Niemczyk T.M. (1988) Acoustic plate mode devices as liquid phase sensors. In: *Proceedings of the IEEE Ultrasonics Symposium*, Hilton Head Island, SC, USA, pp. 23–26.

Ricco A.J. and Martin S.J. (1991) Thin metal film characterization and chemical sensors: monitoring electronic conductivity, mass loading and mechanical properties with surface acoustic wave devices. *Thin Solid Films* **206**, 94–101.

Rinnan R. and Rinnan A. (2007) Application of near infrared reflectance (NIR) and fluorescence spectroscopy to analysis of microbiological and chemical properties of arctic soil. *Soil Biol. Biochem.* **39**, 1664–1673.

Roberts G. (ed.) (1990) *Langmuir-Blodgett Films*. Plenum Press, New York.

Roberts G.G. (1983) Transducer and other applications of Langmuir-Blodgett films. *Sens. Actuators* **4**, 131–145.

Rogers K.R. and Poziomek E. (1996) Fiber optic sensors for environmental monitoring. *Chemosphere* **33**, 1151–1174.

Rouessac F. and Rouessac A. (2007) *Chemical Analysis: Modern Instrumentation Methods and Techniques*. Wiley, Chichester UK.

Sakti S.P., Hauptmann P., Zimmermann B., Buhling F., and Ansorge S. (2001) Disposable HSA QCM-immunosensor for practical measurement in liquid. *Sens. Actuators B* **78**, 257–262.

Sang X.Z., Yu C.X., Yan B.B., Ma J.X., Meng Z.F., Mayteevarunyoo T., and Lu N.G. (2006) Temperature-insensitive chemical sensor with twin bragg gratings in an optical fiber. *Chin. Phys. Lett.* **23**, 3202–3204.

Schierbaum K.D., Weimar U., and Göpel W. (1990) Multicomponent gas analysis: an analytical chemistry approach applied to modified $SnO_2$ sensors. *Sens. Actuators B* **2**, 71–78.

Schoeneberg U., Hosticka B.J., Zimmer G., and Maclay G.J. (1990) A novel readout technique for capacitive gas sensors *Sens. Actuators B* **1**, 58–61.

Schreiter M., Gabl R., Lerchner J., Hohlfeld C., Delan A., Wolf G., Bluher A., Katzschner B., Mertig M., and Pompe W. (2006) Functionalized pyroelectric sensors for gas detection. *Sens. Actuators B* **119**, 255–261.

Seiyama T., Yamazoe N., and Arai H. (1983) Ceramic humidity sensors. *Sens. Actuators* **4**, 85–96.

Senillou Jaffrezic-Renault N., Martelet C., and Griffe F. (1998) A miniaturized ammonium sensor based on the integration of both ammonium and reference FETs in a single chip. *Mater. Sci. Eng. C* **6**, 59–63.

Shamsipur M., Rouhani S., Mohajeri A., and Ganjali M.R. (2000) A bromide ion-selective polymeric membrane electrode based on a benzo-derivative xanthenium bromide salt. *Anal. Chim. Acta* **418**(2), 197–203.

Shen C.Y., Huang C.P., and Huang W.T. (2004) Gas-detecting properties of surface acoustic wave ammonia sensors. *Sens. Actuators B* **101**, 1–7.

Sheppard N.F., Day D.R., Lee H.L., and Senturia S.D. (1982) Microdielectrometry. *Sens. Actuators* **2**, 263–274.

Siesler H.W., Ozaki Y., Kawata S., and Heise H.M. (2002) *Near-Infrared Spectroscopy: Principles, Instruments, Applications*. Wiley-VCH, Weinheim.

Da Silva M.G., Vargas H., Miklós A., and Hess P. (2004) Photoacoustic detection of ozone using a quantum cascade laser. *Appl. Phys. B* **78**, 677–680.

Široký K. (1993) Use of the Seebeck effect for sensing flammable gas and vapours. *Sens. Actuators B* **17**, 13–17.

Somov S.I., Reinhardt G., Guth U., and Göpel W. (2000) Tubular amperometric high-temperature sensors: simultaneous determination of oxygen, nitrogen oxides and combustible components. *Sens. Actuators B* **65**, 68–69.

Streetman B.G. (1990) *Solid State Electronic Devices*, 3rd ed. Prentice Hall, Englewood Cliffs, NJ.

Su P.-G. and Tsai W.-Y. (2004) Humidity sensing and electrical properties of a composite material of nano-sized $SiO_2$ and poly(2-acrylamido-2-methylpropane sulfonate). *Sens. Actuators B* **100**, 425–430.

Sundmacher K., Rihko-Struckmann L.K., and Galvita V. (2005) Solid electrolyte membrane reactors: status and trends. *Catal. Today* **104**, 185–199.

Symons E.A. (1992) Catalytic gas sensors. In: Sberveglieri G. (ed.), *Gas Sensors*. Kluwer Academic, Dordrecht, The Netherlands, pp. 169–185.

Taillades G., Valls O., Bratov A., Dominguez C., Pradel A., and Ribes M. (1999) ISE and ISFET microsensors based on a sensitive chalcogenide glass for copper ion detection in solution. *Sens. Actuators B* **59**, 123–127.

Taylor R. and Schultz J. (1996) *Handbook of Chemical and Biological Sensors*. IOP Publishing, Philadelphia.

Tan O.K., Cao W., Zhu W., Chai J.W., and Pan J.S. (2003) Ethanol sensors based on nano-sized $\alpha$-$Fe_2O_3$ with $SnO_2$, $ZrO_2$, $TiO_2$ solid solutions. *Sens. Actuators B* **93**, 396–401.

Tau S. and Sarma T.V.S. (2006) Evanescent-wave optical Cr VI sensor with a flexible fused-silica capillary as a transducer. *Opt. Lett.* **31**, 1423–1425.

Then D., Vidic A., and Ziegler Ch. (2006) A highly sensitive self-oscillating cantilever array for the quantitative and qualitative analysis of organic vapor mixtures. *Sens. Actuators B* **117**, 1–9.

Tirole N., Choujaa A., Hauden D., Martin G., Blind P., Froelicher M., Pommier J.C., and Cachard A. (1993) Lamb waves pressure sensor using an AlN/Si structure. In: *Proceedings of the IEEE Ultrasonics Symposium*, Hawaii, pp. 371–374.

Trinchi A., Galatsis K., Wlodarski W., and Li Y.X. (2003) A Pt/$Ga_2O_3$-ZnO/SiC Schottky diode-based hydrocarbon gas sensor. *IEEE Sensors J.* **3**, 548–553.

Trinchi A., Woldarski W., and Li Y.X. (2004) Hydrogen sensitive $Ga_2O_3$ Schottky diode sensor based on SiC. *Sens. Actuators B* **100**, 94–98.

Tsai Y.-Y., Lin K.-W., Chen H.-I., Lu C.-T., Chuang H.-M., Chen C.-Y., and Liu W.-C. (2003) Comparative hydrogen sensing performances of Pd- and Pt-InGaP metal-oxide-semiconductor Schottky diodes. *J. Vac. Sci. Technol. B* **21**, 2471–2475.

Tuyen L.T.T., Vinh D.X., Khoi P.H., and Gerlach G. (2002) Highly sensitive $NO_x$ gas sensor based on a Au/n-Si Schottky diode. *Sens. Actuators B* **84**, 226–230.

Urashi T. and Arakawa T. (2001) Detection of lower hydrocarbons by means of surface plasmon resonance. *Sens. Actuators B* **76**, 32–35.

Van Oirschot T.G.J., VanLeeuwen D., and Medema J. (1972) The effect of gases on the conductive properties of organic semiconductors. *J. Electroanal. Chem.* **37**, 373–385.

Varghese O.K., Gong D., Dreschel W.R., Ong K.G., and Grimes C.A. (2003) Ammonia detection using nanoporous alumina resistive and surface acoustic wave sensors. *Sens. Actuators B* **94**, 27–35.

Vlachos D.S., Papadopoulos C.A., and Avaritsiotis J.N. (1997) A technique for suppressing ethanol interference employing Seebeck effect devices with carrier concentration modulation. *Sens. Actuators B* **44**, 239–242.

Vellekoop M.J., Lubking G.W., Sarro P.M., and Venema A. (1994) Integrated-circuit compatible design and technology of acoustic-wave based microsensor. *Sens. Actuators A* **44**, 249–263.

Wang A.W., Kiwan R., White R.M., and Ceriani R.L. (1997) A silicon-based ultrasonic immunoassay for detection of breast cancer antigens. In: *Proceedings of the Ninth International Conference on Solid-State Sensors and Actuators, Transducers'97*, Chicago, pp. 191–194.

Wang A.W., Kiwan R., White R.M., and Ceriani R.L. (1998) A silicon-based ultrasonic immunoassay for detection of breast cancer antigens. *Sens. Actuators B* **49**, 13–21.

Wang H., Feng C.-D., Sun S.-L., Segre C.U., and Stetter J.R. (1997) Comparison of conductometric humidity-sensing polymers. *Sens. Actuators B* **40**, 211–216.

Wenzel S.W. and White R.M. (1989) Flexural plate-wave sensor: chemical vapor sensing and electrostrictive excitation. In: *Proceedings of the IEEE Ultrasonics Symposium*, New York, pp. 595–598.

White R.M., Martin S.J., Ricco A.J., Zellers E.T., Frye G.C., and Wohltjen H. (1997) *Acoustic Wave Sensors: Theory, Design, and Physico-Chemical Application*. Academic Press, San Diego, CA.

Wilson D.M., Hoyt S., Janata J., Booksh K., and Obando L. (2001) Chemical sensors for portable, handheld field instruments. *IEEE Sensors J.* **1**, 256–274.

Wohltjen H. (1984) Mechanisms of operation and design considerations for surface acoustic wave device vapor sensors. *Sens. Actuators* **5**, 307–325.

Wohltjen H., Barger W.R., Snow A.W., and Jarvis N.L. (1985) A vapor-sensitive chemiresistor fabricated with planar microelectrodes and a Langmuir-Blodgett organic semiconductor film. *IEEE Trans. Electron. Dev.* **ED-32**, 1170–1174.

Wolfbeis O.S. (1990) Chemical sensors—survey and trends. *Fresenius J. Anal. Chem.* **337**, 522–527.

Wright J.D., Cado A., Peacock S.J., Rivalle V., and Smith A.M. (1995) Effects of nitrogen dioxide on the surface plasmon resonance of substituted pthalocyanine films. *Sens. Actuators B* **29**, 108–114.

Wu B., Chang C., Sree U., and Chiu K. (2006) Measurement of non-methane hydrocarbons in Taipei city and their impact on ozone formation in relation to air quality. *J. Anal. Chim. Acta* **576**, 91–99.

Yamazoe N. and Shimuzu Y. (1989) Humidity sensors: principle and applications. *Sens. Actuators* **10**, 379–398.

Yotter R.A. and Wilson D.M. (2004) Sensor technologies for monitoring metabolic activity in single cells—Part II: nonoptical methods and applications. *IEEE Sensors J.* **4**, 412–429.

Yuqing M., Jianguo G., and Jianrong C. (2003) Ion sensitive field effect transducer-based biosensors. *Biotechnology Adv.* **21**, 527–534.

Zao Z., Buttner W.J., and Stetter J.R. (1992) The properties and applications of amperometric gas sensors. *Electroanalysis* **4**, 253–266.

Zhang G., Lui J., and Yuan M. (2000) Novel carbon dioxide gas sensor based on infrared absorption. *Opt. Eng.* **39**, 2235.

Zhang G. and Wu X. (2004) A novel $CO_2$ gas analyzer based on IR absorption *Opt. Lasers Eng.* **42**, 219–231.

Zimmer M., Burgmair M., Scharnagl K., Karthigeyan A., Doll T., and Eisele I. (2001) Gold and platinum as ozone sensitive layer in work-function gas sensors. *Sens. Actuators B* **80**, 174–178.

CHAPTER 2

# DESIRED PROPERTIES FOR SENSING MATERIALS

G. Korotcenkov

## 1. INTRODUCTION

The problem of optimization is a key factor in both design and manufacturing of electronic devices. In the case of either chemical sensors or solid-state gas sensors (GS), this problem has some specific peculiarities due to the absence of strict quantitative theory which would describe their operation. Some semiquantitative approaches can be found in the literature (Clifford 1983; Morrison 1987; Zemel 1988; Geistlinger 1993, 1994; Moseley et al. 1991; Brinzari et al. 2000; Barsan and Weimar 2001). However, since the number of physical and chemical parameters that characterize sensor properties is large, and some of these parameters are difficult to control, the problem of optimization is largely empirical and remains a kind of art.

At present, therefore, the field of chemical sensors is characterized by a search for optimal sensing materials and design of adequate theoretical models that may promote optimization of chemical sensors. The term *optimization* here means achieving either necessary or the best available values of sensitivity, selectivity, and response time of chemical sensors at given operating conditions.

For example, according to at least one earlier view of the problem of gas sensor design (Moseley et al. 1991), almost any metal oxide can serve as the basis for a solid-state gas sensor. For this purpose one needs only to prepare this metal oxide as a sufficiently fine dispersed porous substance with properties controlled by the surface state.

However, as requirements for gas sensors became more complex and precise, and understanding of the nature of the gas-sensing effects was became more fundamental, our conception of material compatibility for gas sensor elaboration began to change. We began to understand that for implementation of such requirements as

- High response to the target agent
- Low cross-sensitivity
- Fast and reversible interaction with analytes
- Low sensitivity of the signal to a change in air humidity
- Absence of long-term drift
- Short time to operational status
- Effective low-cost technology
- High reproducibility
- Uniform and strong binding to the surface of the substrate
- Easy connection to control units, etc.

materials for chemical sensors have to possess specific combination of physical and chemical properties, and not every material can fulfill these requirements.

A systematic consideration of the required properties of materials for chemical sensor applications indicates that the key properties which determine a specific choice include:

- Adsorption ability
- Electronic, electrophysical, and chemical properties
- Catalytic activity
- Permittivity
- Thermodynamic stability
- Crystallographic structure
- Interface quality
- Compatibility with current or expected materials to be used in processing
- Reliability, etc.

Many different materials appear to be favorable in terms of some of these properties, but very few materials are promising with respect to all of these requirements.

As confirmation of this statement, let us examine parameters which can determine a material's compatibility for use as a chemical sensor. Our attention will focus mainly on solid-state gas sensors, which are the most complicated in terms of their principles of operation (Kupriyanov 1996; Barsan et al. 1999; Kohl 2001; Barsan and Weimar 2003); however, the information in this chapter is general in nature and can be used as a guide for both the analysis of behavior and the choosing of materials for all types of chemical sensors, including optical sensors, in which the basic sensitive parameter is the coefficient of optical refraction. Even in such devices, the change of the sensing element's parameters takes place through adsorption processes, which will be thoroughly reviewed in this chapter. The same requirements also apply to metal oxides for electrochemical sensors. According to Trasatti (1980), these materials must possess parameters such as large surface area, high electrocatalytic activity, excellent long-term stability, high electrical conductivity, etc.

This brief review cannot cover all promising materials that have been developed for chemical sensors. In addition to binary oxides and various polymers, there are numerous ternary, quaternary, and complex metal oxides, which are of interest for particular applications. For simplicity of presentation, priority has been given to examining binary oxides. Properties of more complicated oxide phases will be discussed in Chapter 5.

Readers who are interested in further information about these metal oxides and polymers may consult relevant reviews and monographs (Sandier and Karo 1974; Samsonov 1973; Jolivet 2000). The bulk properties of simple binary oxides are fairly well understood, and some excellent reviews and books treat the thermodynamics (Johnson 1982), the structures, and nonstoichiometric aspects (Sorensen 1981; Catlow 1997) which are particularly important for oxides. Spectroscopy aspects are described by Hamnett and Goodenough (1984), and mechanical properties by Sorensen (1981). Present knowledge about the surface science of metal oxides has been comprehensively reviewed by Dufour and Nowotny (1988), Freund and Umbach (1993), Henrich and Cox (1994), Szuber and Gopel (2000), and Calatayud et al. (2003). Detailed information about transition metal oxides is given by Krilov and Kisilev (1981), Hamnett and Goodenough (1984), and Cox (1992). Electrochemistry of metal oxides and interface phenomena are reviewed by Gellings and Bouwmeester (1997). The interaction of metal oxide surfaces with agents dissolved in water is discussed in detail by Blesa et al. (2000).

## 2. COMMON CHARACTERISTICS OF METAL OXIDES

### 2.1. CRYSTAL STRUCTURE OF METAL OXIDES

In most cases we do not know exactly where atoms are on the surface of metal oxides, but their bulk crystal structures are known very accurately, thanks to about a century of x-ray crystallography study of various crystalline materials. The most prevalent crystallographic structures of metal oxides used for chemical sensors, with some examples, are listed in Table 2.1 (Samsonov 1973, 1982; Cox 1992; Henrich and Cox 1994). It is necessary to note that the crystal structure of metal oxides is complicated; most metal oxides have large unit cells which include many atoms.

The most important determinants of crystal structure are the stoichiometry, or relative numbers of the different types of atoms present, and the coordination of ions, i.e., the number of ions of one type surrounding another and their geometric arrangement. Commonly, metal oxides consist of several structural units, which may be triangular, tetrahedral, octahedral, or cubic (see Figure 2.1). A consideration of oxide structures suggests that metal-ion coordination numbers and geometries show reasonably systematic features that carry over among different structure types. For example, it has been established that the ratio between the ionic radii of cations and anions (the "ratio rule") serves as a rough prediction of preferred coordination in the unit cells of metal oxides (see Table 2.2) (Pauling 1929). According to Pauling, the size of the $O^{2-}$ anion is 0.14 nm, and the size of cations in metal oxides is in the range from 0.05 to 0.15 nm. As follows from Table 2.1, sixfold octahedral coordination is the most common metal-ion geometry in the metal oxides used for chemical sensors.

Another very important crystallographic parameter of sensing materials is the concentration of point defects, i.e., the degree of deviation of the chemical composition from the ideal chemical formula.

### Table 2.1. Oxide structures of some stable metal oxides

| Formula | Examples | Name | Symmetry | Coordination |
|---|---|---|---|---|
| $M_2O$ | $Cu_2O$, $Ag_2O$ | Cuprite | Cubic | 2: Linear |
| | $Na_2O$ | Antifluorite | Cubic | 4: Tetrahedral |
| MO | MgO, NiO, CoO, CaO | Rocksalt | Cubic | 6: Octahedral |
| | (no oxides) | Zincblende | Cubic | 4: Tetrahedral |
| | ZnO | Wurtzite | Hexagonal | |
| | PdO | Pd O | Tetragonal | 4: Planar |
| | CuO | Tenorite | Monoclinic | |
| | PbO | PbO | Tetragonal | 4: Pyramidal |
| $M_3O_4$ | $Fe_3O_4$ | Magnetite (inverse spinel) | Cubic | Two M6 (oct.); One M4 (tet.) |
| $M_2O_3$ | $\alpha$-$Al_2O_3$, $Rh_2O_3$ $\alpha$-$Ga_2O_3$ $\alpha$-$Fe_2O_3$ $Cr_2O_3$, $V_2O_3$ | Corundum | Hexagonal | 6: Octahedral |
| | $In_2O_3$, $Sc_2O_3$, $Mn_2O_3$ | $Mn_2O_3$ | Cubic | 6: Octahedral |
| $MO_2$ | $ZrO_2$, $UO_2$, $CeO_2$ | Fluorite | Cubic | 8: Cubic |
| | $TiO_2$, $SnO_2$, $RuO_2$ $GeO_2$ | Rutile | Tetragonal | 6: Octahedral |
| $M_2O_5$ | $\alpha$-$V_2O_5$ | $V_2O_5$ | Orthorhombic | 5+1: Distorted |
| $MO_3$ | $ReO_3$, $WO_3$ (distorted variants) | $ReO_3$ | Cubic (pseudo-cubic) | 6: Octahedral |
| | $MoO_3$ | $MoO_3$ | Monoclinic | 6: Distorted |
| | $\alpha$-$MoO_3$ | $\alpha$-$MoO_3$ | Orthorhombic | |
| $AMO_3$ | $SrTiO_3$, $BaTiO_3$ $CaTiO_3$, $NaNbO_3$ | Perovskite | Cubic (or distorted) | M6 (oct.), A12 (cub. oct) |
| | $LiNbO_3$, $FeTiO_3$ | Ilmenite | Trigonal | M and A6 (oct.) |
| $A_2MO_4$ | $La_2CuO_4$, $Nb_2CuO_4$ ($K_2NiF_4$) | Layer perovskite | Tetragonal (or distorted) | M6 (oct.), A12 (cub. Oct.) |

The structure of point defects in metal oxides, especially in transition-metal oxides, has been extensively studied for many years (Mrowec 1978; Murch and Nowick 1984; Nowick 1991; Cox 1992). It has been established that many physical and chemical properties of nonstoichiometric metal oxides depend on the nature, concentration, and mobility of defects in the crystalline lattice of these materials. In particular, the reactivity of solids, as well as their sintering, semiconducting, catalytic, and many other properties, depend to a greater degree on the nature of the defects than on the kind of substance itself.

It is necessary to note that point defects in oxides behave quite differently from point defects in metals and standard semiconductors. According to Nowick (Murch and Nowick 1984; Nowick 1991), the reasons are as follows:

DESIRED PROPERTIES FOR SENSING MATERIALS • 67

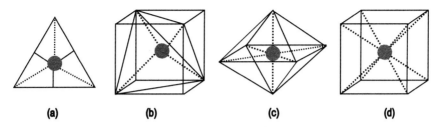

**Figure 2.1.** Basic building blocks of metal oxide crystal structures: (a) triangle; (b) tetrahedron; (c) octahedron; (c) cube.

1. In a binary oxide there are two different sublattices. The diffusion of cations and of anions proceeds each on its own sublattice, involving different point defects. Accordingly, cation and anion diffusion rates can be very different and, in fact, usually are.
2. The exact composition of a large class of oxides is a sensitive function of the ambient partial pressure of oxygen, $P(O_2)$. Oxides can exchange oxygen with the ambient atmosphere. Thus, the stoichiometry, and therefore the concentration, of point defects can be controlled by the annealing of a metal oxide specimen at a specified $P(O_2)$ at a given temperature.
3. The melting temperatures of many oxides ($SiO_2$, $Al_2O_3$, MgO, CaO, $La_2O_3$, etc.) are very high, so, below approximately 1000°C the intrinsic point defect concentrations in these oxides are very small. This implies that the presence of accidentally or intentionally introduced impurities strongly affects the defect concentrations in such metal oxides. Therefore, unintentional and uncontrolled defects may dominate the transport behavior.
4. Most important, oxides are highly ionic, so that, in general, point defects carry an effective charge. This implies that the behavior of point defects is controlled by charge compensation that results in the simultaneous introduction of point defects of opposite charge to maintain the condition of charge neutrality. The existence of charge-compensation defects on opposite sublattices and the corresponding equilibrium conditions through mass action equations means that the defect concentrations in the two sublattices are coupled—i.e., the introduction of defects in one sublattice can profoundly affect the defect structure in the other. Furthermore, the presence

**Table 2.2. Relationship between cation coordination number, shape of the coordination polyhedron, and cation-to-anion size ratio**

| CATION COORDINATION | COORDINATION POLYHEDRON GEOMETRY | RATIO OF IONIC RADII ($R_c/R_a$) |
|---|---|---|
| 3 | Triangle | 0.155–0.225 |
| 4 | Tetrahedron | 0.225–0.414 |
| 6 | Octahedron | 0.414–0.732 |
| 8 | Cube | 0.732–1.00 |

*Source:* Data from Pauling 1929.

of oppositely charged point defects results in the formation of associated pairs or larger clusters, especially when the temperature is not very high.

Because of the potential problem of impurities controlling the defect structure, it is desirable to deliberately control the introduction of defects. Probably the two best ways to accomplish this are (1) to introduce aliovalent impurities and (2) to take the material off stoichiometry through appropriate $P(O_2)$ annealing. However, it is necessary to remember that at high concentration of point defects the mutual interaction among point defects can be accompanied by formation of complexes and defect clusters called extended defects (Mrowec 1978). These extended defects may become further ordered, leading to superstructure ordering and to formation of intermediate phases. In some cases, point defects may be eliminated by the process of crystallographic sheer, which is connected to formation of a whole series of intermediate phases.

It is important to note that, in principle, most of the point defects mentioned above can also be present at free surfaces of a crystal, but their energies are very different from those of the defects in the bulk sample (Gellings and Bouwmeester 2000). This means that the defect concentrations at the surface also differ from those in the bulk. Depending on the sign of the energy difference, this can lead either to increased concentration, also called positive defect segregation to the surface, or the reverse, called negative (or reverse) segregation. Due to the changes in band energies near the surface in semiconducting oxides, the so-called band bending or band curvature, the electron or electron-hole concentrations in the conduction or valence bands, respectively, are different close to the surface from those in the bulk.

## 2.2. ELECTRONIC STRUCTURE OF METAL OXIDES

Metal oxides exhibit a very wide range of electrophysical properties. Their electrical behavior ranges from the best insulators (e.g., $Al_2O_3$ and MgO) through wide-band-gap and narrow-band-gap semiconductors ($TiO_2$, $SnO_2$, and $Ti_2O_3$, respectively) to metals ($V_2O_3$, $Na_xWO_3$, and $ReO_3$) and superconductors (including reduced $SrTiO_3$). The range of electronic structures of oxides is so wide that metal oxide compounds have been divided into two categories: (1) transition-metal oxides ($Fe_2O_3$, NiO, $Cr_2O_3$, etc.) and (2) non–transition-metal oxides, which include (a) pre–transition-metal oxides ($Al_2O_3$, etc.) and (b) post–transition-metal oxides (ZnO, $SnO_2$, etc.).

The common feature of the non–transition-metal oxides is that the valence orbitals of the metal atoms have $s$ and $p$ symmetry. In transitional-metal oxides, however, the $d$ atomic orbitals assume crucial importance. Many of the complications in using transition-metal oxides stem from this difference, because of the different bonding properties associated with $d$ orbitals. These complexities include the existence of variable oxidation states, the frequent failure of the band model, and the crystal-field splitting of the $d$ orbitals (Henrich and Cox 1994). Table 2.3 shows the electron configuration of $3d$ transition-metal oxides.

Henrich and Cox (1994) gave the following explanation of the difference in behavior between non–transition and transition-metal oxides. The non–transition-metal oxides contain elements that with some exceptions have only one preferred oxidation state. Other states are inaccessible because too

### Table 2.3. Electron configurations for 3d transition-metal oxides

| Electron configuration | Metal oxides |
|---|---|
| $3d^0$ | $Sc_2O_3$, $TiO_2$, $V_2O_5$, $CrO_3$ |
| $3d^1$ | $VO_2$, $Ti_2O_3$ |
| $3d^2$ | $CrO_2$, $V_2O_3$, $TiO_x$ |
| $3d^3$ | $\beta$-$MnO_2$, $Cr_2O_3$, $VO_x$ |
| $3d^4$ | $Mn_2O_3$, $Mn_3O_4$ |
| $3d^5$ | $\alpha$-$Fe_2O_3$, $MnO$, $Fe_3O_4$ |
| $3d^6$ | $FeO$, $Co_3O_4$ |
| $3d^7$ | $CoO$ |
| $3d^8$ | $NiO$ |
| $3d^9$ | $CuO$ |
| $3d^{10}$ | $Cu_2O$, $ZnO$ |

*Source:* Data from Krilov and Kisilev 1981.

much energy is needed to add or remove an electron from the cations when they are coordinated with $O^{2-}$ ligands. Transition-metal oxides behave differently in that the energy difference between a cation $d^n$ configuration and either a $d^{n+1}$ or $d^{n-1}$ configuration is often rather small. The most obvious consequence is that many transition elements have several stable oxides with different compositions. It is also much easier than with non–transient-metal oxides to create defects that have different electron configurations. As a result of high defect concentrations, the bulk and surface chemistries of transition-metal oxides are very complicated.

Trends in the stability of different oxidation states are very important in surface chemistry, as they control both the types of defect that may be formed easily and the type of chemisorption that may take place. The $d^0$ configuration represents the highest oxidation state that can ever be attained; thus, pure $TiO_2$, $V_2O_5$, etc., cannot gain any more oxygen, although they can lose oxygen to form defects or other bulk phases. On the other hand, $d^n$ oxides with $n \geq 1$ are potentially susceptible to oxidation as well as reduction. The stability of high oxidation states declines with increasing atomic number across a given series.

This big difference between the behavior of non–transition-metal and transition-metal oxides means that transition-metal oxides are more sensitive to a change of external conditions. It seems, therefore, that these oxides might be preferable for use in chemical sensors, and gas sensors particularly. However, in practice, simple transition-metal oxides are not being used for chemical sensor design. Structure instability, and nonoptimality of other parameters that are important for chemical sensors, such as $E_g$ and electroconductivity, considerably limit their use.

Only transition-metal oxides with $d^0$ and $d^{10}$ electronic configurations find real application. As we know, the post–transition-metal oxides, such as $ZnO$ and $SnO_2$, have cations with the filled $d^{10}$ configuration. The $d^0$ configuration is found in binary transition-metal oxides such as $TiO_2$, $V_2O_5$, $WO_3$, and perovskites such as $ScTiO_3$, $LiNbO_3$, etc., as well. These compounds share many features with the non–transition-metal oxides. They have a filled valence band of predominantly $O2p$ character and a gap between this and an empty conduction band. Typical band gaps are 3–4 eV. Stoichiometric $d^0$ oxides are therefore good isolators, diamagnetic, and have no electronic excitations at energies less than the band gap.

So, the main differences between transition-metal oxides and non–transition-metal oxides are that (1) the lower part of the conduction band is based on metal $d$, rather than $s$, orbitals, and (2) many transition-metal oxides are relatively easily reduced to form semiconducting or metallic phases.

The band model predicts that most oxides that have a partially filled $d$ band—i.e., for $d^n$ with $0 < n < 10$—should be metallic. These expectations are frequently not fulfilled because of the intervention of various types of electron–electron and electron–lattice interactions. Nevertheless, "simple" metallic behavior is found with a number of oxides of elements in the $4d$ and $5d$ series ($ReO_3$, $RuO_2$). Some oxides of the $3d$ series ($Ti_2O_3$, $VO_2$, $Fe_2O_3$) also have high conductivity in the metallic range.

However, limited use of pure transition-metal oxides for conductometric gas sensor fabrication does not mean that transition-metal oxides are not of interest to designers of chemical sensors. On the contrary, unique surface properties, plus high catalytic activity, make them very attractive for various sensor applications, such as modification the properties of more stable and wide-band-gap oxides, and formation of more complicated nanocomposite materials (Kanazawa et al. 2001). For example, for optical wave-guide sensors, where the change of optical refraction coefficient is more important than the change of electroconductivity, the most attractive metal oxides are transition-metal oxides such as $WO_3$ (for $H_2$ and alcohol detection), $Mn_2O_3$, $Co_3O_4$, and NiO (for CO detection) (Dakin and Culshaw 1988).

## 2.3. ROLE OF THE ELECTRONIC STRUCTURE OF METAL OXIDES IN SURFACE PROCESSES

In spite of many experimental and theoretical studies (Henrich and Cox 1994; Noguera 1995; Lantto et al. 2000; Maki-Jaskar and Rantala 2002), many surface properties of metal oxides are still not well understood. It is established, however, that adsorption on metal oxides involves very rich chemistry.

To analyze the adsorption and sensing properties of metal oxides, one can sometimes use the concepts of the acid/base properties of the metal oxide surface. In this case, the behavior of different oxides can be rationalized to some extent in terms of electrostatic and chemical bonding modes (Noguera 1995). Henrich and Cox (1994), however, are skeptical of the direct application of Lewis acid/base properties to explain the adsorption properties of real surfaces. Nevertheless, they think that it is possible to apply these ideas to predict some adsorption properties of metal oxides (Horiuchi et al. 1998) as well as surface activity in solutions (Trasatti 1987).

Henrich and Cox (1994) gave the following explanation of Lewis acid/base properties of metal oxides. Lewis acidity should depend on the existence of exposed metal cations at the surface, having empty orbitals and positive charges that can interact with the filled orbitals and/or negative charges or dipoles of donor molecules. Lewis acid strength is expected to depend on factors such as the ion charge, the degree of coordinative unsaturation, and the availability of empty orbitals (i.e., the band gap). For example, the stable (100) surfaces of alkaline-earth oxides should show very weak acidity because of the relatively low cation charge and the large band gap. Stronger acidity is expected with more highly charged ions such $Ti^{4+}$, and also with post–transition-metal ions as in ZnO, where the band gap is smaller and bonding with empty Zn $4s$ and $4p$ orbitals is possible.

Lewis basicity on oxide surfaces is related to the availability of a pair of $2p$ electrons associated with oxygen ions. One again, coordination is expected to be important. However, the charge of the metal cation should now play a role inverse to that expected in acid behavior. Low cation charge and large cation radius lead to weaker bonding and hence more basic $O^{2-}$ ions. Thus, the most basic of the oxides considered in this chapter is BaO, which is easily corroded by acidic adsorbates such as $CO_2$ (see Figure 2.2). These arguments show that not only the composition but also the particular crystal surface under consideration are important in determining the acid/base properties. However, the reported behavior does not always follow simple expectations. One reason is that defects may often be important, even on surfaces supposedly prepared to be defect-free. Surface defects create sites where surrounding ions have lower coordination and thus may be more active than acidic or basic centers.

In the frame of the approach discussed, Dimitrov and Komatsu (2002) tried to classify simple metal oxides. For this purpose they used the values of refractive index, the energy of the band gap, ion polarizability, cation polarizability, bulk basicity, $O1s$ binding energy determined by x-ray photoelectron spectroscopy (XPS) measurements, metal (or nonmetal) binding energy, and the Yamashita-Kurosawa interaction parameter of the oxides. Some of these parameters are presented in Table 2.4.

Analysis of these electronic properties of metal oxides allowed Dimitrov and Komatsu (2002) to propose a classification of the oxides, as shown in Table 2.5. According to this classification, the simple oxides can be separated into three groups. The first group of *semicovalent,* predominantly acidic oxides includes BeO, $B_2O_3$, $P_2O_5$, $SiO_2$, $Al_2O_3$, $GeO_2$, and $Ga_2O_3$, which have low oxide ion polarizability, high $O1s$ binding energy, low cation polarizability, high metal (or nonmetal) outermost binding energy, comparatively low bulk basicity, and strong interionic interaction, leading to the formation of strong covalent bonds.

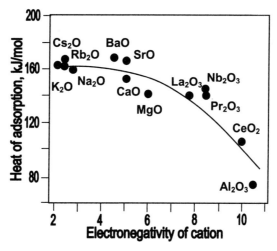

**Figure 2.2.** Heat of $CO_2$ adsorption versus electronegativity of the metal oxide cation. (Reprinted with permission from Horiuchi et al. 1998. Copyright 1998 Elsevier.)

### Table 2.4. Electronic properties of simple metal oxides

| Element | Oxide | Cation | Orbital | Binding energy (eV) element | (O1s) | Optical basicity |
|---|---|---|---|---|---|---|
| (a) | | | | | | |
| Be | BeO | $Be^{2+}$ | $1s^2$ | 113 | — | 0.375 |
| B | $B_2O_3$ | $B^{3+}$ | $1s^2$ | 191 | 533.2 | 0.425 |
| P | $P_2O_5$ | $P^{5+}$ | $2p^6$ | 133 | 533.5 | 0.33 (0.40) |
| Si | $SiO_2$ | $Si^{4+}$ | $2p^6$ | 102 | 532.8 | 0.50 |
| Al | $Al_2O_3$ | $Al^{3+}$ | $2p^6$ | 74 | 531.2 | 0.60 |
| Mg | MgO | $Mg^{2+}$ | $2p^6$ | 51 | 530.9 | 0.68 |
| Ge | $GeO_2$ | $Ge^{4+}$ | $3d^{10}$ | 31 | 531.3 | 0.70 |
| Ga | $Ga_2O_3$ | $Ga^{3+}$ | $3d^{10}$ | 20 | 530.6 | 0.755 |
| (b) | | | | | | |
| Li | $Li_2O$ | $Li^+$ | $1s^2$ | 56 | — | 0.87 |
| Ca | CaO | $Ca^{2+}$ | $3p^6$ | 25 | 529.8 | 1.00 |
| Sc | $Sc_2O_3$ | $Sc^{3+}$ | $3p^6$ | 31 | — | 0.87 |
| Ti | $TiO_2$ | $Ti^{4+}$ | $3p^6$ | 37 | 529.7 | 0.97 |
| V | $V_2O_5$ | $V^{5+}$ | $3p^6$ | 40 | 530.0 | 1.04 |
| Mn | MnO | $Mn^{2+}$ | $3d^5$ | — | 529.8 | 0.95 |
| Fe | $Fe_2O_3$ | $Fe^{3+}$ | $3d^5$ | — | 530.0 | 1.02 |
| Co | CoO | $Co^{2+}$ | $3d^7$ | — | 529.9 | 0.98 |
| Ni | NiO | $Ni^{2+}$ | $3d^8$ | — | 530.0 | 0.915 |
| Cu | CuO | $Cu^{2+}$ | $3d^9$ | — | 530.3 | 1.10 |
| Zn | ZnO | $Zn^{2+}$ | $3d^{10}$ | 10 | 530.3 | 1.08 |
| Y | $Y_2O_3$ | $Y^{3+}$ | $4p^6$ | 25 | 529.3 | 0.99 |
| Zr | $ZrO_2$ | $Zr^{4+}$ | $4p^6$ | 29 | 529.9 | 0.825 |
| Nb | $Nb_2O_5$ | $Nb^{5+}$ | $4p^6$ | 35 | — | 1.05 |
| Mo | $MoO_3$ | $Mo^{6+}$ | $4p^6$ | 38 | 530.4 | 1.07 |
| In | $In_2O_3$ | $In^{3+}$ | $4d^{10}$ | 19 | 530.1 | 1.07 |
| Sn | $SnO_2$ | $Sn^{4+}$ | $4d^{10}$ | 25 | 530.1 | 0.85 |
| Te | $TeO_2$ | $Te^{4+}$ | $5s^2$ | 14 | 530.5 | 0.93 |
| Ce | $CeO_2$ | $Ce^{4+}$ | $5p^6$ | 18 | 529.1 | 1.01 |
| Ta | $Ta_2O_5$ | $Ta^{5+}$ | $4f^{14}$ | 25 | — | 0.94 |
| W | $WO_3$ | $W^{6+}$ | $4f^{14}$ | 34 | 530.2 | 1.045 |
| (c) | | | | | | |
| Na | $Na_2O$ | $Na^+$ | $2p^6$ | 31 | 529.7 | — |
| Sr | SrO | $Sr^{2+}$ | $4p^6$ | 20 | 529.0 | 1.14 |
| Cd | CdO | $Cd^{2+}$ | $4d^{10}$ | 11 | 528.6 | 1.115 |
| Sb | $Sb_2O_3$ | $Sb^{3+}$ | $5s^2$ | 7 | — | 1.18 |
| Cs | $Cs_2O$ | $Cs^+$ | $5p^6$ | 12 | 529.4 | — |
| Ba | BaO | $Ba^{2+}$ | $5p^6$ | 16 | 528.2 | 1.22 |
| Pb | PbO | $Pb^{2+}$ | $6s^2$ | 3 | 529.7 | 1.18 |
| Bi | $Bi_2O_3$ | $Bi^{3+}$ | $6s^2$ | 8 | — | 1.19 |

*Source:* Reprinted with permission from Dimitrov and Komatsu 2002. Copyright 2002 Elsevier.

**Table 2.5. Classification of simple oxides (group of oxides) according to their electronic properties**

| OXIDES | OXIDE ION POLARIZABILITY | O1s BINDING ENERGY | CATION POLARIZABILITY | METAL-BINDING ENERGY | BULK BASICITY | INTERACTION PARAMETER | BONDING |
|---|---|---|---|---|---|---|---|
| **Semicovalent** (predominantly acidic oxides) $BeO, B_2O_3, P_2O_5, SiO_2, Al_2O_3, MgO, GeO_2, Ga_2O_3$ | Low | High | Low | High | Acidic | Strong interionic interaction | Large overlap between $O2p$ and valence metal orbitals; Strong covalent bonds |
| **Ionic** (basic) $Li_2O, CaO, Sc_2O_3, TiO_2, V_2O_5, MnO, Fe_2O_3, CoO, NiO, CuO, ZnO, Y_2O_3, ZrO_2, Nb_2O_5, MoO_3, In_2O_3, SnO_2, TeO_2$ | High | Medium range | High | Low | Basic | Mainly weak interionic interaction | Smaller overlap between $O2p$ and metal valence orbitals; Bonds with increased ionicity |
| **Very ionic** (very basic) $Na_2O, SrO, CdO, Sb_2O_3, Cs_2O, BaO, PbO, Bi_2O_3$ | Very high | Low | Very high | Very low | Very basic | Very weak interionic interaction | Small overlap between $O2p$ and metal valence orbitals; Very ionic bonds |

*Source:* Reprinted with permission from Dimitrov and Komatsu 2002. Copyright 2002 Elsevier.

The second main group includes so-called ionic or basic oxides such as CaO, $In_2O_3$, $SnO_2$, and $TiO_2$, and most of these transition-metal oxides show relatively high oxide ion polarizability, $O1s$ binding energy in a very narrow medium range, high cation polarizability, and low metal (or nonmetal) binding energy. Their bulk basicities vary in a narrow range and are close to that of CaO.

The third group of very ionic or very basic oxides includes CdO, SrO, and BaO, as well as PbO, $Sb_2O_3$, and $Bi_2O_3$, which possess very high oxide ion polarizability, low $O1s$ binding energy, very high cation polarizability, and very low metal (or nonmetal) binding energy. Their bulk basicities are higher than that of CaO, and the interionic interaction is very weak, giving rise to the formation of very ionic chemical bonds.

It is necessary to note that this metal oxide classification is very productive from our point of view, and can be applied for understanding the behavior of metal oxides in interacting with analytes that have different properties. For example, the results presented in Figure 2.2 may be understood in the frame of this approach.

"Redox" properties (ionosorption processes) is another characteristic which can be used to describe the surface behavior of metal oxides during their interaction with gas and liquid surroundings. "Redox" reactions involve an electron transfer, either directly or through the removal or addition of an oxygen atom in the metal oxide lattice. Pre–transition-metal oxides (e.g., MgO) are expected to be quite inert, since they can neither be reduced nor oxidized easily. In terms of electronic structure, this property is related to the large band gap, which means that neither electrons nor holes can be formed easily.

On the other hand, the post–transition-metal oxides ZnO, $In_2O_3$, and $SnO_2$, as well as most transition-metal oxides, are active in redox reactions, since the electron configuration of the solid may be altered. Stoichiometric ZnO, $SnO_2$, and $d^0$ transition-metal oxides may be reduced but not oxidized. Reaction with oxidizing species such as $O_2$ is therefore expected only with samples that have been bulk reduced or in which the surfaces have been made oxygen-deficient. The detailed behavior with reducing molecules may be different for different metal oxides. On ZnO, the reduction leads to the formation of free carriers, which greatly increase the surface conductivity, a fact that is crucial for sensor applications.

Reduction of transition-metal oxides, by contrast, is more likely to lead to local changes in electron configuration, with electrons trapped at specific ions in the surface region. For $SnO_2$ there is the possibility of both localized formation and free carriers. In this case it is necessary to remember that the $d^n$ transition-metal oxides may, in principle, be both reduced and oxidized, so that their redox chemistry is likely to be particularly complex.

For solid-state conductometric gas sensors, it is important that the change in electroconductivity takes place exactly through ionosorption processes, because ionosorption means charge carrier exchange with the bands of the bulk solids. Local bonding in principle does not affect the conductivity of the semiconductor, because such injection or capture of electrons is not implied. Local bonding of an adsorbate to a surface stats may affect the conductivity only indirectly, because the surface state may exchange electronic carriers with the bands of the semiconductor (Henrich and Cox 1994).

For stoichiometric metal oxides in which the metals are in their highest oxidation state, the metal atoms have lost all their valence electrons. Then they can be understood as insulators. The valence band has a predominant bonding character, and the crystal orbitals are mainly localized on $O^{2-}$. The conduction band, on the contrary, has a predominantly antibonding character, and the crystal orbitals are localized mainly on $Me^{n+}$. The most stable surfaces do not show intrinsic surface states in the band gap (Henrich and Cox 1994; Calatayud et al. 2003). However, when the oxide deviates from the stoichiometric state because of the presence of defects such as vacancies or adatoms (atoms that lie on the surface and therefore can be thought of as the opposite of surface vacancies), the oxidation state of the surface atoms varies and the electron count does not correspond to that of an insulator. All the atoms are not in their highest oxidation state. If oxygen vacancies are present, the metal oxide is reduced, some electrons filling the bottom of the conduction band or levels in the gap. If oxygen adatoms are present, the valence band is not completely filled (and some O atoms have an oxidation state of $-1$ instead of $-2$).

According to Post et al. (1999), if initially there are perfect surfaces, adsorption through an acid–base mechanism is the best way to maintain adsorbed species. Electron-rich molecules (Lewis bases) will interact at the cationic site ($Me^+$), and electron-poor ones (Lewis acids) will interact at the anionic site ($O^{2-}$). MgO and $TiO_2$ surfaces clearly appear to be predominantly acidic, and molecules that do not dissociate generally bind to the metal cation. Sites of low coordination are in general more reactive than sites of high coordination. The former is more common, and metal oxide surfaces are predominantly acidic surfaces when organic molecules are adsorbed without dissociation (Minot 2001). The adsorption on the surface of oxygen atoms involves only the cations that result from heterolytic cleavage, the metal atoms and a few specific adsorbates. Acid–base reactions do not modify the oxidation states of the atoms. In a molecular orbital description, the occupied orbitals of the base are stabilized and the vacant orbitals of the acid are destabilized. It follows that stoichiometric oxides remain insulating. The electron count for the whole system is determining.

The reactivity at surfaces that are not stoichiometric differs from that on ideal surfaces. The difference originates not only from the modification of the coordination number but also from the electron count. In this case (a defective surface), the most favorable adsorption scheme is the redox mechanism. It is thus generally believed that the presence of defects, adatoms or oxygen vacancies, creates active sites and enhances the surface activity. Oxygen vacancies modify the adsorption scheme by decreasing the coordination of the surface atoms and by modifying the electron count. The first factor enhances the activity of the acidic sites close to the vacancy. The second factor, the electronic factor, can operate in both ways, either improving or diminishing the reactivity. When the oxide is irreducible, the electron pair is trapped in an $F$ center that is extremely basic. The geometry of the oxide does not change extensively, the electron pair replacing the anion and preserving the structure of alternating charges. However, Calatayud et al. (2003) believe that the acid–base mechanism of adsorption at defective surface is nevertheless still possible. The deviation from the stoichiometry in that case modifies the acidic and basic properties of the surface atoms.

According to a results presented by many researchers, molecules that adsorb without dissociation always bind to one or several metal cations (Minot et al. 1995; Minot 2001). The oxygen ions, anionic sites with a formal charge of $-2$, are therefore less reactive. They should adsorb Lewis acids, but we have seen that $CO_2$ is not a strong enough Lewis acid for that. Adsorption at the $O^{2-}$ site is the exception. On terraces, it takes place as a secondary interaction complementing the main interaction with the metal cation, or it happens at very uncoordinated oxygen sites. Basic properties are revealed when the molecules dissociate and when the cationic fragment adsorbs at the oxygen surface sites. However, the adsorption on $O^{2-}$ of bases or radicals requires electron transfer, either a pair or a single electron, to the cations of the metal oxide—i.e., the adsorption takes place through a redox mechanism. Calatayud et al. (2003) have shown that this is easier for an electron-acceptor adsorbate and more difficult for an electron-donor one. The adsorption of a radical species is incompatible with the electron count being maintained in the stoichiometric metal oxide. A first possibility is to couple the electrons and form two opposite ions adsorbed on the two surface sites, as for $H_2$/MgO, using an acid–base mechanism. Another possibility for adsorbing radicals is via an electron transfer to/or from the metal oxide (a redox mechanism). Adsorptions with donor-like or acceptor-like adsorbates implies a redox mechanism and an electron transfer from (to) the adsorbate to (from) the metal oxide. After accepting electrons, acceptor-like moieties behave like Lewis bases and bind to the metal cations. When the adsorbate is an electropositive group (a donorlike adsorbate such as NO), the unpaired electron can be transferred to a reducible metal cation of the metal oxide. When the radical is an electronegative atom (an acceptor-like adsorbate such as Cl), it can capture an electron from the metal oxide, provided that it is in a reduced form. Then, the first adsorption corresponds to a Lewis acid (such as $NO^+$) on the oxygen atom, and the second one to that of a Lewis base (such as $Cl^-$) on the metal atom.

$SO_2$ chemisorption on the surface of metal oxides may also be understood in the frame of Lewis acid/base properties (Ziolek et al. 1996). The amount of $SO_2$ chemisorbed on metal oxides such as MgO, $Al_2O_3$, $ZrO_2$, $TiO_2$, and $CeO_2$ corresponds to the Lewis basicity strength. On all metal oxides that have been studied by IR spectroscopy, $SO_2$ chemisorption leads to the creation of Brønsted acidic sites. Results of catalysis have shown that the strength of these acid sites depends on the metal oxide: It is quite weak on MgO, more pronounced on $\gamma$-$Al_2O_3$, $ZrO_2$, and $TiO_2$ anatase, and high on rutile. The creation of such sites on ceria is accompanied by a drastic modification of its redox properties.

Adsorption of metal atoms on the oxide surface has been extensively studied during the last decade. It has been established that a metal cation has acidic properties and binds to the oxygen anions of the surface. The adsorption of neutral metal atoms, on the other hand, is weak. The charge transfer from the metal to the cations that would equilibrate the charges is not very important. This charge transfer also depends on the difference in electronegativity and is thus weak in first approximation. Electron transfer is significant for K and Ca only when the metal oxide is oxidized. Considering the first row of the transition metals, the curve of the heats of adsorption as a function of the atomic number (Figure 2.3) resembles that of the cohesive energies. They are small for the alkali and the noble metals. They are larger for the transition metals, with a depression for Cr due to the high stability of the atom in the high-spin state. Heats of adsorption are from Krilov and Kisilev (1981).

It was shown by Calatayud et al. (2003) that even though interaction of the metal atom with $O^{2-}$ is weak, it splits the metal atomic levels. The $\sigma$ levels, $4s$, and $3d_{z^2}$ (or $4p_z$) are mainly nonbonding and antibonding M–O orbitals; it is important that the latter remains unoccupied. When it is occupied by two electrons (as in Zn), the heat of adsorption is negligible. When it is occupied by one electron (as in Cr, Mn, and Cu), the heat of adsorption is weak. The heat of interaction is large for Ti, since there are no antibonding occupied orbitals, while the $\pi$ levels are occupied for V–Co (high-spin state). The trend from K to Ti is due to the variation of electronegativity. The lower the metal atomic levels are, the stronger is the interaction with the oxygen orbitals. This also explains the high heat of adsorption for Ni and Cu. Since we qualified the metal adsorption as a physisorption, the effect is of small amplitude.

Finally, it is necessary to note that the reaction medium plays an important role in adsorption processes. Clean surfaces exist in dry conditions when the surface is exposed at low gas pressure. In hydrated conditions, when the metal oxide surface is covered by water, the surface sites are not available for other molecules. As a consequence, adsorption at a hydroxylated surface is possible only in two situ-

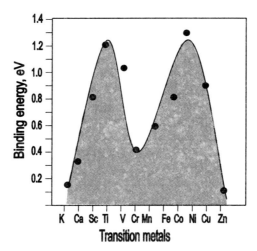

**Figure 2.3.** Binding energies of the transition metals on the MgO(100) surface at low coverage as a function of their atomic numbers.

ations: Either the adsorption is strong enough to allow the desorption of the water molecules that are bound directly to the clean surface (or a substitution of hydroxyl groups), or it occurs directly on these groups through hydrogen bonds. For example, experimental research reported by Leblanc et al. (2000) has shown that the presence of hydroxyl groups at the surface of $SnO_2$ strongly modifies the nature and the content of species formed by $NO_2$ adsorption. In the presence of hydroxyl groups, an additional interaction is observed, with the formation of hydrogenonitrate species. These groups appear to be more stable that the unidentate nitrato complexes that form in the absence of OH groups.

## 3. SURFACE PROPERTIES OF SENSING MATERIALS

### 3.1. ELECTRONIC PROPERTIES OF METAL OXIDE SURFACES

The work function, $W$, is one of the most important properties of a surface. The work function is defined as the energy necessary to remove an electron from the Fermi level in a material and put it at rest in vacuum an infinite distance away from the material. Averaged work functions measured for clean surfaces of some metal oxides are presented in Table 2.6.

The work function is extremely sensitive to the state of the surface (Henrich and Cox 1994). In fact, for metal oxides, $W$ is so sensitive that its absolute value has little significance. However, this is not to say that work functions of metal oxides are entirely useless. Relative changes in $W$ reflect variation of surface parameters and contain important information about the surface state. Therefore, measurement of these changes is often used both in studies of adsorption effects (De Fresart et al. 1982; Barsan et al. 1999; Barsan and Weimar 2003) and for design of room-temperature gas sensors (Doll et al. 1998). It should be noted that, in selecting a material for gas sensors, the absolute value of the surface potential is not as important as its change upon chemisorption.

**Table 2.6. Average work functions of some metal oxides**

| Oxide | Work function (eV) |
|---|---|
| BaO | 1.6–2.7 |
| CaO | 2.5–2.8 |
| ZnO | 4.2–4.9 |
| $SnO_2$ | 4.3–4.7 |
| $TiO_2$ | 5.1–5.9 |
| $ZrO_2$ | 5.0–6.6 |
| $SrTiO_3$ | 4.2–5.2 |
| $V_2O_3$ | 4.9 |
| $Fe_2O_3$ | 5.4 |
| NiO | 4.3–4.65 |

Concentration of native surface states is another metal-oxide surface parameter that is important for chemical sensors, especially solid-state gas sensors. To achieve effective operation of solid-state gas sensors, the concentration of these states must be minimized, because in this case the surface Fermi level will not be pinned. A low concentration of surface states creates a possibility for modulating the surface potential of a semiconductor with a change of surrounding atmosphere, because the charge of the native surface state becomes commensurable with, or less than, the charge of the chemosorbed particles.

A similar correlation with the concentration of the native surface state was observed for Schottky barrier heights at the metal–semiconductor interface (Shaw 1985). The height of the Schottky barrier at the semiconductor surface can be represented as

$$U_S = K(W_{Me} - W_S) \qquad (2.1)$$

where $W_{Me}$ and $W_S$ are the work functions for emission of an electron from a metal and a semiconductor, respectively, and $K$ is a chemical parameter that depends on the nature of the semiconductor. The parameter $K$ is defined as a coefficient of a linear dependence relating $U_S$ and $(W_{Me} - W_S)$, and can be interpreted as a demonstration of the sensitivity of the electronic properties of a semiconductor to the state of its surface. Figure 2.4 shows the parameter $K$ in relation to the difference in electronegativy $\Delta X$ (by Pauling) between the anion and cation that form the semiconductor. The abrupt change in the $K$ value at $\Delta X = 0.8$ corresponds to a transition from materials with covalent bonding (Si, Ge, GaAs) to those in which ionic bonds predominate (ZnO, $SiO_2$, $SnO_2$).

If we employ this correlation for gas sensing, the greatest sensitivity to changes in the concentration of molecules adsorbed onto the surface and, consequently, to changes in the gas-phase composition, will be exhibited by materials with predominantly ionic bonding—e.g., materials such as CdS, ZnS, and $SiO_2$. However, standard ionic semiconductors such as ZnS and CdS have low chemical and thermal stability, which limits their possible application in the design of chemical sensors. Metal oxides, which also have low concentrations of native surface state, have much higher thermal and temporal stability, which allows their successful use in chemical sensors. Furthermore, this statement is true only for conducting oxides. Oxides that are characterized by exceedingly high resistivity, such as $SiO_2$ and $Al_2O_3$, are not used as sensing materials in resistive gas sensors because of the difficulties encountered in measuring electrical conductivity. Some fluorides, such as $LaF_3$, are more thermodynamically stable than oxides in air at high temperature, but they often are more volatile.

Brinzari et al. (1999a, 2000, 2001) illustrated the influence of electronic parameters of metal oxide surfaces on gas-sensing effects, for example, CO detection by $SnO_2$ gas sensors. Unfortunately, reliable data about $SnO_2$ surface properties, which would be necessary for theoretical simulation, are practically nonexistent in the literature, so this example can serve only as preliminary guidance for experimental confirmation and further development.

Brinzari et al.'s (2001) results for $U_s(P_{CO})$ dependencies are presented in Figure 2.5. Here we can see that the density of native surface states, $N_{ss}$, really is a factor in determining the behavior of $U_s$: The slope of the $U_s(P_{CO})$ curve decreases with an increase of $U_s$. Moreover, it was shown that when $N_{ss} \rightarrow N^*$, where $N^*$ is the concentration of adsorption sites, the surface charge associated with $SnO_2$ native defects and adsorbed charged species ($Q = Q_{SS} + Q^*$) is not affected by the surrounding ambient gas. In other words, the surface potential is pinned and is not sensitive to a change in gas atmosphere. Further, the increase of density of the native surface states also leads to an increase in the sensor's threshold of sensitivity.

## 3.2. ROLE OF ADSORPTION/DESORPTION PARAMETERS IN GAS-SENSING EFFECTS

Much research has shown that for effective chemisorptions, sensor materials should have particular combinations of adsorption/desorption parameters for oxygen and detecting gases (Brinzari et al.

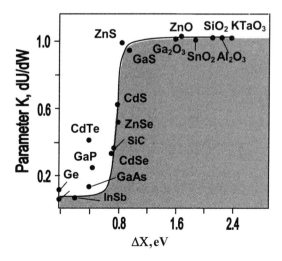

**Figure 2.4.** Influence of electronegativity on the value of $K$ in $U_S = K \Delta X$. (Adapted with permission from Kurtin et al. 1969. Copyright 1969 American Physical Society.)

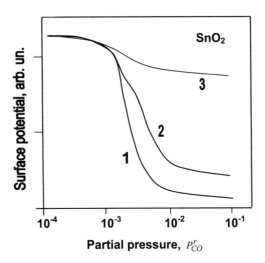

**Figure 2.5.** Influence of the concentration of surface states on the surface potential dependencies of undoped $SnO_2$ films on relative CO pressure ($P_{CO}/P_{O2}$): $P_{O2} = 2 \times 10^4$ Pa; $\beta_{CO}/\beta_{RO} = 10^{-3}$; $\alpha_{CO}/\beta_{RO} = 1–10^{-5}$; $N_d = 10^{19}$ cm$^{-3}$; $N^* = 10^{15}$ cm$^{-3}$; $T_{oper} = 300°C$: (1) $N_{ss} = 0$; (2) $N_{ss} = 6 \times 10^{12}$ cm$^{-2}$; and (3) $N_{ss} = 10^{13}$ cm$^{-2}$ (Adapted with permission from Brinzari et al. 2001. Copyright 2001 Elsevier.)

1999a, 2000, 2001). It is known that the smaller the activation energy of chemisorption and the higher the activation energy of desorption, the greater is the gas-sensing effect of adsorption-type sensors (Lundstrem 1996; Brinzari et al. 2000). At the same time, we have to take into account that at excessively large activation energy of adsorbed species, the desorption might lead to a considerable increase of recovery time after changing the surrounding atmosphere, which is not acceptable for practical applications. That is why for chemisorptions-type chemical sensors a material with optimal activation energy of desorption for the given work temperature is needed. Otherwise, to reduce recovery time, it would be necessary to increase the working temperature, which could lead to a sharp drop of the sensor's reliability and durability. For instance, simulations carried out by Brinzari et al. (2000) showed that at an operating temperature of 300°C, an activation energy for oxygen desorption of ~1.0 eV is optimal.

Brinzari et al. (2000) also considered the influence of some adsorption/desorption parameters on the surface potential ($\Delta U_s$) and $SnO_2$ conductivity ($\Delta G$) during CO detection. The pattern of the influence of the main physical-chemical parameters of a reducing gas (R) on $\Delta U_s$ and $\Delta G$ is presented in (2.2), where ↑ indicates an increase and ↓ indicates a decrease:

$$\alpha_R/\alpha_O\uparrow, \beta_4\uparrow, \beta_{RO}\uparrow, N^*\downarrow, N_{ss}\downarrow, \beta_3\downarrow, \beta_R\downarrow \Rightarrow \Delta U_s\uparrow, \Delta G\uparrow \quad (2.2)$$

where $\alpha_R$ and $\alpha_O$ are the coefficients of R and $O_2$ adsorption; $\beta_R$ and $\beta_{RO}$ are the coefficients of R and RO desorption; $\beta_3$ and $\beta_4$ are the coefficients of charging and neutralization of RO; and $N^*$ and $N_{ss}$ are the total number of adsorption sites and sites originating from native (biographic) surface charge. The scheme shows the directions of adsorption/desorption parameters changes, which are necessary to improve sensor response.

According to the model of solid-state gas sensors presented by Brinzari et al. (2000, 2001), the temperature dependence of gas sensitivity for undoped metal oxide films ($SnO_2$) is determined by the following processes:

- Dissociative adsorption of oxygen (molecular oxygen does not interact with reducing gas). This process begins at $T > 170°C$.
- Adsorption of R molecules (by impact or another mechanism) and conversion to RO (catalytic oxidation). This process depends on the type of reducing gas—for example, in the case of CO molecules, it begins at $T > 250°C$; the difference between CO and $H_2$ adsorption results from an associative mechanism for $H_2$ adsorption.
- Decrease of conversion rate of R (reverse desorption of R).
- Desorption of products of catalytic reactions.
- Desorption of chemisorbed oxygen.

The first two processes determine the drop in sensitivity at low operating temperatures, the last three determine it at high temperatures (see Figure 2.6). Competition among these processes determines both the temperature of maximum sensitivity and the half-width of the $S(T_{oper})$ sensitivity curves.

Brinzari et al. (2000) simulated the influence of some adsorption/desorption (A/D) parameters on $S(T)$ dependence. The influence of some A/D parameters on $S(T)$ dependence is shown in Figure 2.7. Here $E_{CO}$ is the activation energy of CO adsorption, and $q_{CO}$ is the activation energy of CO desorption. One can see that the decrease of activation energy of CO and $O_2$ adsorption processes may really change the sensor sensitivity and shift greatly the temperature of the maximum sensor response. In addition, this process may be accompanied by a considerable decrease of response and recovery times.

Brinzari et al. (2000) found that the position of the $U_s(P_{CO})$ curve along the $P_{CO}$ axis depends on the $\alpha_{CO}/\alpha_O$ ratio and $N^*$. The $U_s(P_{CO})$ curve shifts in the direction of lower $P_{CO}$ with an $\alpha_{CO}/\alpha_O$ increase

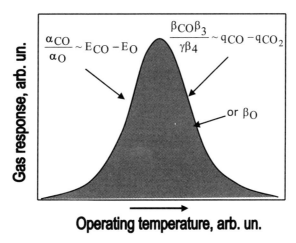

**Figure 2.6.** Adsorption/desorption parameters controlling temperature dependence of metal-oxide sensor response to CO.

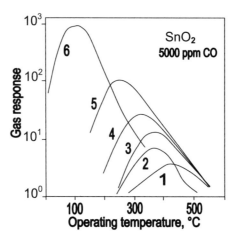

**Figure 2.7.** Simulation of the influence of adsorption/desorption parameters on temperature dependence of $SnO_2$ conductivity response to CO. $E_{CO}$: 1, 0.9 eV; 2, 0.85 eV; 3, 0.8 eV; 4, 0.7 eV; 5, 0.6 eV. $q_{CO}$: 1, 2, 3, 4, 1.6 eV; 5, 1.4 eV. (Reprinted with permission from Korotcenkov 2007. Copyright 2007 Elsevier.)

and an $N^*$ decrease (curve 3 in Figure 2.5). This means that indicated changes of adsorption/desorption parameters promote an increase in sensor response. However, in this case we need to find the optimum value for $N^*$, because with a decreased number of adsorption sites, a decrease of sensor response is observed.

The main methods of influencing the electronic parameters of adsorbed species are changing the composition of the metal oxide film, i.e., change from simple binary to multioxide films, and/or doping the metal oxide surface by adding catalyst particles (Yamazoe 1983; Korotcenkov et al. 2003). For example, doping with metal catalyst additives (Pd, Pt) seems to result in a decrease of $\alpha_{CO}/\alpha_O$, at least on metal catalyst particles, because modifying the surface with noble metals increases adsorption of dissociative oxygen. The impact adsorption of CO (by Redeal-Elley mechanism [Masel 1996]) is less affected by doping. However, due to the competition of molecular and atomic forms of oxygen on the $SnO_2$ surface, there is an "apparent" decrease of $\alpha_O$ and increase of $\alpha_{CO}/\alpha_O$ with surface doping by these noble metals. The result is shifts of both the $S(T)$ and $S(P_{CO})$ curves at lower values of $T_{oper}$ and $P_{CO}$. Simulations of $S(P_{CO})$ dependencies for doped $SnO_2$:Pd films are shown in Figure 2.8.

Analysis of gas detection reactions indicates that materials for gas sensors should also be stable to surface poisoning, i.e., they should have acceptable desorption energy of catalytic reaction products. Otherwise these products may accumulate on the surface of sensitive elements (see Figure 2.9), and parameters of interest may worsen. "Sulfur poisoning" (Somorjai 1981; Satterfield 1991) is one such type of poisoning. One can also consider species such as lead tetraethylene, metal hydride vapors, chlorides, and metal organics (Madou and Morrison 1987) as reagents which could poison active sites of the sensors.

In addition, it is desirable that surface coking does not take place. This process is one of the most important reasons for a decrease of catalytic activity during use (Somorjai 1981; Satterfield 1991). Another source of poisoning is other compounds that may be reducible to metals and elements under

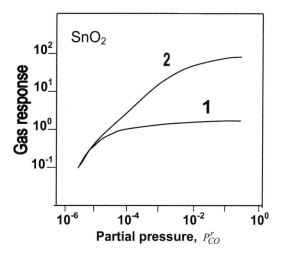

**Figure 2.8.** Simulation of the influence of metal catalysts on sensor response dependencies on CO pressure $P^r_{CO} = P_{CO}/P_{O2}$; $P_{O2} = 2\text{--}10^{-4}$ Pa; $N_{ss} = 6 \times 10^{12}$ cm$^{-2}$; $N_d = 10^{19}$ cm$^{-3}$; $T_{oper} = 300°$C; $N^* = 2 \times 10^{13}$ cm$^{-2}$: 1, undoped film; 2, modified film [$\alpha_O$(doped)/$\alpha_O$(undoped) = $10^5$; $\alpha_{CO}$(doped)/$\alpha_{CO}$(undoped) = $10^2$]. (Adapted with permission from Brinzari et al. 2001. Copyright 2001 Elsevier.)

reaction conditions. Elements such as As, Fe, P, etc., may alloy with the catalytically active metals and metal oxides, reducing their effectiveness. Arsenic is present in trace amounts in various feedstocks. Iron is a ubiquitous material of construction and can be a serious poison to platinum-group catalysts. Phosphorus compounds are typically used in lubricating oils for pumps, blowers, fans, and other machinery. In this context, one should note that metal oxides are more resistant to certain types of poisoning (especially halogens, As, Pb, and P) than noble metals. These effects of poisoning have been discussed in terms of occupancy (site blocking) and of electronic effects. It is now quite clear that strong electronic effects play a fundamental role in these changes (Ustaze et al. 1998).

Analogous requirements exist for materials for adsorption sensors, for example, such as SAW and cantilever sensors, where the change in weight of the sensing element is a determining factor (Houser

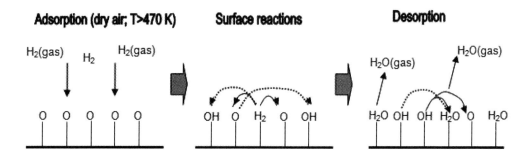

**Figure 2.9.** Schematic illustration of surface reactions during $H_2$ detection.

2001). It is known that the role of the sensing material in such devices is selectively and reversibly to sorb an analyte of interest from sampled air or liquid, and to concentrate it to achieve lower concentration detection capabilities. Therefore, maximum and reversible sorption of specific analytes or classes of analytes, with rapid sorption kinetics and minimal sorption of interferents, are key aims in the development of a successful chemoselective coating for SAW and work-function sensors (Thomson and Stone 1997).

As an example, the capability of various metal oxides for adsorption of isopropanol and methanol is shown in Table 2.7. These measurements were conducted by Wachs and co-workers (Badlani and Wachs 2001; Kulkari and Wachs 2002). The active surface site densities were calculated from the gain in weight of the samples after adsorption of isopropanol and methanol for 1 h at 110°C and were found to be 0.2–4 $\mu$mol m$^{-2}$ on the majority of metal oxide surfaces Some of the active metal oxides (MgO, $La_2O_3$, $Cr_2O_3$, $Sb_2O_3$) possessed higher active site density, which gives those oxides an advantage for the design of adsorption sensors sensitive to isopropanol and methanol. $Cr_2O_3$, $WO_3$, and BaO show somewhat lower active surface site density. $SiO_2$ is extremely unreactive and has a low $N_s$ in spite of having a high surface area.

However, we have to note that this conclusion cannot be considered to hold in all cases. Every technical task requires individual solution, considering both the nature of the gas tested and the operating conditions. For example, $CO_2$ adsorption (see Figure 2.2) is entirely different.

## 3.3. CATALYTIC ACTIVITY OF SENSING MATERIALS

In many chemical sensors, the sensitivity is determined by the effectiveness of catalytic reactions with the gas being detected at the surface of the sensing material. Results of experiments directed toward simultaneous control of sensor response and efficiency of conversion of the gas detected confirm this conclusion. Therefore, high catalytic reactivity of the surface, and especially selectivity of this reaction to detected gas, are important advantages for a sensor material. As a result, control of the catalytic activity of a material is often the main method used in preliminary evaluation of the material's suitability as a gas sensor, and in determining operating temperatures. As a rule, maximum sensor response is observed at a temperature corresponding to 50% conversion of detected gas (Yamazoe et al. 1983; Li et al. 1999; Cabot et al. 2002) (see Figure 2.10).

It is necessary to note that the maximum catalytic activity to different gases may be observed at different temperatures. This is in fact a favorable property for gas-sensing materials, because by changing the operating temperature we may be able to influence the selectivity of gas sensors. For example, the peak in sensitivity (oxidation) for methane is often at higher temperatures than for CO and other hydrocarbons, suggesting that a higher temperature may be desirable for methane-selective sensors, while a lower temperature will be desirable for CO-selective ones.

However, in spite of the obvious correlations between chemical sensing and heterogeneous catalysis, the choice of a material for gas sensor applications is not determined just by catalytic activity: Catalytic activity is an important parameter, but not a determining one; in confirmation, see Figures 2.11 and 2.12. Figure 2.12 (Krilov and Kisilev 1981) presents results on the relative catalytic activity of metal oxides in logarithmic units for different reactions.

**Table 2.7. Number of active surface sites on metal oxides during isopropanol and methanol adsorption**

| Oxide | Number of active sites ($\mu$mol m$^{-2}$) | |
|---|---|---|
| | Isopropanol adsorption | Methanol adsorption |
| MgO | 8.9 | 22.5 |
| CaO | 2.7 | 5.4 |
| SrO | — | 4.3 |
| BaO | 0.7 | 3.8 |
| $Y_2O_3$ | 3.1 | 4.9 |
| $La_2O_3$ | 5.3 | 34.1 |
| $TiO_2$ | 3.2 | 3.7 |
| $ZrO_2$ | 2.6 | 1.1 |
| $HfO_2$ | 3.1 | 2.6 |
| $CeO_2$ | 2.5 | 4.2 |
| $V_2O_5$ | 1.6 | 0.7 |
| $Nb_2O_5$ | 2.0 | 2.6 |
| $Ta_2O_5$ | 3.6 | 4.6 |
| $Cr_2O_3$ | 0.7 | 12.4 |
| $MoO_3$ | 1.2 | 0.8 |
| $WO_3$ | 0.7 | 2.3 |
| $Mn_2O_3$ | 2.6 | 1.6 |
| $Fe_2O_3$ | 7.9 | 3.7 |
| $Co_3O_4$ | 3.9 | 2.8 |
| $Rh_2O_3$ | 2.7 | 8.1 |
| NiO | 2.8 | 6.5 |
| PdO | 8.3 | 9.9 |
| PtO | 2.1 | 7.2 |
| CuO | 4.3 | 8.4 |
| $Ag_2O$ | 5.8 | 12.0 |
| $Au_2O_3$ | 36.5 | — |
| ZnO | 1.7 | 0.3 |
| $Al_2O_3$ | 2.8 | 5.6 |
| $Ga_2O_3$ | 3.4 | 4.1 |
| $In_2O_3$ | 2.0 | 2.7 |
| $SiO_2$ | 0.5 | 0.2 |
| $SnO_2$ | 2.1 | 1.6 |
| $Bi_2O_3$ | 2.0 | 2.1 |
| $P_2O_5$ | — | 3.6 |
| $Sb_2O_3$ | — | 11.3 |
| $TeO_2$ | — | 4.1 |

*Source:* Data from Badlani and Wachs 2001; Kulkame and Wachs 2002.

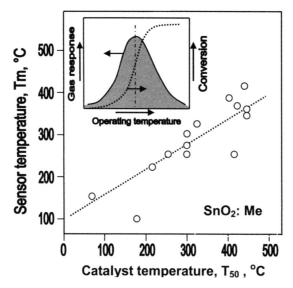

**Figure 2.10.** Correlation between catalysis temperature of 50% conversion ($T_{50}$) and temperature of $SnO_2$ sensor response maximum ($T_m$). (Reprinted with permission from Korotcenkov 2007. Copyright 2007 Elsevier.)

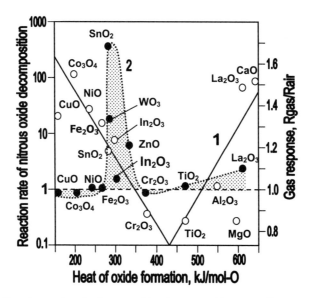

**Figure 2.11.** (1) Reaction rate of nitrous oxide decomposition at 773 K in oxygen-containing atmosphere (8% $O_2$, 1% $N_2O$) as a function of the heat of oxide formation. (2) Sensitivity to 300 ppm $N_2O$ for the various single metal oxides. (Experimental data from Satsuma et al. 2001 and Kanazawa et al. 2001.)

Numerous experiments conducted by various authors have indicated that, as a rule, oxides with electron configurations of $d^3$ ($Cr_2O_3$, $MnO_2$) and $d^{6-8}$ ($Co_3O_4$, NiO) are the most catalytically active. Minimum activity is observed for oxides with electron configurations of $d^5$ ($Fe_2O_3$, MnO), $d^0$ (CaO, $Sc_2O_3$, $TiO_2$), and $d^{10}$ (ZnO, $Cu_2O$), although the activity of oxides with an electron configuration of $d^5$ is much higher than the activity of oxides with $d^0$ and $d^{10}$ configurations. In practice, however, as we have mentioned before, metal oxides with electron configurations such as $d^0$ ($TiO_2$), $d^{10}$ (ZnO, $SnO_2$, $Cu_2O$, $Ga_2O_3$), and more rarely $d^5$ ($Fe_2O_3$), which are least active in terms of catalysis, are being considered today as the most promising gas-sensing materials. Therefore, catalytic activity, in spite of the temperature of maximum sensitivity being equal to the temperature of 50% conversion of detecting gas, cannot explain the choice of $d^{10}$ and $d^0$ oxides as base materials for conductometric gas sensors. Selection is in fact determined by the totality of the material properties, not just one. For example, based on the data presented by Krilov and Kisilev (1981) (see Figure 2.12), one can conclude that oxygen bond energy at the surface of transition-metal oxides of the fourth period is a parameter that is more useful than catalytic activity in terms of determining the adaptability of particular metal oxides for solid-state gas sensor design.

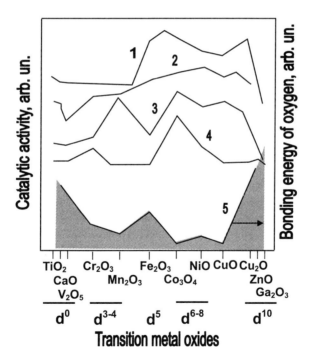

**Figure 2.12.** Relative catalytic activity of metal oxides in different reactions (1–4) and bonding energy of oxygen on the surface of transition metal oxides (5): 1, $N_2O$ decomposition; 2, isopropanol dehydration; 3, CO oxidation; 4, reaction of $H_2$–$D_2$ exchange. (Experimental data from Krilov and Kisilev 1981.)

At the same time, it is true that the choice of a metal oxide as an additive to modify the properties of another metal oxide is often connected to the catalytic properties of the oxides (Yamazoe et al. 1983; Oelerich et al. 2001). For example, the catalytic activity toward a particular gas is the most important parameter for applications in membranes and is used to improve the selectivity of sensor response. Room-temperature (RT) chemical sensors is another field of application of catalytically active oxides. Doll et al. (1998) found that RT work-function sensors based on catalytically active oxides such as CeO, $Fe_2O_3$, and NiO show good operating parameters.

# 4. STABILITY OF PARAMETERS IN SENSING MATERIALS

## 4.1. THERMODYNAMIC STABILITY

Materials for chemical sensors that must work at high temperature have to possess high thermodynamic stability. The better a material's thermodynamic stability, the higher will be the temperatures at which chemical sensors incorporating this material will work, especially in the presence of reducing gases.

Sensor materials with high thermodynamic stability should also have better temporal stability. This condition can be attained by suppressing grain-size increases during use. In this case, the opportunity to use materials with small crystallites exists, which is necessary to achieve both high sensitivity and a good rate of response. Both high heat of formation and high melting temperature characterize such materials. Table 2.8 (data from Samsonov 1973, 1982; Jonhnson 1982; Lamereaux et al. 1987; Weast et al. 1988; Badlani and Wachs 2001; etc.) lists relevant parameters of the metal oxides that are most widely used in design of chemical sensors. Here the heats of formation are normalized to the number of oxygen atoms required for reaction. Oxides in air are in the lowest free-energy state of almost all metals in the Periodic Table, which leads to their high thermodynamic stability.

The data in Table 2.8 can be considered as parameters that characterize the reactivity of elements with respect to oxygen. Table 2.9 shows some other relevant data (Dean 1972; Samsonov 1973; Henrich and Cox 1994). The elements are ordered according to the heat of formation of the most stable oxide, avaluated as $\Delta H_f$ in kJ mol$^{-1}$ of oxygen atoms.

The more reactive metals are those with a more negative heat of oxide formation, lower down in Table 2.8. They should be able to reduce the oxides of metals above them. There are, however, various reasons why the predictions of bulk thermodynamics may not be followed. There is the possibility that surface phases formed may have thermodynamic stability different from those of bulk oxides, but it is also important to remember that surface reactions of this kind require extensive migration of atoms, a process that may have a high activation energy (Henrich and Cox 1994). Such reactions are therefore more likely at higher temperatures.

The thermal program reduction (TPR) technique may be used to probe the stability of different metal oxides. In this method, diluted hydrogen is used to reduce metal oxides. Similar to methanol chemisorption, hydrogen reduction of a metal oxide proceeds through dissociative adsorption of $H_2$, which reacts with lattice oxygen to form surface hydroxyl species. Subsequently, $H_2O$ leaves the surface by eliminating the surface hydroxyl species. The TPR threshold temperatures of the metal oxides, which reflect the reducibility of the metal oxides, are shown in Table 2.8. The initial reduction temperatures

## Table 2.8. Thermodynamic stability of metal oxides

| Material | Melting temperature (°C) | $\Delta H_f$ for metal oxide formation per oxygen atom, $-\Delta H_f$ @ 298 K (kJ mol$^{-1}$) | Temperature-programmed reduction, TPR (°C) | Thermal stability in oxygen atmosphere |
|---|---|---|---|---|
| SiC | 2700 | | | |
| MgO | 2800–2820 | 601.7 | N.R. | Thermally stable (T.S.) |
| CaO | 2587–2620 | 635.1 | 300 | T.S. |
| SrO | 2430–2650 | 590.7 | 326 | T.S. |
| BaO | 1923–2015 | 553 | 330 | $T > 500°C \rightarrow BaO_2$ |
| $Y_2O_3$ | | 586.2 | 325 | T.S. |
| $La_2O_3$ | 2300 | 699.7 | 468 | T.S. |
| $TiO_2$ | 1855 | 470.8 | N.R. | T.S. |
| $ZrO_2$ | 2690 | 547.4 | N.R. | T.S. |
| $HfO_2$ | 2790 | 556.8 | N.R. | T.S. |
| $CeO_2$ | 2727 | 544.6 | 594 | T.S. |
| $V_2O_5$ | 690 | 311.9 | 550 | $T > 700°C$, evaporates with partial dissociation |
| $Nb_2O_5$ | 1512 | 381.1 | N.R. | T.S. |
| $Ta_2O_5$ | 1879 | 409.9 | 340 | T.S. |
| $Cr_2O_3$ | 2300–2435 | 380.0 | 219 | T.S. |
| $MoO_3$ | 795 | 251.7 | 575 | $T > 650°C$, sublimates |
| $WO_3$ | 1470 | 280.3 | 544 | $T > 1000°C$, sublimates |
| $Mn_2O_3$ | 1347 | 323.9 | 184 | $T > 750°C$, decomposes |
| $Fe_2O_3$ | 1347 | 247.7 | 200 | $T > 1400°C$, dissociates |
| $Co_3O_4$ | 1562 | 202.3 | 288 | $T > 900°C \rightarrow CoO$ |
| $Rh_2O_3$ | 1115 | 95.3 | 100 | T.S. |
| NiO | 1957 | 245.2 | 278 | T.S. |
| PdO | 877 | 85.0 | 90 | |
| PtO | 507 | 71.2 | 345 | Decomposes |
| CuO | 1336 | 157.0 | 268 | $T > 800°C$, decomposes |
| $Ag_2O$ | 187 | 30.6 | 200 | $T > 200°C$, decomposes |
| $Au_2O_3$ | | 26.9 | | $T > 160°C$, decomposes |
| ZnO | 1800–1975 | 348 | N.R. | T.S. |
| $Al_2O_3$ | 2050 | 558.4 | N.R | T.S. |
| $Ga_2O_3$ | 1740–1805 | 360 | 320 | T.S. |
| $In_2O_3$ | 1910–2000 | 308.6 | 350 | T.S. |
| $SiO_2$ | 1720 | 429.1 | N.R | T.S. |
| $SnO_2$ | 1900–1930 | 290.5 | 500 | T.S. |
| $Bi_2O_3$ | 817 | 192.6 | 400 | |
| $P_2O_5$ | 563 | 306.3 | N.R. | $T > 359°C$, sublimates |
| $Sb_2O_3$ | 655 | 233.2 | 563 | Easily sublimates |
| $TeO_2$ | 2127 | 162.6 | 355 | $T > 450°C$, sublimates |

N.R., no reduction detected between 150 and 700°C.
*Source:* Reprinted with permission from Korotcenkov 2007. Copyright 2007 Elsevier.

**Table 2.9. Reactivity of elements toward oxygen**

| Heat of oxide formation, $\Delta H_f$ (kJ per mole O) | Elements |
|---|---|
| 0–50 | Au, Ag, Pt |
| 50–100 | Pd, Rh |
| 100–150 | — |
| 150–200 | Ru, Cu |
| 200–250 | Re, Co, Ni |
| 250–300 | Na, Fe, Mo, Sn, Ge, W |
| 300–350 | Rb, Cs, Zn, V |
| 350–400 | K, Cr, Nb, Mn |
| 400–450 | Si, Ta |
| 450–500 | Ti |
| 500–550 | U, Ba, Zr |
| 550–600 | Al, Sr, La, Ce |
| 600–650 | Mg, Th, Ca, Sc |

*Source:* Data from Dean 1972; Henrich and Cox 1994.

vary in the range from 100°C ($Rh_2O_3$) to more than 700°C ($CeO_2$). Most of the bulk metal oxides exhibit multiple peaks in their TPR profiles due to their multiple oxidation states when they are extensively reduced. Note, however, that only the onset reduction temperatures are reported in Table 2.8.

For some metal oxides, such as $HfO_2$, MgO, ZnO, $TiO_2$, $Al_2O_3$, $SiO_2$, $ZrO_2$, $P_2O_5$, and $Nb_2O_5$, no detectable $H_2$ consumption was observed in the temperature range of 150–700°C. Some of these samples ($TiO_2$, $Al_2O_3$, $ZrO_2$, and $Nb_2O_5$) probably experienced slight surface reduction because their color changed after a TPR run and the color quickly disappeared when the sample was exposed to ambient conditions (Badlani and Wachs 2001; Kulkari and Wachs 2002). It can be seen that the initial reduction temperature is in full accordance with the heat of oxide formation, which characterizes the oxide's thermodynamic stability. The less stable the metal oxide is, the more easily the surface is reduced to form oxygen adsorption sites. The higher the energy of a stable oxide's formation is, the higher is the initial reduction temperature. Therefore, sensors based on such oxides will have better parameter stability in reducing atmosphere.

As has been mentioned, the initial reduction temperature is a very important parameter for chemical sensors, because if this temperature is exceeded, the metal oxide may be reduced to the metal during interaction with a reducing gas. So, the applicability of such oxides as $Cr_2O_3$, $Mn_2O_3$, $Fe_2O_3$, and NiO for high-temperature solid-state gas sensor design is very limited, because their working temperature in a reducing atmosphere cannot exceed 200°C.

For more complex oxides there are also some correlations which may predict their thermodynamic stability (Kreuer 1997). For example, for simple perovskites of the $ABO_3$ type with alkaline-earth metals on the A site, it was established that the thermodynamic stability of oxides is determined mainly by the choice of the B cation. In accordance with an increasing perovskite tolerance factor $R_A/R_B$ (where $R_A$ and $R_B$ are the ionic radiuses of the A and B cations), one observes increasing stability in the order cerates → zirconates → titanates. Even higher stability of perovskites, apparently, can be expected upon introduction of $Nb^{+5}$ an the B position, due to a more advantageous perovskite tolerance factor.

There are also stability criteria for polymers. According to Sandier and Karo (1974), a polyamide with high resistance to degradation should have the following properties: (1) high melting/softening point, (2) low weight loss as determined by thermogravimetric analysis, and (3) a structure that is not susceptible to degradative chain scission or intra- or intermolecular bond formation. Intensive research is being carried out in this area. For example, when a segment of the aliphatic polymer's main chain is replaced by a ring segment, the melting temperature and hence the thermal stability increases due to the decrease in the flexibility of the polymer chain (Bhuiyan 1984). However, accomplishing this is quite complicated, because it is necessary to attain high stability while maintaining the high activity of the polymer.

In addition, there are some fundamental restrictions to achieving the required thermal stability with polymers. Figure 2.13 shows that the melting point of the polymer decreases as the chain length increases. This means that a complicated polymer will inevitably have a lower melting temperature and therefore lower stability. Figure 2.13 shows relevant data for nylon-type polymers (Bhuiyan 1984).

Results given by Ryabtsev et al. (1999) show the importance of high thermal stability. $Fe_2O_3$:Pt-based sensors operating at $T_{oper} \approx 200$–$250°C$ had maximum sensitivity to acetone compared to $SnO_2$-, CdO-, and $Nb_2O_5$-based sensors. However, $Fe_2O_3$:Pt-based sensors were not used in the instrument prototype for acetone vapor analysis, due to the strong dependence of the long-term stability on operating temperature. Increasing $T_{oper}$ to higher than 250°C resulted in a sharp worsening of gas-sensing characteristics. It is important to note that this behavior of $Fe_2O_3$-based sensors corresponds to data presented in Table 2.8.

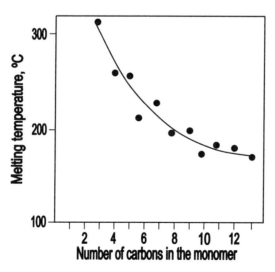

**Figure 2.13.** Influence of the number of carbon atoms in the monomer on the approximate melting temperature of nylon-type polymers. (Experimental data from Bhuiyan 1984.)

## 4.2. CHEMICAL STABILITY

Sensing materials should be characterized by high chemical stability. This property means the absence of corrosion during interaction with gases and solutions, i.e., an opportunity to work in corrosive media. The specificity of interaction of some materials with chemical reagents is given in Table 2.10 (data from Samsonov 1973, 1982; Jolivet 2000; etc.).

The chemical activity of sensor materials is also important in medical applications. Sensing elements, as well as sensor construction elements, often contact patients' blood, so prevention of infection is critical in the application of chemical sensors in medicine.

When polymers were first considered for chemical sensors, it was feared that they might not provide the necessary durability in organic solvents. However, results presented by Matsuguchi et al. (1998) showed that this need not be a problem. Polymers such as cross-linked polyamide (C-PI) and cross-linked fluorinated polyamide (C-FPI) can have high stability against many strong solvents. After interaction with solvents such as saturated acetone vapor, formadehyde, ethylene oxide gas, chlorine-type sterilizer, and ampholytic surfactant, the change in resistance of C-PI and C-FPI polymers did not exceed 2–4%.

At the same time, however, excessive chemical inactivity of a material may actually create difficulties during sensor design, when one wants to localize the sensing material on the surface of the chemical sensor platform, i.e., when creating the sensor's configuration. Such widely used oxides as tin oxide and aluminum oxide have this increased chemical resistance. To create the necessary surface configuration, one must use passive masks, dry etching, or elaborate special technological methods for local deposition of these materials in required spots (Majoo et al, 1996; Semancik et al. 1996).

On the other hand, the chemical activity of some materials with respect to certain reagents can be used to advantage. As an example, two-phase systems such as $SnO_2$-$CuO$ and $SnO_2$-$AgO$ may be used in gas sensors that will be sensitive to $H_2S$ (Tamaki et al. 1998; Jianping et al. 2000). The high sensitivity of sensors based on these materials is a consequence of the following reactions:

$$CuO + H_2S \Rightarrow CuS + H_2O\uparrow \qquad CuS + O_2 \Rightarrow CuO + SO_2\uparrow \qquad (2.3)$$

Or

$$AgO + H_2S \Rightarrow AgS + H_2O\uparrow \qquad AgS + O_2 \Rightarrow AgO + SO_2\uparrow \qquad (2.4)$$

These reactions lead to changes in chemical composition and physical properties of material, forming intercrystallite interlayers in the gas-sensing matrix.

The same principle is used in designing solid-state chemical sensors for $CO_2$ detection, for example, using the $La_2O_3/Li_2CO_3$ (1/10) system (Bogue 2002). In the presence of $CO_2$, the $La_2O_3$ is converted to lanthanum carbonate, which alters the sensor's conductivity.

$$La_2O_3 + 3CO_2 \Rightarrow La_2(CO_3)_3 \qquad (2.5)$$

It is important to note here that the use of such materials for sensor design is possible only when the reactions are completely reversible and take place at an acceptable rate.

## Table 2.10. Chemical activity of metal oxides

| Metal oxide | Reagent | Reaction |
|---|---|---|
| $MgO$ | Dilute acids | Dissolves |
| | $H_2O$ | Interacts with hydroxide formation |
| $Al_2O_3$ | Acids and alkalis | Does not interact |
| $SiO_2$ | HF | Interacts slightly |
| | Molten alkalis | Dissolves |
| $CaO$ | $H_2O$ | Interacts with $CaCO_3$ formation |
| | $CO_2$ | Dissolves |
| $SrO_2$ | Acids | Does not interact |
| $TiO_2$ | $H_2SO_4$ (conc.) | Interacts slowly during heating |
| $V_2O_5$ | $H_2SO_4$ (conc.) | Interacts during heating |
| | Alkalis | Dissolves |
| $Cr_2O_3$ | Acids and alkalis | Dissolves slightly |
| $Mn_2O_3$ | $H_2SO_4$ | Interacts |
| | HCl | Interacts |
| $Fe_2O_3$ | $H_2SO_4$ (conc.) | Interacts |
| | HCl | Interacts |
| $Co_3O_4$ | Acids | Dissolves slowly |
| $NiO$ | $H_2SO_4$ (conc.) | Dissolves |
| $Cu_2O$ | Ammonia | Dissolves |
| | HCl | Dissolves |
| | Dilute $H_2SO_4$, $HNO_3$ | Dissolves |
| $ZnO$ | Dilute acetic acid | Dissolves |
| | Dilute mineral acids (HCl, $H_3PO_4$) | Dissolves |
| | Ammonia, ammonium carbonate | Interacts |
| | Fixed alkalis | Dissolves |
| $Ga_2O_3$ | Alkalis | Dissolves |
| | Acids | Dissolves |
| $ZrO_2$ | $H_2SO_4$ (conc.) | Dissolves during heating |
| $Nb_2O_5$ | HCl | Dissolves |
| | $H_2SO_4$ | Dissolves |
| $In_2O_3$ | Acids | Dissolves |
| $SnO_2$ | Acids and alkalis | Does not interact |
| $Sb_2O_3$ | Alkalis | Dissolves |
| | HF | Dissolves |
| $CeO_2$ | Acids | Dissolves |
| $HfO_2$ | HF | Dissolves |
| $WO_3$ | Acids and alkalis | Dissolves with difficulty |
| $ReO_2$ | $HNO_3$ | Interacts |
| | HCl | Interacts |
| $La_2O_3$ | Mineral acids | Dissolves |
| $Bi_2O_3$ | Acids | Dissolves |

## 4.3. LONG-TERM STABILITY

Well-designed chemical sensors should provide long-term use regardless of the operating conditions. In general, it is required that, for example, any gas-sensing device should exhibit stable and reproducible signal for a period of at least 2–3 years (17,000–26,000 h). Therefore, high temporal stability of bulk and surface properties of sensing materials, even in corrosive media, is very important for sensor applications. This is not possible for standard semiconductors, such as Si, InP, GaAs, and GaP, because surface oxidation takes place in oxygen atmosphere, which inevitably leads to changes in the electronic, adsorption, and catalytic properties of the surface. Gas sensors based on ionic compounds, such as CuBr, have unstable parameters as well (Bendahan et al. 2002).

Organic materials, polymers, and especially biological systems also do not have the necessary temporal parameter stability. Receptors based on biological recognition elements have excellent selectivity for some reagents, but the temporal and thermal stability of these materials is very poor. For example, according to some estimates, electrochemical sensors, using the organic polymer Nafion may retain the capability for work for up to 1 year. However, to achieve this result, the Nafion must be wetted using a wicking system to a reservoir (Pasierb et al. 2004). Polymer sensors used for environmental control have another big problem: their sensitivity to ultraviolet (UV) radiation and the presence of oxidizing gases. It has been reported that ozone and other oxidizing components ($NO_x$) of the polluted atmosphere of industrial centers may be initiators or accelerators of the photochemical destruction of polymers (Razumovskii and Zaikov 1982; Heeg et al. 2001). Because of either polymerization or destruction, their properties change irreversibly within a fairly short period of time. As a result, gas sensors based on such materials have short life spans, especially in normal atmosphere containing water and active gases. Among other polymers, undoped PPY is fairly stable toward UV irradiation (Fang et al. 2002). However, the stability of PPY against UV irradiation depends on the type of dopant present in the polymer and the power density of the UV irradiation (Rabek 1995). Moreover, UV irradiation may change the thickness and surface roughness of even PPY films (Fang et al. 2002). As a result, in spite of the wide range of gas sensor prototypes designed on polymer films, very few have found their way to the market. Even those that show excellent analytical qualities are often not suitable for industrial fabrication, because of low technological effectiveness of the fabrication process as well as insufficient reliability and stability.

All recognize that to realize the advantages of polymers that have a rare combination of electrical, electrochemical, and physical properties, it is very important to increase their processability and their environmental and thermal stability (Sandler and Karo 1974; Kumar and Sharma 1998). Intensive research is being carried out in this area. For example, when a segment of an aliphatic polymer's main chain is replaced by a ring segment, the melting temperature, and hence the thermal stability, increases due to the decrease in flexibility of the polymer chain (Razumovskii and Zaikov 1982; Bhuiyan 1984). However, this task with reference to gas sensor design is rather complicated, because it is necessary to attain high stability while maintaining the polymer's activity. This again confirms that stability and reliability are determinants for practical use of any sensing material.

The same situation is observed for silicon-based chemical sensors, such as field-effect transistors (FETs) or humidity sensors. No Si-based humidity sensors are currently produced commercially because of the technological and fundamental problems of reproducibility and stability. Crystalline porous Si creates a problem because PSi is very reactive and is easily oxidized. Chemical sensors based on Si generally need an aging treatment for them to have reliable and repeatable sensitivity. Even then, lifetimes of chemical sensors based on standard semiconductors (InP, GaAs, GaP), and especially on porous Si, can be short (Han et al. 2001).

Only metal oxides and wide-band semiconductors such as SiC and GaN with a dielectric covering have the necessary stability of surface and bulk properties in both oxygen atmosphere and water environment to make them of wide practical use in real devices for long-term use. Thus the long-term stability of signal of chemical sensors is one of the most important factors determining the practical use of such devices.

Zirconia-based solid electrolytes have high stability as well: Zirconia-based solid electrolytes retain their electroconductivity even at $T > 1000°C$. However, as established by Badwal et al. (Badwal 1992; Badwal et al. 2000), to achieve such results it is necessary to use special additives, such as $Sc_2O_3$ or $Y_2O_3$ (6–10 mol%) to stabilize the parameters of zirconia-based ceramics. Note that, with respect to other solid electrolytes, stabilized zirconia ceramics have a minimal electronic contribution to total conductivity in the oxygen partial pressure $p(O_2)$ range from approximately 100–200 atm down to $10^{-25}$–$10^{-20}$ atm. This is important for practical applications.

Potentiometric sensors based on carbonates also have yet not achieved the required durability. However, recent research has indicated progress in this direction. Stable behavior of potentiometric sensors was observed for a eutectic phase of Li:K:Na carbonates used as an auxiliary electrode material (Seo et al. 2000). Another example of progress in improving sensor stability and performance was exact definition of reference electrode influence on sensor parameters. The use of two-phase systems of $Na_2Ti_6O_{13}$–$TiO_2$, $Na_2Ti_3O_7$–$Na_2Ti_6O_{13}$, and $Na_2SnO_3$–$SnO_2$ (Maier et al. 1994; Holzinger et al. 1996) or $LiCoO_2$–$Co_3O_4$ (Zhang et al. 1997) allowed experimenters to obtain sensors with stable signal, in which no shift of signal value was observed over several weeks. This research shows that all elements in chemical sensors play their parts in achieving acceptable operating parameters for practical applications.

Pasierb et al. (2004) have shown that solid-state potentiometric $CO_2$ sensors of the type $M_2CO_3$–$BaCO_3$ | Nasicon | -X (where M = Li or Na, and X is a $Na_2O$ activity buffer) with a $Na_2Ti_6O_{13}/Na_2Ti_3O_7$ reference electrode had stable signal over 1000 h at $T_{oper} = 748$ K. This is a very good result for potentiometric sensors working at high operating temperatures.

These results are in accordance with the conclusions of Meixner and Lampe (1996) that the main reasons for long-term instability of solid-state gas sensors are the following:

- A change in metal oxide parameters caused by (1) a change in the crystallite size, a consequence of insufficient preaging by tempering; (2) irreversible reactions with the gas phase—i.e., reduction during interaction, or reactions with active gases such as $SO_2$, $Cl_2$, etc., that create new phases; and (3) reactions with the substrate
- A change of metalization (sensor heating) elements, contacts for metal oxides
- Instability of the wire contacts
- Interaction with an insutable sensor casing

An additional source of temporal drift might be ionic drift, which can modify electrophysical and surface properties of metal oxides (Madou and Morrison 1987).

This strong influence of both contacts and substrate shows that the problem of stability of chemical sensors is not an easy one to resolve, and all the affecting factors need to be considered. Attempts to improve stability should be targeted to removing all possible causes of instability. High-temperature devices such as solid-state gas sensors do not have secondary elements (components). These devices operate in extreme conditions, and degradation of any component may be responsible for long-term instability.

To see how important this is, consider the results of Esch et al. (2000). In analyzing the behavior of Ti–Pt layers destined for the heater elements of chemical sensors, Esch et al. (2000) established that during use the Pt surface lost its metallic luster, and at a temperature of 650°C, hillocks appeared on the heater surface. It was found that this thermal treatment induces diffusion and oxidation of Ti, especially for annealing times up to 2 h at 450°C. Longer heat treatments did not further affect the chemical composition. Annealing at 650°C gave rise to hillock formation and strong adhesion problems.

Chemical sensors should work in both water solutions and in atmospheres containing water vapors. As we know, the relative humidity of the surrounding atmosphere can reach 100%. One can judge the importance of water vapor influence on sensor parameters by analyzing the results of research by Skouras et al. (1999). They established that adsorption of water is a dominant factor in the formation of surface characteristics, both with respect to adsorption of other species and to surface catalysis. For example, hydroxylation of a $SnO_2$ surface was found to inhibit sorption for all gas mixtures ($CH_4$, $CO$, $CO_2$, $O_2$) examined.

A hydroxylated surface is formed on an oxide by the chemisorption of a monolayer of water. Water may also catalyze reactions taking place on the surface of a chemical sensor. The adsorption of water also has an effect on the electronic properties of semiconducting metal oxides, usually acting as a donor. Morrison (1982) has shown that hydroxyiation is an intermediate stage in the interaction of water with the metal oxide. It is intermediate between hydration of the surface and physical adsorption of water. Long exposure can lead to hydration of the surface layer, and correspondingly to drift of chemical sensor characteristics. Therefore, a low tendency to hydration is an important requirement for a material intended for practical use. Only this property can provide stable operation in wet atmospheres. For example, early research in the field of humidity sensors showed that ceramic humidity sensors had progressive drift in resistance, which was caused by the gradual formation of stable chemisorbed OH groups on the oxide surface after prolonged exposure to humid environments (Kohl 1990). Given the ionic-type humidity-sensing mechanism, proton hopping was adversely affected by the surface presence of hydroxyl ions instead of water molecules, thereby resulting in a decrease in surface conductivity (Traversa et al. 1996). As a result, most commercial humidity sensors based on ceramic sensing elements were equipped with a heater for regeneration before each operation (Nitta 1980, 1981). Unfortunately, this makes it necessary to expend energy for the recovery of sensitivity of the porous ceramics, and during the cleaning operation the sensor is unable to give information about humidity.

Eventually, however, this problem was solved by using materials with different humidity-sensing mechanisms (Yanagida 1990). For example, Arshak et al. (2002), because of a happy choice of components ($SiO_2/In_2O_3$ = 75%/25%) and, probably, of parameters of thermal stabilization, succeeded in obtaining very good use parameters and high temporal stability with metal oxide conductometric humidity sensors operating at room temperature. A humidity sensitivity of 0.25%/RH% was achieved. The samples exhibited low drift over a 1-year time span (0.0013 RH%/year), low hysteresis (0.34 RH%), good linearity (±2 RH%), and a reasonably fast time response (18 s). This stability was achieved without using any additional thermal treatments.

Experiments carried out by Matsuguchi et al. (1998) showed that proper choice of polymer allows considerable improvement of temporal stability of capacitance-type humidity sensors operating at room temperatures as well. Long-term stability tests of cross-linked polyamide (C-PI) and cross-linked fluorinated polyamide (C-FPI) sensor exposed to hot and wet atmospheres showed that C-FPI sensors are quite stable after exposure to hot and wet atmospheres for long times. During these tests, sensor elements were placed in a vessel in which the atmosphere was maintained at 40°C and 90% RH. The humidity dependence of capacitance was measured over a humidity cycle of 10–90% RH at 25°C. It was established that the drift of capacitance is very small.

The reason for the drift of sensor output in hot and wet atmospheres is not clear at the present time. Hydrolysis of the film is one possible explanation. Another possibility is a change in the amount and the state of sorbed water with time, caused by morphology (distribution of microvoids) changes in the film. It is well known that a fluorinated polymer is water-repellent (Matsuguchi et al. 1998). The introduced fluorine atom also reduces the cohesive forces between the polymer segments. These characteristics of the fluorinated polymer reflect the stability of the sensor in hot and wet atmospheres. Because the melting points of cross-linked polymers are high and they are heat-resistant, cross-linked polymer sensors can have good stability even under high-temperature conditions.

All the work discussed in this section testifies that the problem of temporal stability of chemical sensors parameters, in spite of its complexity, is a temporary problem, which will be resolved in time.

## 5. ELECTROPHYSICAL PROPERTIES OF SENSING MATERIALS

### 5.1. OXYGEN DIFFUSION IN METAL OXIDES

In the signal-determining elementary interaction processes in oxide-based chemical sensors, one may distinguish between thermodynamically controlled chemisorptions and kinetically controlled catalytic reactions of the molecules to be detected, as well as between thermodynamically controlled bulk point defect equilibria (Gopel et al. 1989; Mosely 1997). Semiconductor oxides are in general nonstoichiometric, and the oxygen vacancies are the main bulk point defects. This means that changes in oxygen partial pressure at operating temperatures may cause changes in bulk conductance of metal oxides. For example, the oxygen vacancies can diffuse from the interior of the grains to the surface and vice versa, and the bulk of the oxide has to reach an equilibrium state with ambient oxygen. So, the coefficient of oxygen diffusion, which controls the equilibration time between the concentration of bulk point defects in metal oxides and the surrounding gas, is an important parameter, just like the other physical-chemical properties we have discussed.

Considering this, one can conclude that, depending on the type of solid-state gas sensor, materials may be needed that have extreme properties, e.g., very high coefficients of bulk diffusion of oxygen and point defects, or very low ones. The first type of material is necessary for chemical and gas sensors whose function depends on a change of the bulk properties of materials. In such sensors the change in bulk conductivity is a reflection of the equilibrium between the oxygen activity in the oxide and the oxygen content (oxygen partial pressure, $P_{O_2}$) of the surrounding atmosphere. Usually, in explaining their behavior, the following equation is used (Sberveglieri 1995; Wang et al. 1998):

$$G = G_O \exp\left(-\frac{E_a}{kT}\right) P_{O_2}^{\pm 1/n} \tag{2.6}$$

where $G_o$ is a constant and $E_a$ is the activation energy for conductivity. The value and sign of $1/n$ are determined by the type of dominant bulk point defect involved in the equilibration process. Positive and negative signs of $1/n$ correspond to $p$-type and $n$-type conduction, respectively. The sensitivity of a semiconducting gas sensor is determined by the value of $1/n$—the higher the value of $1/n$, the greater the sensitivity of the sensor. A high diffusion coefficient in such devices decreases both operating temperature and response time.

The main application of this kind of solid-state gas sensor is in the measurement of oxygen partial pressure as required in combustion-control systems, particularly the feedback control of the air/fuel ratio of automobile engine exhaust gases near the λ point, in order to improve fuel economy and reduce the harmful emission of gases such as CO, $NO_x$, and hydrocarbons (Gopel et al. 1989; Moseley 1997). Normally, electrochemical cells based on solid-state electrolytes such as $ZrO_2$ are used as λ sensors.

At lower temperatures, the change in ambient gas concentration does not necessarily lead to equilibration of bulk properties of metal oxides and the surrounding gas. The surrounding gas affects electrical properties through surface reactions. Therefore, for chemosorption-type sensors, the diffusion of oxygen in the bulk of crystallites is a source of temporal drift of parameters (Jamnik et al. 2002). This means that such sensors should utilize materials in which the coefficient of oxygen diffusion is minimized.

However, according to Fleischer and Meixner (1997), for any metal oxide sensor, there are three temperature regions for operation under specific conditions. At high temperatures, the kinetics of the rate processes are fast enough that equilibrium is rapidly established between the partial pressure of oxygen in the surrounding gas and the composition of the oxide. At intermediate temperatures, we can observe reaction of the gas with the metal oxide lattice, but because of the small coefficient of bulk oxygen diffusion, the chemical composition of the material does not reach equilibrium during the time of gas detection. This is the so-called "redox" (reduction/reoxidation) mechanism. At still lower temperatures, chemisorption (adsorption/desorption) processes can dominate in surface reactions. For these modes of operation, there are no fixed temperature borders (Figure 2.14)—they are fairly diffuse and can be shifted by exchanging one metal oxide for another. One can only say that the first border is in the temperature range 200–500°C, while the second border occurse at temperatures of 400–700°C.

It is impossible to say which mode of operation is preferable for practical use. Every mode has advantages and disadvantages. Some of them, relating to solid-state gas sensors, are given in Table 2.11. $SnO_2$- and $In_2O_3$-based sensors, operating in the temperature range 200–450°C, are an example of low-temperature sensors, while $TiO_2$-, $WO_3$-, $SrTiO_3$-, and $Ga_2O_3$-based sensors, functioning at temperatures higher than 450–500°C, are high-temperature sensors.

The sensing mechanisms of low-temperature gas sensors based on the widely used $SnO_2$ oxide are in general well investigated (Fleischer and Meixner 1997, 1998; Barsan et al. 1999; Brinzari et al. 2000). This sensor material is characterized by a grain boundary controlled sensing mechanism and a tendency

## DESIRED PROPERTIES FOR SENSING MATERIALS

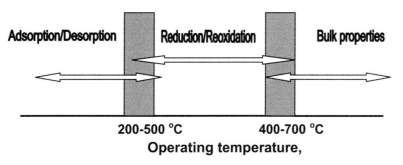

**Figure 2.14.** Processes controlling conductivity response of metal-oxide gas sensors. (Reprinted with permission from Korotcenkov 2007. Copyright 2007 Elsevier.)

to suffer from surface contamination due to a low operating temperature range, 100–400°C. This temperature is not high enough to completely burn out organic deposits or desorb certain adsorbates. This problem limits long-term stability of the electrical output signal. However, low-temperature chemosorption sensors could be easily adapted into modern microelectronics, having sufficient limitations in temperature of use. Using these, it is easier to maintain good selectivity and to create sensor arrays for design of an "electronic nose" (Gardner and Bartlett 1992, 1999).

On the other hand, high-temperature sensors are better for conducting *in-situ* control of many high-temperature technological processes, including control of combustion engines (Fleisher and Meixner 1997, 1998). The sensing behavior of such devices is mostly explainable and predictable and is based

**Table 2.11. Advantages and disadvantages of sensors operating in different modes**

| Operating temperature range | Advantages | Disadvantages |
|---|---|---|
| Low operating temperatures ($T_{oper} < 400°C$) | Low dissipated power of sensor<br>Low threshold of sensitivity<br>Long lifetime<br>Wide choice of sensitive materials<br>Good compatibility with micromachining technology | Strong dependence on relative air humidity<br>Long response and recovery times<br>Necessity of prolonged aging before start of exploitation |
| High operating temperatures ($T_{oper} > 500°C$) | Weak dependence on air humidity signal<br>Good signal reproducibility<br>Short response time<br>Fast recovery of initial state | High dissipated power<br>Lower reliability<br>Lower sensitivity<br>Strict requirements for sensing material and sensor construction<br>Poor compatibility with standard silicon technology |

*Source:* Adapted with permission from Korotcenkov 2007. Copyright 2007 Elsevier.

on well-established thermodynamic principles (Brynn and Tseung 1979). In addition, during a study of semiconducting metal oxides operated at high temperatures (400–900°C), it was established that for some materials, there is a grain boundary–independent conduction mechanism and self-cleaning effects on the sensor surfaces, thus indicating progress toward stability and reproducibility (Fleisher and Meixner 1997).

As we noted earlier, virtually all oxides can function as high-temperature gas sensors (Wang et al. 1998). In practical applications, however, the usefulness of a metal oxide for this temperature range is determined by parameters such as material stability, response time, $E_g$, type of conductivity, etc. Many semiconducting oxides have been investigated; $BaTiO_3$, $SrTiO_3$, $Ga_2O_3$, $WO_3$, $Nb_2O_3$, $CoO$, $MoO_3$, $CeO_2$, and $BaSnO_3$ are some examples. All these oxides are stable enough (see Table 2.8) and can provide sensors that operate with more or less effectiveness right up to 900°C.

For example, for $Ga_2O_3$ sensors, the optimal operating temperature is $T_{oper} \approx 600-800°C$, whereas for oxygen sensors based on $SrTiO_3$, operating temperature is 1000°C. It is necessary to note that considerable attention to complex metal oxides such as $SrTi_{0.65}Fe_{0.35}O_{3-\delta}$ is determined by unusual temperature-independent conductivity of these materials above 700°C and $P_{O2} > 1$ Pa (Litzelman et al. 2005; Rothschild et al. 2005). It was established that at an intermediate composition of $X = 0.35$, the band-gap energy is such that the Fermi energy lies just far enough above the valence band to compensate for the temperature dependence of mobility, yielding a zero temperature coefficient of resistance (TCR) from the product of the free carrier (hole) concentration and mobility terms (Rothschild et al. 2005). Strong sensitivity to oxygen partial pressure variation and negligible cross-sensitivity to temperature fluctuations make these metal oxides promising candidates for oxygen sensors in lean-burn engines (Litzelman et al. 2005; Rothschild et al. 2005).

If one considers that the presence of structural vacancies in the lattice promotes an increase in the coefficient of oxygen bulk diffusion, it is possible to assume that for design of sensors in which the appearance of diffusion processes worsens exploitation parameters, materials which do not contain structural vacancies are preferable. At the same time, for design of sensors in which bulk diffusion controls the sensor's parameters, materials with native structural vacancies are preferable.

The perovskite materials, investigated intensively in recent years for high-temperature sensors, have these very structural properties. The structures of the perovskite phases can be viewed as distortions arising from ordering of oxygen vacancies in the cubic lattice of $SrFeO_3$. Post et al. (1999) established that as the content of oxygen drops, the phase changes from cubic perovskite ($x \approx 0.5$) to tetragonal ($x \approx 0.35$), to orthorhombic ($x \approx 0.25$), to the brownmillerite structure ($x \approx 0$). These phases can be represented as a repeating sequence of octahedral and tetrahedral layers. The $SrFeO_{2.5+x}$ compounds are normally prepared and operated as sensors at high temperatures, where the oxygen-ion mobility is reasonably high. The ability to prepare these phases at room temperature was interpreted in terms of high mobility of oxygen anions along extended defects in the material as it formed (Post et al. 1999). The mobility along defects is several orders of magnitude greater than in an ordered crystal. It was postulated that the defects responsible for the increased mobility are the stacking faults in the repeating sequence of octahedral and tetrahedral layers. Rapid mobility of the oxygen anions or vacancies along the stacking faults would be followed by diffusion over very small distances within the ordered domains, on the order of some nanometers. Thus, the overall process is accelerated as a result of high diffusion coefficients over long distances and small diffusion coefficients over short distances.

Another interesting conclusion regarding requirements for materials based on dual-phase composites was made by Chena et al. (1996). On the basis of a mixed oxides study, it was established that for large oxygen permeability, the following requirements must be met: (1) percolation of the electronic conducting metal phase in the oxide matrix, allowing electrons to pass through; (2) large ionic conductivity of the oxide phase, allowing oxygen ions to migrate through; and (3) high catalytic activity of the metal phase toward surface oxygen exchange. Bismuth oxide–silver composite fulfils these requirements, showing the best oxygen permeability. For example, for a 1.60-mm-thick bismuth oxide–silver composite, an oxygen flux of $1.19\ 10^{-7}$ mol cm$^{-2}$s$^{-1}$ was observed at 800°C under conditions of $P_{O2}(h)$ = 0.21 atm and $P_{O2}(l)$ = 0.026 atm. For comparison, zirconia-based composites had oxygen permeability at the same conditions almost two orders less. Such dual-phase composites are promising for applications in oxygen separation, oxygen-enriched combustion, and catalytic membrane reactors.

It is necessary to note that considerable uncertainty exists in the choice of material for low-temperature sensors, for a temperature range lower than 450°C. The gas response of such sensors can be controlled by either chemisorption processes or "redox" processes (Korotcenkov et al. 2004). Sensors based on tin dioxide, the most studied example, are sensors of the first type, whereas $In_2O_3$-based sensors, studied intensively earlier, are sensors of the second type. However, the fact that $SnO_2$-based sensors have been better studied and are more widely used is not the basis for concluding that the chemisorption mechanism of sensitivity has advantages over the "redox" mechanism for sensor design. $SnO_2$-based sensors have better sensitivity to reducing gases, and better stability during operation in reducing atmospheres. However, $In_2O_3$-based sensors have better sensitivity to oxidizing gases, and show less dependence of parameters on changes in air humidity (Korotcenkov et al. 2004a, 2004b, 2004c). Which is better, high response to reducing gases, or to oxidizing ones? The answer to this question depends on the user's requirements.

One can make a similar comparison for other pairs of metal oxides, for example, $SnO_2$–CTO, and $SnO_2$–$WO_3$ (see Table 2.12).

All of the above indicates that the choice of materials and operation modes is determined by such factors as the type of gas sensor (see Tables 2.9–2.11), the objective (apparatus, device) for which the sensor is being designed, and the construction (structure) chosen for the sensor's fabrication. However, any competition between sensing materials can be ignored if the devices are intended for incorporation into an "electronic nose." Different behavior during interaction with the same gas is one of the most important requirements for sensors designed for this application (Gardner and Bartlett 1992, 1999).

## 5.2. CONDUCTIVITY TYPE

Metal oxides can have either $n$- or $p$-type conductivity, a difference which is very important in terms of their application. For materials with $p$-type conductivity, the conductivity increases with an increase in oxygen pressure; whereas for $n$-type oxides, the conductivity decreases with an increase in oxygen pressure (see Table 2.13).

If one considers oxides in their main valence states, there is a regular change of conductivity type depending on the metal's position in the Periodic Table. The stable metal oxides are shown in Figure 2.15. There is some information in the literature about their type of conductivity. Oxides of transition

Table 2.12. Advantages and disadvantages of gas-sensing materials

| Material | Advantages | Disadvantages |
|---|---|---|
| $SnO_2$ | High sensitivity<br>Good stability in reducing atmosphere | Low selectivity<br>Dependent on air humidity |
| $WO_3$ | Good sensitivity to oxidizing gases<br>Good thermal stability | Low sensitivity to reducing gases<br>Dependent on air humidity<br>Slow recovery process |
| $Ga_2O_3$ | High stability<br>Possibility to operate at high temperatures | Low selectivity<br>Average sensitivity |
| $In_2O_3$ | High sensitivity to oxidizing gases<br>Fast response and recovery<br>Low sensitivity to air humidity | Low stability at low oxygen partial pressure |
| CTO (CrTiO) | High stability<br>Low sensitivity to air humidity | Average sensitivity |

*Source:* Reprinted with permission from Korotcenkov 2007. Copyright 2007 Elsevier.

elements at the beginning of big periods have $n$-type conductivity, while the ones at the end usually have $p$-type conductivity.

According to Krilov and Kiselev (1981), the metal oxides of highest valence states as a rule are disposed to partial reduction, which leads to an appearance of stoichiometric excess of metal in the lattice of an $n$-type semiconductor. Oxides of lowest degrees of oxidation are disposed to partial oxidation, and both stoichiometric excess of oxygen and $p$-type conductivity is a consequence of this fact. For example, both $CrO_2$ and $CrO_3$ are oxides with $n$-type conductivity, while CrO is a $p$-type semiconductor. $Cr_2O_3$ may have either $n$-type or $p$-type conductivity, depending on oxygen pressure.

If one considers the main gas-sensing materials according to their conductivity type, it turns out that all the most effective chemisorptions-type sensors are based on materials with $n$-type conductivity, such as $SnO_2$, $TiO_2$, $WO_3$, ZnO, and $In_2O_3$, providing the opportunity for oxygen chemisorption. Previous research has shown that, in general, all $n$-type metal oxides are thermally stable and have low oxygen availability in comparison with $p$-type metal oxides (Gordon et al. 1996). It is known that inter-

Table 2.13. Expected resistance responses for reducing and oxidizing gases on $n$- and $p$-type semiconducting oxides

| Material | Decreased $P_{O2}$ | Reducing gases | Oxidizing gases |
|---|---|---|---|
| $n$-Type | − | − | + |
| $p$-Type | + | + | − |

+, Resistance rises; −, resistance falls.

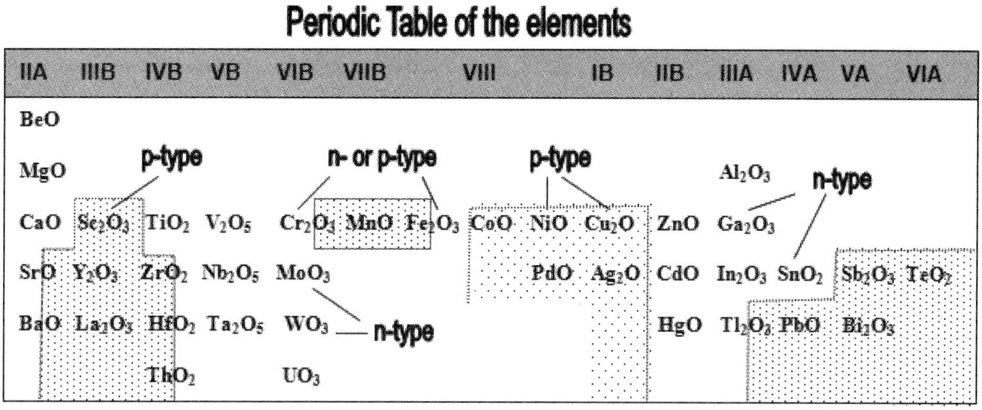

### DESIRED PROPERTIES FOR SENSING MATERIALS • 103

**Figure 2.15.** Dependence of the metal oxide's type of conductivity on the position of metals in the Periodic Table. Hatched, metal oxides with a predominance of *p*-type conductivity. Unhatched, metal oxides with a predominance *n*-type conductivity.

action with reducing gas decreases the resistance of *n*-type metal oxides. This is the preferred direction for detection of reducing gases, leading to simpler compatibility with peripheral measuring devices, and probably better reproducibility. In addition, Madou and Morrison (1987) showed that many *p*-type oxides are relatively unstable because of their tendency to exchange lattice oxygen easily with air.

However, this does not mean that *p*-type materials are not applicable for sensor design. For example, the metal oxide $Cr_{2-x}Ti_xO_3$ ($x < 0.4$) (CTO), a prospective for gas sensors, is a *p*-type material (Williams and Pratt 1997, 1998). According to Williams and Pratt (1997, 1998), CTO is in many ways an excellent material. It shows minimal effects from humidity, unlike $SnO_2$, for which humidity can dominate overall response. This may be attributed to metal oxides having separate surface sites for catalytic combustion, electrical and water vapor response, as discussed by Williams and Pratt (1998) for $SnO_2$. In addition, labile bridging oxygens, characteristic of the rutile structure of $SnO_2$, are absent in the corundum crystal structures of $Cr_2O_3$ and CTO, allowing a longer time constant for the surface modification of CTO. This, in turn, may explain the superior resilience of CTO to surface poisoning (Williams and Pratt 1997).

Stable perovskites, for example, $ReCoO_3$ (Re = La, Nd, etc.), promising candidates for chemical sensor application, are also *p*-type semiconductors. It was demonstrated on the example of $LaCoO_3$ that these materials have high activity in the oxidation of CO, reduction of NO (Voorhoeve 1972), and reduction of $SO_2$ in the presence of CO (Brynn and Tseung 1979). Concerning the catalytic mechanism, it has been clarified that a fundamental role is played by the oxygen present in the structure (bulk) and by the one adsorbed on the surface. The rate of the catalytic reactions may depend on the surface reaction and/or the oxygen bulk diffusion. In particular, at high temperatures, when the oxygen is fairly mobile, the contribution of bulk oxygen becomes more relevant. Moreover, the cations exposed to the surface have some dangling bonds and these, together with their incomplete *d* shells, favors chemisorption and electron transfer between the solid phase and the interacting molecules (Kosima et al. 1983; Chan et al. 1994).

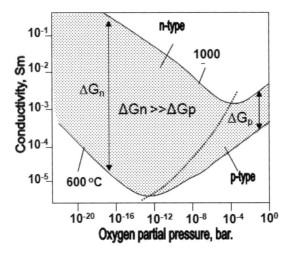

**Figure 2.16.** Typical view of the influence of oxygen partial pressure on the conductivity of perovskite-type metal oxides with *p*-type conductivity. (Adapted with permission from Moseley 1997. Copyright 1997 IOP.)

The fact that perovskites can have *p*-type conductivity gives them additional advantages for applications in high-temperature oxygen sensors. It was found that the temperature dependence of conduction at high temperatures is considerably less in the *p*-type range than in the *n*-type one; i.e., in the *p*-type range, the isothermal data lines are very much closer together than in the *n*-type range (see Figure 2.16).

Parameters of the dependence of conductivity on both oxygen partial pressure and temperature for some metal oxides are shown in Table 2.14 (data from Lee and Park 1990; Sberveglieri 1995; Moseley 1997; Menesklou et al. 2000).

Moseley (1997) gives the following explanation: In general, as the temperature is raised at a fixed oxygen partial pressure, the oxides of transition metals lose oxygen; if they are in the *p*-type range of oxygen partial pressure, then conductivity will tend to decrease. This effect is opposite to the thermal

**Table 2.14. Dependence of high-temperature conductance on oxygen partial pressure and temperature for some metal oxides**

| MATERIAL | OXYGEN SENSITIVITY ($n$) | ACTIVATION ENERGY (eV) | CRYSTAL STRUCTURE |
|---|---|---|---|
| $TiO_2$ | −4, −6 | 1.5 | Rutile |
| $Ga_2O_3$ | −4 | 1.2–1.9 | Corundum |
| $SnO_2$ | −4, −6 | 1.3–2.0 | Casseterite |
| $CeO_2$ | −4, −6 |  | Fluorite |
| $SrTiO_3$ | −4, −6 | 1.24 | Perovskite |
|  | +4 | ~0 |  |
| $BaFe_{0.8}Ta_{0.2}O_3$ | +5 | ~0 | Perovskite |

promotion of charge carriers, so the possibility arises of doping the oxide so that the two effects cancel and the effect of a temperature change is to leave the conductivity little altered. Two points need to be considered here. First, if such a balance of thermal effects is achieved, it need not alter the effect of changes in oxygen partial pressure at a fixed temperature. Second, such a balance is only possible for oxides in the $p$-type regime, where the thermal activation of charge carriers and the stoichiometry effect are opposed. The data in Figure 2.16 show that for the materials tested, the thermal activation and stoichiometric effects are not quite in balance, so the activation energy is small but not zero. A family of doped perovskite structure ferrates has been found in which balance is achieved, and the activation energy is almost zero over a considerable temperature range (Moseley 1997). Materials such as $BaFe_{0.8}Ta_{0.2}O_3$ (see Table 2.14) retain substantial oxygen sensitivity and thus should be useful in the construction of oxygen sensors without the need for additional temperature control or compensation elements. $Sr(Ti_{0.65}Fe_{0.35})O_3$ has the same properties (Menesklou et al. 2000).

CuO and ferrum oxides also have $p$-type conductivity. These oxides are effective additives to both tin and indium oxides for the formation of nano-composite-based sensors with extremely high conductivity response to $H_2S$ and series of other specific gases (Ivanovskaya et al. 2003; Kotsikau et al. 2004). In addition, materials with $p$-type conductivity are being used successfully in adsorption electrochemical sensors.

Metal oxides of $p$-type conductivity also show specific catalytic properties. Metal oxides of $p$-type are generally active oxidation catalysts. For example, $p$-type semiconducting bulk oxides ($Co_3O_4$, $MnO_2$, $MoO_3$, and $V_2O_5$) have been found to promote catalytic oxidation of carbonaceous residues through the thermal-redox cycle by releasing lattice oxygen to initiate combustion under pyrolytic conditions in the 450–600°C temperature range (Gordon et al. 1996). It turns out that ozone dissociation takes place with greater effectiveness on the surface of $p$-type oxides (Dhandapani and Oyama 1997). Metal oxides of $n$-type are generally not active oxidation catalysts. For example, Dhandapani and Oyama (1997) found that catalytic activity of unsupported oxides based on equal volume gave the following order of reactivity in terms of ozone decomposition:

$$Ag_2O > NiO > Fe_2O_3 > Co_3O_4 > CeO_2 > Mn_2O_3 > CuO$$
$$> Pb_2O_3 > Bi_2O_3 > SnO_2 > MoO_3 > V_2O_5 > SiO_2 \qquad (2.7)$$

Unfortunately in this progression, surface areas have not been considered. Later research yielded a different progression:

$$MnO_2 > Co_3O_4 > NiO > Fe_2O_3 > Ag_2O > Cr_2O_3 > CeO_2 > MgO > V_2O_5 > CuO > MoO_3 \qquad (2.8)$$

According to Gordon et al. (1996), $n$-type bulk oxides do not exhibit redox activity to any great extent under these conditions of temperature and reaction environment, because surface oxygen mobility is too low.

It is possible that distinction of catalytic properties of $p$-type conductivity oxides is a consequence of a particular surface's materials properties. Kohl (1990) notes that oxygen cannot be chemisorbed on undoped stoichiometric $n$-type oxides. Oxygen ions can only be adsorbed if their negative charge is compensated by ionized bulk donors in a space-charge layer. In addition, for thermodynamic reasons, only a small fraction of the oxygen monolayer can be chemisorbed on the surface of $n$-type oxides (the Veisz limitation). In contrast, on $p$-type semiconductors, a full monolayer of oxygen ions typically occurs, because the metal ions of the lattice can be oxidized into a higher oxidation state. However, it is necessary to note that, as a rule, with a rise in temperature, in such oxides the probability of transition of adsorbed oxygen into lattice oxygen due to its incorporation into the metal oxide lattice increases sharply. As a result, one can observe the following distinction in the behavior of $n$- and $p$-type metal oxides (Sprivey 1987): If $n$-type metal oxides have lost oxygen upon heating in air, $p$-type metal oxides gain oxygen during such thermal treatments. Numerous researchers have identified that the multiple stable oxidation states and high concentration of positive "holes" in $p$-oxides stimulate surface oxygen mobility to a greater extent than $n$-type oxides. Each "hole" provides a vacancy where the free electrons of mobile surface oxygen species can be stabilized (Gordon et al. 1996).

## 5.3. BAND GAP

The electronic structure of metal oxides is much more complex than that of most metals and standard semiconductors. For example, the bulk electronic structure of the $3d$ transition-metal oxides lies somewhere between itinerant and localized, and neither theoretical approach—the former of which has been developed for metals and semiconductors and the later that is used to describe molecules—is entirely appropriate. The great number of atoms per unit cell in many structures, coupled with the reduced symmetry at the surface and the necessity to calculate many unit cells to separate surface and bulk effects, means that first-principles electronic structure calculations still require a great deal of computer time (Maki-Jaskar and Rantala 2001, 2002). Therefore, in this section we will limit our consideration to the band gap, which may be determined experimentally.

Fairly large band gap ($E_g$) and small activation energy of the centers, responsible for metal oxide conductivity, is an optimal combination of parameters for materials designed for semiconductor solid-state chemical sensors. This combination of activation energies is necessary in order to avoid sensor operation in the region of self-conductance. In this case the influence of the surrounding temperature on the sensor's parameters is reduced. At that, as a rule, the higher the operating temperature, the larger should $E_g$ be (see Figure 2.17). The curve in Figure 2.17 was built on the basis of results presented by Bogue (2002a, 2000b). As follows from experimental results, for solid-state gas sensors operating at temperatures exceeding $T > 300°C$, the optimal band gap must be larger than 2.5 eV. For sensors working at room temperature, $E_g$ can be considerably smaller. Moreover, for sensors that function at room temperature, a small $E_g$ may be an advantage. Doll and Eisele (1996, 1998) have shown that the average work function shifts in an atmosphere of dry oxidizing gases ($Cl_2$, $NO_2$, $SO_2$), increasing when the energy band gap of metal oxides decreased.

Note that the ability to operate at higher temperatures is an important advantage of solid-state chemical sensors, because this allows one to reduce considerably the influence of air humidity on gas-sensing characteristics. It has been established that, as a rule, the lower the operating temperature, the greater is the sensitivity of the sensor's parameters to relative air humidity (Korotcenkov et al. 2003).

Analyzing the data in Table 2.15 (data from Samsonov 1973; Kindery et al. 1976; Walton 1990; Cox 1992; Gordon et al. 1996; Kumar and Sharma 1998; Sol and Tilley 2001; Colladet et al. 2004; Wallace 2004; etc.), one can see that most known metal oxides satisfy this requirement. Note that low dielectric constant is required for design of humidity sensors.

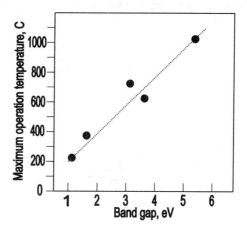

**Figure 2.17.** Influence of band gap on the maximum operating temperature of semiconductor devices. (Data from Bogue, 2002a, 200b.)

Many polymers also have fairly large band gaps. Table 2.15 lists only a range of possible $E_g$ values, because there are so many known polymers. For example, Harsanyi (1994, 2000) lists more than 10 types of polymers that are used for gas sensors, including PMODS [poly(methyl-octadecyl siloxane)], polychloroprene, polyvinylpyrrolidone, polyepichlorohydrin, polycaprolactone, DEGA (diethylene glycol adipate), PEVA (18% VA) [polyethylene-co-vinyl acetate (18% vinyl acetate)], polystyrene-butadiene, PEVA (45% VA), and polyethylene oxide. Taking into account the polymers' limited temperature range, however, they need not necessarily have wide band gaps. It is also necessary to know that the $E_g$ for polymers is only a conditional value, which may change within a certain range depending on the mode of polymerization. For example, the $E_g$ for thienylene-PPV polymers may vary within the range of 1.4 to 2.4 eV (Colladet et al. 2004).

A large band gap is also an advantage for ion semiconductors, because in this case the contribution of electron conductivity to that of the entire material, especially at high operating temperatures, is reduced.

## 5.4. ELECTROCONDUCTIVITY

Certainly a large variety of metal oxides can be used in many types of chemical sensors. However, this does not mean that metal oxides have no limitationss in application. For example, for a chemisorptional conductometric gas sensor, the sensing material should be conducting,, i.e. the concentration of point defects in the metal oxide should be fairly high. Experimental results have indicated that the optimum lies in the range $10^{17}$–$10^{19}$ cm$^{-3}$, which corresponds to an electroconductivity of the metal oxide of $10^{-2}$–$10^{1}$ Sm/cm. Too high a concentration of point defects, i.e., high electroconductivity, reduces the influence of the surface on the bulk concentration of charge carriers in the grains. In this case the effects on the surface are screened and cannot be observed. As a result, metals are usually not used for conductometric sensors.

However, results have shown that metals can be used for these purposes if they are in the form of superthin films with thickness less than 20–40 nm (Najafi et al. 1994; Johnson et al. 1994; Galdikas et

Table 2.15. Electrophysical parameters of sensing materials

| Material | Band gap (eV) | Type of conductivity | Resistance (ohm m$^{-1}$) | Dielectric constant | Coefficient of refraction |
|---|---|---|---|---|---|
| **Semiconductors** | | | | | |
| Si | 1.1 | $n, p$ | $10^3$–$10^{-5}$ | 11.8 | 3.4–3.55 |
| InP, GaAs | 1.35–1.41 | $n, p$ | $10^8$–$10^{-5}$ | 12 | |
| SiC | 3.27 | $n$ | | 9.7 | 3.4 |
| GaN | 3.6 | $n$ | | | |
| Diamond | 5.4 | $n$ | $10^{13}$ | 5.5 | |
| **Polymers** | | | | | |
| Polymers | 0.3–3.5 | $n, p$ | $10^{-18}$–$10^4$ | | |
| *trans*-Polyacetylene | 1.4 | $n, p$ | $10^{-12}$–$10^4$ | | |
| Polyphenylene | 3.4 | $n, p$ | $10^{-16}$–$10^3$ | | |
| Polypyrrole | 2.7–3.0 | $p$ | $10^{-14}$–$10^3$ | 8 | |
| Polythiophene | 2.0 | $p$ | $10^{-14}$–$10^2$ | | |
| **Metal oxides** | | | | | |
| MgO | 7.7–7.8 | $n$ | $10^{13}$–$10^{14}$ | 3–3.5 | 1.74 |
| CaO | 6.9 | $n$ | >$10^{10}$ | 3 | 1.84 |
| SrO | 5.3 | | $10^2$ | | 1.87 |
| BaO | 4.4 | | >$10^6$–$10^8$ | 4 | 1.98 |
| Y$_2$O$_3$ | 5.5–5.6 | $p$ | >$10^9$ | 14–15 | |
| La$_2$O$_3$ | 4.3 | $p$ | >$10^{10}$ | 20–30 | |
| TiO$_2$ | 3.0–3.4 | $n$ | $10^{11}$ | 50–86 | 2.00–2.70 |
| ZrO$_2$ | 4.5–5.8 | $n$ | $10^4$–$0^8$ | 17–25 | 2.10 |
| HfO$_2$ | 5.7 | $n, p$ | $10^9$–$10^{11}$ | 25 | 1.98–2.02 |
| CeO$_2$ | 3.4–5.4 | $p$ | >$10^6$–$10^{10}$ | 21 | |
| V$_2$O$_5$ | 2.2–2.45 | $n$ | $10^1$–$10^5$ | 13–15 | |
| Nb$_2$O$_5$ | 3.3–3.9 | $n$ | | 11–40 | 2.0 |
| Ta$_2$O$_5$ | 3.5–4.0 | $n$ | >$10^3$ | 21–27 | 2.5 |
| Cr$_2$O$_3$ | 1.8–3.3 | $n, p$ | >$10^2$–$10^6$ | 9 | 2.1–2.5 |
| MoO$_3$ | | $n$ | $10^5$–$10^8$ | | |
| WO$_3$ | 2.4–2.85 | $n$ | >$10^1$–$10^4$ | 20 | |
| Mn$_2$O$_3$ | | $p$ | $10^1$–$10^6$ | | 2.35 |
| Fe$_2$O$_3$ | 1.9–2.2 | $n, p$ | $10^2$–$10^6$ | | |
| Co$_3$O$_4$ | 1.0–1.1 | $p$ | $10^2$–$10^4$ | | |
| Rh$_2$O$_3$ | | | | | |
| ReO$_2$ | | | $10^{-6}$–$10^{-3}$ | | |
| NiO | 2.0–4.3 | $p$ | $10^9$–$10^{11}$ | 10 | 2.23–2.37 |
| PdO | 1.0 | $p$ | | | |
| PtO | | | | | |
| CuO | 1.2–1.4 | $n, p$ | $10^{-1}$–$10^1$ | 10 | 2.63–2.84 |

(*Continued on following page*)

**Table 2.15. (Continued)**

| MATERIAL | BAND GAP (eV) | TYPE OF CONDUCTIVITY | RESISTANCE (ohm m$^{-1}$) | DIELECTRIC CONSTANT | COEFFICIENT OF REFRACTION |
|---|---|---|---|---|---|
| Metal oxides (*Continued*) | | | | | |
| $Ag_2O$ | | p | $10^{-1}$ | | |
| $Au_2O_3$ | | | | | |
| ZnO | 3.2–3.4 | n | $10^{-4}$–$10^2$ | 8–10 | 2.00–2.02 |
| $Al_2O_3$ | 6.2–8.7 | n | $>10^{12}$–$10^{13}$ | 8–9 | 1.6–1.70 |
| $Ga_2O_3$ | 4.2–4.4 | n | $>10^6$–$10^8$ | | |
| $In_2O_3$ | 3.2–3.7 | n | $10^{-2}$–$10^2$ | | |
| $SiO_2$ | 8.9 | | $>10^{12}$–$10^{14}$ | 3.5–4.0 | 1.461–1.54 |
| $SnO_2$ | 3.6 | n | $1$–$10^4$ | 14 | 1.99–2.09 |
| $Bi_2O_3$ | | p | $10^6$–$10^{10}$ | | 2.42 |
| $P_2O_5$ | | | | | |
| $Sb_2O_3$ | 1.6 | p | | | 2.087–2.18 |
| $TeO_2$ | ~6–7 | p | | | |
| $Si_3N_4$ | 5.1 | | | 7 | |

al. 1996, 1998). Prototypes of such gas sensors were produced using thin films of various metals such as Pt, Au, and Ni. Reversible resistance response was obtained for the new sensors in air contaminated with various gases. The sensitivity and selectivity of the new sensors are comparable to those of the metal-oxide gas sensors that have already been commercialized. However, the stability of the parameters has to be studied thoroughly before these new sensors can be used for practical applications. The results of first experiments carried out by Galdikas et al. (1998) indicated that the electric parameters are stable for the metal sandwiches within fixed intervals of working temperatures. The high-temperature limit depends on the metal used for the bottom layer of the sandwich. The upper limit of temperature also depends on the amount of contaminating gases in some sandwiches. Overheating the sensors leads to limited drift of the parameters, probably caused by oxidation of the metal films. Tin oxide coating protects metal film sensors against oxidation. The coating also stabilizes the electrical parameters and extends the limits for stable operation of the sandwiches.

For conductometric sensors, excessively low concentration of free charge carriers ($n < 10^{16}$ cm$^{-3}$, i.e., $\sigma < 10^{-4}$–$10^{-5}$ Sm/cm) is also not acceptable. In nanosize structures it reduces the modulation limits of the position of the Fermi level and leads to a sharp increase in the resistance of the sensing material.

For other chemical sensors, however, such as sorptional sensors, fiber-optic sensors, sensors utilizing the fluorescence effect, etc., where conductivity is not a controlled parameter, there is no need to impose restrictions on the electroconductivity of the materials used. The materials may be either insulators or have metal-type conductivity. For example, metals may be used successfully in devices such as metal-insulator-semiconductor sensors and work-function sensors, which exploit the catalytic properties of metals (Eisele et al. 2001). Similarly, such insulators as $Al_2O_3$ are good materials for humidity sensors. Materials for high-temperature sensors ($T > 800°C$) may be insulators at room temperature. Conductivity in such materials may become apparent only at sufficiently high temperatures.

Table 2.15 provides data on the electroconductivity of various oxides used in chemical sensors (data from Samsonov 1973; Kindery et al. 1976; Harsanyi 1994; Gellinger and Bouwmeester 1997; etc.). Applicability of metals, semiconductors, and dielectrics for the most common types of sensors is given in Table 2.16.

Note that for electrochemical sensors, in contrast to metal-oxide conductometric gas sensors, high electroconductivity is an advantage. Therefore, metal oxides with metallic-type conductivity, such as $RuO_2$, $Co_3O_4$, $PbO_x$, and $MnO_2$, may be used successfully in electrochemical technology as so-called activated electrodes (Trasatti 1987).

Figure 2.18 shows the possible range of electroconductivity change for some polymers (data from Kumar and Sharma 1998; Skotheim et al. 1998; etc.). One can see that polymers cover a much wider range of conductivity than other materials, demonstrating that polymers can satisfy any requirement for electroconductivity. Polymers can exhibit conductivity across a range of some 15 orders of magnitude; however, the conductivity may involve different mechanisms within different ranges. Conductivity of polymers is influenced by a variety of parameters, including polaron length (the number of rings or double bands spanned), the conjugation length (the extent of delocalization along the chain before a defect is encountered), the overall chain length (which also affects other physical properties of the polymer), and, importantly, charge transfer to adjacent molecules. This latter involves interchain hopping within a fibril or polymer particle and finally interparticle hopping (Walton 1990).

Also, it should be noted that since conducting polymers are generally insoluble and able to interact, they are not amenable to conventional methods of purification and characterization, and some discrepancy and apparent irreproducibility may originate from changes in preparation procedures that are sufficient to alter the exact nature of the polymer under study. In addition, the mechanism of sensitivity—for example, the sensitivity of polymer materials to gases—differs from the mechanism of the sensitivity of metal oxides to gases. Therefore, the limitations described above for metal oxide sensors do not apply to polymer sensors.

## 5.5. OTHER IMPORTANT PARAMETERS FOR SENSING MATERIALS

Taking into account that at present a large variety of optical methods may be used in chemical sensors, including ellipsometry, luminescence, fluorescence, phosphorescence, Raman spectroscopy, interferometry, surface Plasmon, etc., one can conclude that for chemical sensor design based on these methods, param-

**Table 2.16. Applicability of various materials for chemical sensor design**

| | Type of chemical sensor | | | | | | |
|---|---|---|---|---|---|---|---|
| Material | Conductometric gas sensor | HT sensor | Optic sensor | Fiber optic sensor | Membrane | RT sensor (SAW, work function) | Electrochemical cell |
| Metal | + | − | − | + | + | + | + |
| Semiconductor | + | + | + | + | + | + | + |
| Insulator | − | − | + | + | + | + | − |

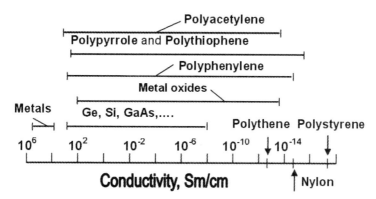

**Figure 2.18.** Comparison of conductivity ranges for several conducting polymer systems compared with conventional materials.

eters such as refractive index, absorbance and fluorescence properties of analyte molecules or chemo-optical transducing elements will have definitive importance. For example, as noted earlier, in order to shift the operating point of surface plasmon resonance gas sensors toward an aqueous environment, a thin, high-refractive-index dielectric overlayer can be employed (Koudelka-Hep 2000, 2001). Use of an overlayer with high refractive index allows for a thinner overlayer and potentially better sensor sensitivity. Analysis of data presented in Table 2.17 indicates that tantalum pentoxide, which has high refractive index and good environmental stability, may be used for this purpose.

However, metal oxides such as $Co_3O_4$, NiO, $Mn_3O_4$, CuO, and $WO_3$ are more suitable for use in optochemical sensors based on optical absorption change during interaction with analyte (Ando et al. 1997, 2001). The reversible absorbance change in the visible–near IR range and relatively fast response make these oxides potential candidates for optical detection of CO, $H_2$, and air humidity. $Co_3O_4$-based optochemical sensors can operate at room temperature (Ando et al. 1997). Of course, these sensors

**Table 2.17. Refractive indices of some sensing materials**

| Material | Refractive index |
|---|---|
| **Semiconductors** | |
| Si, InP, GaAs | 3.4–3.55 |
| **Metal oxides** | |
| $Al_2O_3$, $SiO_2$ | 1.4–1.7 |
| MgO, CaO, SrO | 1.7–2.0 |
| BaO, $ZrO_2$, $HfO_2$, $Nb_2O_5$ | ~2–2.1 |
| ZnO, $SnO_2$, $Sb_2O_3$ | 2.1–2.5 |
| $Cr_2O_3$, $Fe_2O_3$, NiO, $Bi_2O_3$, $TiO_2$, $Ta_2O_5$, CuO | >2.5 |

*Source:* Reprinted with permission from Korotcenkov 2007. Copyright 2007 Elsevier.

do not possess as high sensitivity as standard semiconductor gas sensors. However, this type of sensor has some unusual features which can be utilized in real applications. The advantages of optochemical sensors over conventional electricity-based gas sensors are higher resistivity to electromagnetic noise, compatibility with optical fibers, and the potential for multiple-gas detection using differences in the intensity, wavelength, phase, and polarization of the output light signals (Ando et al. 2001).

Dielectric constant is another important parameter of materials intended for chemical sensor applications (see Table 2.18) and should be considered when selecting materials for capacitance-type gas sensors (Ishihara and Matsubara 1998). Capacitive-type sensors have good prospects given that the capacitor structure is so simple, enabling miniaturization and achieving high reliability at low cost. In addition, application of capacitance is easily performed by oscillator circuits and thus, capacitive-type sensors enable sensitive detection. In addition, oscillator circuits consist of only a standard resistor and a sensor capacitor. Therefore, the signal treatment circuit is also very simple and low in cost. The humidity sensor is the most well known capacitive-type sensor. Since water has an abnormally high dielectric constant, the adsorption of water into porous metal oxide changes the relative permittivity of the gas-sensing matrix. In relation to other types of humidity sensors, the capacitive type has the advantage of high sensitivity over a wide humidity range. Porous $Al_2O_3$ is the best-known metal oxide for using in such sensors (Ishihara and Matsubara 1998).

## 6. STRUCTURAL PROPERTIES OF SENSING MATERIALS

It has been established that materials in different structural states can be used for chemical sensors. These states include the amorphous-like, glass-like, nanocrystalline, polycrystalline, and single-crystalline

**Table 2.18. Dielectric constants of some sensing materials**

| Material | Dielectric constant |
|---|---|
| **Metal oxides** | |
| MgO, CaO, BaO, $SiO_2$ | 3–5 |
| $Cr_2O_3$, NiO, CuO, ZnO, $Al_2O_3$ | 5–10 |
| $Y_2O_3$, $ZrO_2$, $V_2O_5$, $WO_3$, $SnO_2$ | 10–20 |
| $La_2O_3$, $HfO_2$, $CeO_2$, $Nb_2O_5$, $Ta_2O_5$ | 20–50 |
| $TiO_2$ | >50 |
| **Semiconductors** | |
| Si, InP, GaAs | 11.8–12 |
| SiC | 9.7 |
| Diamond | 5.5 |
| **Polymers** | |
| Polypyrrole | 8 |

*Source:* Reprinted with permission from Korotcenkov 2007. Copyright 2007 Elsevier.

states. Each state has its own unique properties and characteristics that can affect sensor performance. However, in practice, nanocrystalline and polycrystalline materials have found the greatest application in solid-state chemical sensors (Gopel et al. 1989–1995; Sberveglieri 1992a; Ihokura and Watson 1994; Schierbaum et al. 1992; Henrich and Cox 1994; Barsan et al. 1999; Kohl 2001; Korotcenkov 2007, 2008b). Nanocrystalline and polycrystalline materials have the optimal combination of critical properties for sensor applications, including high surface area due to small crystallite size, cheap design technology, and stability of both structural and electrophysical properties

Typically, amorphous-like and glassy materials are not stable enough for gas-sensing applications, especially at high temperature (Korotcenkov 2007a, 2008b; Tsiulyanu et al. 2004). Single-crystalline and epitaxial materials have maximum stability, and therefore the use of materials in these states for gas sensors may improve the temporal stability of the sensor. Unlike polycrystalline materials, devices based on epitaxial and single-crystalline materials will not be plagued with the problem of instability of grain size. However, the high cost and technological challenges associated with their deposition limit their general use in gas sensors.

One-dimensional structures, which are single-crystalline materials, can be synthesized using simple, inexpensive technology (Barsan et al. 1999; Dai et al. 2002; Lu et al. 2006). Wide use of one-dimensional structures is, however, impeded by the great difficulties encountered in their separation and manipulation (Deb et al. 2007; Huang and Choi 2007). During the synthesis of one-dimensional structures, one may observe a considerable diversity of geometric parameters. In polycrystalline and even nanocrystalline material we work with averaged grain size, whereas in using one-dimension structures, each sensor is characterized by the specific geometry of the one-dimensional crystal. Therefore, reproducibility of performance parameters for sensors based on one-dimensional structures depends on the uniformity of those structures. Unfortunately, the problem of separation, sizing, and manipulation of one-dimensional structures is not yet resolved. To achieve uniform sizing and orientation, advanced new technologies will need to be implemented, and these will probably be expensive and not accessible for wide use. There are a few interesting proposals for controlling one-dimensional structures (Heo et al. 2004), but they require further improvement for practical implementation.

Thus, chemical sensors based on individual one-dimensional structures are not yet readily available commercially. Further, the manufacturing cost of sensors based on one-dimensional structures would far exceed that of polycrystalline devices. Therefore, in the near future, polycrystalline materials will probably remain the dominant platform for solid-state gas sensors. However, nano- and polycrystalline materials are very complicated objects to study, because a lot of structural parameters influence on their properties and therefore control operating characteristics of chemical sensors based on these materials.

## 6.1. GRAIN SIZE

At present, either the "grains" model or the "necks" models (see Figure 2.19) are applied to rationalize the electrophysical properties of polycrystalline materials, which depend strongly on their microstructure (Xu et al. 1991; Barsan et al. 1999; Brynzari et al. 1999, 2000; Barsan and Weimar 2001; Rothschild and. Komem 2004). The grain size and the width of the necks are also the main parameters that control

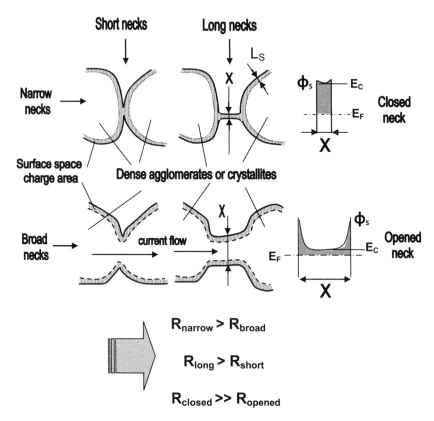

**Figure 2.19.** Diagram illustrating the role of necks in the conductivity of a polycrystalline metal oxide matrix and the potential distribution across the neck. (Reprinted with permission from Korotcenkov 2008. Copyright 2008 Elsevier.)

gas-sensing properties in metal oxide films. Moreover, in the frame of modern gas-sensor models, the influence of grain size and neck size on sensor response may be attributed to the fundamentals of gas-sensor operation (Yamazoe 1991; Xu et al. 1991; Barsan et al. 1999; Brynzari et al. 1999, 2000a; Gas'kov and Rumyantseva 2001; Rothschild and. Komem 2004; Schierbaum et al. 1991). Usually the relationship is expressed through the so-called dimension effect, i.e., a comparison of the grain size ($d$) or neck width ($X$) with the Debye length ($L_D$):

$$L_D = \sqrt{\frac{\varepsilon k T}{e^2 N}} \tag{2.9}$$

where $k$ is the Boltzmann constant, $T$ is the absolute temperature, $\varepsilon$ is the dielectric constant of the material, and $N$ is the concentration of charge carriers.

In brief, as an explanation of the dimension effect on the gas-sensing effect, we offer the following argument (see Figure 2.19) (Schierbaum et al. 1991; Barsan and Weimar 2001). For large crystallites with grain size diameter $d \gg 2L_S$, where $L_S$ is the width of the surface space charge $L_S = L_D\sqrt{eV_S/kt}$, and for a small neck width ($d < L_S$), the conductance of both the film and ceramics usually is limited by Schottky barriers ($V_S$) at the grain boundary. In this case the gas sensitivity is practically independent of $d$.

In the case of $d \approx 2L_S$, every conductive channel in necks between grains is overlapped (see Figure 2.19). If the number of long necks is much larger than the intergrain contacts, they control the conductivity of the gas-sensing material and define the size dependence of the gas sensitivity.

If $d < 2L_S$, every grain is fully involved in the space-charge layer and the electron transport is affected by the charge at the adsorbed species. Ogawa et al. (1982) demonstrated that when the grain size becomes comparable to twice the Debye length, a space-charge region can develop in the whole crystallite. The latter case is the most desirable, since it allows achievement of maximum sensor response. More detailed descriptions of models used for explanation of both grain size and neck influence on gas-sensing effects may be found in the literature (Xu et al. 1991; Yamazoe 1991; Yamazoe and Miura 1992; Wang et al. 1995; Barsan and Weimar 2001).

Note that the applicability of the "grains" or "necks" model depends strongly on the technological routes used for metal-oxide synthesis or deposition and the sintering conditions (Yamazoe 1991). Usually, the appearance of necks is a result of high-temperature annealing ($T_{an} > 700$–$800°C$). Considering processes which take place at intergrain interfaces during high-temperature annealing, one can assume that the forming of long necks in intergrain spaces is a consequence of mass transport from one grain to another. According to Xu et al. (1991), for metal oxide samples after high-temperature annealing, the neck size ($X$) is proportional to the grain size ($d$) with a proportionality constant ($X/d$) of $0.8 \pm 0.1$. However, this constant depends on the sintering parameters and may vary. For thin metal oxide films and ceramics which are not subject to high-temperature treatments, the gas-sensing matrix is formed from separately grown grains. Therefore, in such metal oxides, the necks between grains are very short or are absent. This means that we can use the "grains" model to describe of their gas-sensing properties. According to this model, there are Schottky-type contacts between grains with the height of the potential barrier depending on the surrounding atmosphere. In this approach, the grain boundary space charge or band bending on intergrain interfaces are the main parameters controlling the conductivity of nanocrystalline metal oxides. The adequacy of this model was estimated from results obtained during the impedance spectroscopy of metal oxides. (Bose et al. 2006; Kaur et al. 2005; Labeau et al. 1995; Labidi et al. 2005; Vasiliev et al. 2006).

Regarding experimental confirmation of the influence of grain size on sensor response, a dramatic increase in sensitivity for metal oxides with grain size smaller than a Debye length has been demonstrated many times for various materials, such as $SnO_2$ (Suzuki and Yamazaki 1990; Xu et al. 1991; Yamazoe and Miura 1992; Brinzari et al. 2001; Korotcenkov, 2005, 2008), $WO_3$ (Kanda and Maekawa 2005; Gillet et al. 2005), and $In_2O_3$ (Gurlo et al. 1997; Korotcenkov et al. 2004, 2007).

This effect is illustrated in Figures 2.20 and 2.21 for $In_2O_3$ and $SnO_2$-based sensors. Shimizu and Egashira (1999) established that the sensitivity of sensors based on tin oxide nanoparticles increased dramatically when the particle size was reduced to 6 nm. Below this critical grain size, sensor sensitivity decreased rapidly. As the calculated Debye length of $SnO_2$ is $L_D = 3$ nm at $250°C$ (Ogawa et al. 1982), the greatest sensitivity was actually reached when the particle diameter was 2 nm.

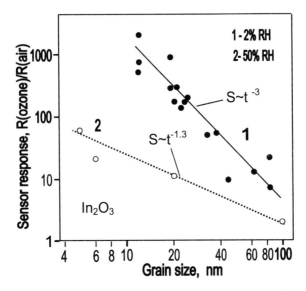

**Figure 2.20.** Influence grain size on sensor response to (1) ozone (~1 ppm) and (2) $NO_2$ (~1 ppm) of undoped $In_2O_3$-based devices. (1) Thin-film technology. Films deposited by spray pyrolysis at $T_{pyr}$ = 475°C ($T_{oper}$ = 270°C; RH = 1–2%). (2) Thick-film technology. $In_2O_3$ powders were synthesized by the sol-gel method ($d \approx 200$ nm; $T_{oper}$ = 150°C; RH = 50%). (Experimental data from Korotcenkov et al. 2004a and Gurlo et al. 1997.)

However, it is necessary to note that the threshold value of grain sizes determined by Shimizu and Egashira (1999) is not an invariable constant of $SnO_2$. This parameter must be dependent on the material's properties and the doping [see Eq. (2.9)]. For example, the Debye length in $SnO_2$ (~3 nm), estimated by Ogawa et al. (1982), corresponds to a donor concentration equal to $n_d = 3.6 \times 10^{24}$ m$^{-3}$. The decrease in concentration of free charge carriers—e.g., by improving the metal oxide stoichiometry through annealing or doping, can considerably increase this threshold value (Xu et al. 1991). Thus, Al-doped $SnO_2$ (see Figure 2.21, curve 8) shows high sensitivity with increasing grain size even at $d > 20$ nm, while Sb-doped $SnO_2$ (see Figure 2.21, curve 6) is insensitive in the whole $d$ region (see Figure 2.19). If $d \gg L_D$, the increase of sensor response with grain size decrease is not so significant.

It is important to note that the relationship to grain size is dependent on the type of metal oxide, the detection mechanism, and the gas being analyzed. For example, in $In_2O_3$-based gas sensors, the size of the crystallites is appreciably less important for the detection of reducing gases in comparison with oxidizing gases (Korotcenkov et al. 2004). Korotcenkov et al. (2004b) found that for certain deposition conditions, the sensor signal to the reducing gases can either increase or decrease with increasing grain size. In other words, for reducing gases, the $In_2O_3$ grain size is less important than for $SnO_2$-based gas sensors (Korotcenkov et al. 2004b).

For $In_2O_3$, it was also found that sensors fabricated using thin films with minimal crystallite size had minimal response time (see Figure 2.22) in addition to maximum sensor response (see Figure 2.20). For example, a decrease in grain size from 60–80 nm to 10–15 nm decreased $\tau_{res}$ during ozone detec-

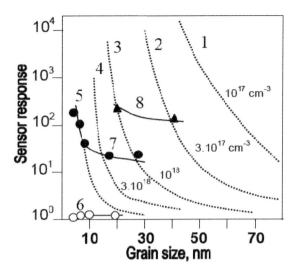

**Figure 2.21.** (1–5) theoretical and (6–8) experimental dependencies of $SnO_2$ sensor response to reducing gas on grain size for (7) undoped $SnO_2$ ceramics and ceramics doped by (6) Sb and (8) Al: 1, $N_d = 10^{17}$ cm$^{-3}$; 2, $3 \times 10^{17}$ cm$^{-3}$; 3, $10^{18}$ cm$^{-3}$; 4, $3 \times 10^{18}$ cm$^{-3}$; 5, $10^{19}$ cm$^{-3}$. Porous sensor elements were fabricated on an alumina tube with Pt wire electrodes. Elements were sintered at 700°C for 4 h. $SnO_2$ powders were synthesized by conventional hydrolysis of $SnCl_4$. Foreign oxides (5 at.%) were added by an impregnation method. Sensitivity to $H_2$ (800 ppm) was estimated at 300°C. (Data from Xu et al. 1991 and Korotcenkov et al. 2001. Reprinted with permission from Korotcenkov 2008. Copyright 2008 Elsevier.)

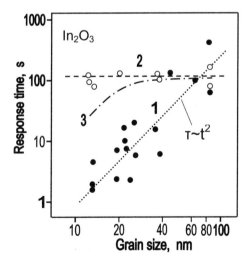

**Figure 2.22.** Influence of grain size on (1, 2) the response and (3) recovery times of the undoped $In_2O_3$-based sensors during ozone detection: $T_{pyr} = 475$°C (0.2 M), $T_{oper} = 270$°C, 1 and 2, RH = 1–2%; 3, RH = 45–50%. (Adapted with permission from Korotcenkov 2004a. Copyright 2004 Elsevier.)

tion in dry air by a factor of 50–100 times. In a humidified atmosphere, however, the correlation was significantly weakened (Korotcenkov et al. 2004a, 2004c).

On the other hand, the decrease in grain size cannot be unlimited. At some critical dimension, the number of free electrons in the grain can become zero even at $V_s = 0$. This leads to a grain resistance that does not depend on changes in the surrounding atmosphere. For charge-carrier concentrations in metal oxides of $10^{21}$ cm$^{-3}$, the critical crystallite size is 1 nm.

The use of finely dispersed small crystallites can also have a deleterious effect on the temporal stability of the sensor (Korotcenkov 2005; Korotcenkov et al. 2005). An excessive decrease of grain size leads to a loss of structural stability (Korotcenkov 2005; Korotcenkov et al. 2005b) and, as a consequence, to changes in both the surface properties and the catalytic properties of the material (Rao et al. 2002). For example, it was shown that for $SnO_2$ with a grain size of about 1–4 nm, the grain growth process already begins at temperatures of ~200–400°C (see Figure 2.23) (Shek et al. 1999; Korotcenkov et al. 2005b). In contrast, $SnO_2$ crystallites with average sizes ranging from 1.7 to 4.0 μm were stable up to 1050°C.

It thus becomes clear that the presence of a finely dispersed fraction with a grain size smaller than 2–5 nm will lead to some structural instability of the metal oxide matrix at moderate operating temperatures ($T < 600°C$). This is true even for films with average grain sizes greater than 100 nm. Therefore, future design methods for nano-scale devices, which assure grain size stabilization during long-term operation at high temperature, will gain priority over the design of methods that produce nano-scaled materials with minimal grain size. The production of both metal oxide powders and deposition of

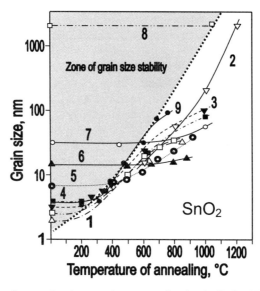

**Figure 2.23.** Influence of annealing temperature on grain size in $SnO_2$ thin films, thick films, and ceramics fabricated using different manufacturing methods. (Reprinted with permission from Korotcenkov 2005b. Copyright 2005 Elsevier.)

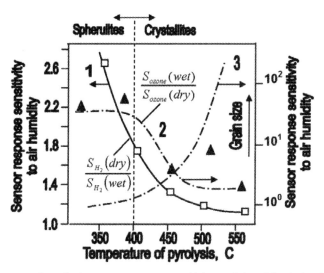

**Figure 2.24.** Influence of pyrolysis temperature on sensitivity to air humidity during (1) $H_2$ (5000 ppm) detection by undoped $In_2O_3$ sensors ($T_{oper}$ = 370°C) and (2) $O_3$ (1 ppm) detection by undoped $SnO_2$ sensors ($T_{oper}$ = 270°C), and (3) on the grain size of $In_2O_3$- and $SnO_2$-based sensors. Dry air ~1% RH, wet air ~35–45% RH. (Adapted with permission from Korotcenkov 2008. Copyright 2008 Elsevier.)

films with a small dispersion of grain sizes is an effective method for improving the temporal stability of solid-state gas sensors as well. Utilizing thin films in gas sensors also leads to improved stability of the gas-sensing matrix. Research has shown that the grain size during the annealing process changes considerably less in thin films than in thick films. Therefore, unless there is a pressing need to improve detection limits (sensitivity), it is not advisable to design sensors with excessively reduced grain sizes. This problem is especially serious for operation in atmospheres of reducing gases.

As the grain size decreases, there is one more interesting effect. With decreased grain size, a significantly increased sensitivity to humidity was observed (see Figure 2.24). The same effect was observed experimentally by Williams and Coles (1998b). Williams and Coles found that alumina-based humidity sensors also increased their sensitivity by an order of magnitude when they were prepared from 13-nm particles as compared to 300-nm particles. This means that the humidity influence of sensor parameters is structure-dependent and may be controlled through structural engineering of the metal oxides used. It has been confirmed that water adsorption on the surface of metal oxides depends on the crystallographic structure (Emiroglu et al. 2001; Golovanov et al. 2005; Batzill et al. 2005, 2006a). The difference in water attachment on different crystallographic $SnO_2$ planes appeared not only in concentration, but also in the type of bonds between hydroxyl groups and the $SnO_2$ surface. The OH groups participate in gas-detection reactions, and water adsorption/desorption processes can control the kinetics of gas response (Korotcenkov et al. 1998, 2004d, 2004e; Michel et al. 1998; Barsan et al. 1999). Thus, if minimum sensitivity to humidity is required, the size of the crystallites in the sensor material should be the maximum permissible.

Mechanical properties of materials are connected directly to their crystal structure as well. Zhang et al. (2003) showed that with a decrease in grain size, the multiplication and mobility of the dislocations are hindered, and the hardness of materials increases according to the Hall-Petch relationship, $H(d) = H_O + Kt^{1/2}$. This effect is especially prominent for grain size ($t$) down to tens of nanometers (see Figure 2.25). However, dislocation movement, which determines the hardness and strength in bulk materials, has little effect when the grain size is less than approximately 10 nm. At this grain size, further reduction in grain size brings about a decrease in strength because of grain boundary sliding. Softening caused by grain boundary sliding is attributed mainly to a large amount of defects in grain boundaries, which allows fast diffusion of atoms and vacancies under stress.

## 6.2 CRYSTAL SHAPE

At present, the effects of grain shape and faceting of crystallites on gas sensing has not been studied. However, there are reasons to believe that the role of these parameters is undeservedly understated because, depending on the external form of the nanocrystallites, a nanostructured gas-sensing matrix will have a unique combination of structural, electronic, and adsorption/desorption process parameters (Henrich and Cox 1994; Lucas et al. 2001), these parameters control the operating characteristics of all types of chemical sensors (Brinzari et al. 2001, 2002; Korotcenkov 2005; Golovanov et al. 2005; Batzill et al. 2005; Korotcenkov et al. 2005c; Batzill, 2006d). So, the determination of crystallographic planes with optimal combinations of adsorption/desorption and catalytic parameters, and the development of methods for metal-oxide deposition or synthesis with desired faces of grains can be considered an important contemporary goal for thin-film technology as applied to metal-oxide gas sensors. Current research results (Lucas et al. 2001; Kawabe et al. 2001; Brinzari et al. 2002; Batzill et al. 2005) can be considered as the first step in this direction. For example, research carried out by Batzill et al. (2005)

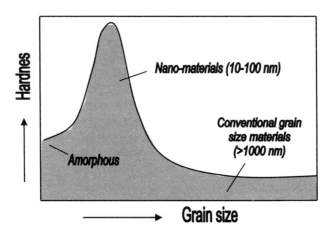

**Figure 2.25.** Schematic diagram of the influence of grain size on the hardness of the covering. (Adapted with permission from Zhang et al. 2003. Copyright 2003 Elsevier.)

has shown that the (110), (110), and (101) $SnO_2$ planes have different surface energies, different phase-transition conditions, and different energy spectra of surface states generated during their reduction. Different crystallographic planes also have different peculiarities of interaction with water (Golovanov et al. 2005; Batzill, 2006a). Kawabe et al. (2001) established that $SnO_2$ films with different texturings have different catalytic activity for selective oxidation of $CH_4$. In these investigations, one type of $SnO_2$ film had predominant orientation in the (110) crystallographic direction and another in the (211) and (301) directions. Lucas et al. (2001) observed that monodentate adsorption on MgO and CaO is preferred on the edge/corner sites. This effect is unique to small nanocrystallites (spherulites). In contrast, bidentate adsorption is favored by flat planes, which are more prevalent on the larger nanocrystals. This feature of surface species adsorption indicates that the control of surface roughness and grain shape can be exploited for improvement of gas-sensing characteristics such as absolute magnitude and selectivity. Such control may be as effective as the use of catalytic additives. This last factor is important for improving the stability of solid-state gas sensors (Korotcenkov 2005a).

Detailed studies (Brinzari et al. 2002; Korotcenkov 2005a; Korotcenkov et al. 2005a, 2005cc) conducted on $SnO_2$ films deposited by a spray pyrolysis method confirmed the appropriateness of this approach for modifying gas-sensing properties. For example, it was established that the growth of grains, especially in the range from nanometers to micrometers, during which transition from spherulites to nanocrystallites and from nanocrystallites to nanocrystals and crystals takes place, is accompanied by changes in both the size and the external shape of the crystallites (Brinzari et al. 2002; Korotcenkov et al. 2005). Possible morphologies of $SnO_2$ films deposited by spray pyrolysis are shown in Figure 2.26.

A proper choice for crystallite deposition technology or synthesis with a necessary grain facet can be also one method to decrease humidity effects in gas sensors. For example, Golovanov et al. (2005), using a Mulliken population analysis, have shown that the chemisorption of OH groups on the (110) face is accompanied by the localization of negative charge to a greater extent than the chemisorption of OH groups at the (011) surface of $SnO_2$. This means that adsorption/desorption processes and surface reactions with water vary with different $SnO_2$ crystallographic planes.

Understandably, the preparation of polycrystalline metal oxides with necessary grain faceting is difficult to control, but it is achievable for one-dimensional sensors and should be a high-priority area of research. One-dimensional structures are crystallographically perfect and have clear faceting with a fixed set of planes, which may be modified via control of synthesis parameters (Liang et al. 2001; Dai et al. 2003; Kong et al. 2003; Li et al. 2003).

Semiconducting one-dimensional metal oxide structures with well-defined geometry and perfect crystallinity might represent a perfect model material family for systematic experimental study and theoretical simulation of gas-sensing effects in metal oxides. Many parameters used for characterizing polycrystalline materials lose meaning for one-dimensional structures, because they are single-crystalline materials. These parameters include film thickness, porosity, grain size, grain network, grain boundary, agglomeration, and texturing, i.e., all the parameters that we have considered that influence the metal oxide sensor response. The main structural and morphologic parameters that characterize one-dimensional structures are geometric size, characterizing the profile of one-dimensional structures, and crystallographic planes, framing these one-dimensional structures (see Figure 2.27).

The minimal distance between faceting planes in one-dimensional structures plays the same role in gas-sensing effects as grain size plays in polycrystalline material. Undoubtedly, the decreased number

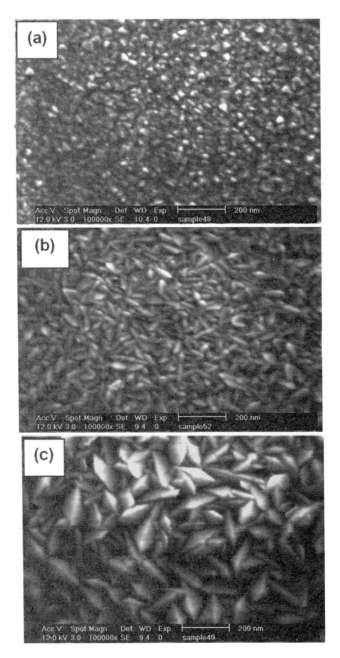

**Figure 2.26.** SEM images of SnO$_2$ films with different thicknesses ($T_{pyr}$ = 450–475°C): (a) $d \approx$ 40–50 nm; (b) $d \approx$ 80–90 nm; (c) $d \approx$ 310–380 nm. (Reprinted with permission from Korotcenkov 2005a. Copyright 2005 Elsevier.)

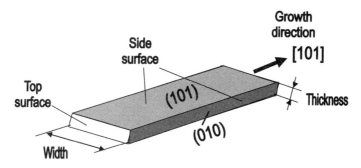

**Figure 2.27.** Schematic diagram of the geometric configuration of $SnO_2$ nanobelts. (Reprinted with permission from Korotcenkov 2008. Copyright 2008 Elsevier.)

of parameters which control the sensor response of one-dimensional structures should facilitate better understanding of the nature of the observed effects. Theoretical simulation of gas-sensing effects can become significantly simpler. As noted above, only two planes usually participate in gas-sensing effects in one-dimensional structures. This means that in simulations, one-dimensional structures should be considered as single crystals with limited sizes. On the other hand, in one-dimensional gas sensors the role of contacts increases because of their small area and therefore greater specific resistance.

It should be noted that the indicated simplified approach to consideration of gas-sensing effects in one-dimensional structures is possible only when sensors are formed based on single one-dimensional structures with invariability along the length parameters. In the presence of segmented morphology, small-diameter segments may act as necks between particles in traditional thin-film gas sensors, but with the significant advantage of greater morphological integrity and stability. Studies conducted by Dmitriev et al. (2007) have shown that the responsiveness of these structurally modulated nanowires to gases is improved over that of straight nanowires of the same average diameter. The better tolerance of such nanostructures toward contact effects is another advantage of such chemiresistors based on quasi-one-dimensional single-crystal nanostructures.

Sensors based on nanowire arrays behave similarly to sensors based on the usual nanoparticle films. Impedance spectroscopy studies have showed that the gas-sensing mechanism for sensors based on networked nanowire thin films involves changes in both the nanowire and the internanowire boundary resistances (Deb et al. 2007). Thin films containing nanowires in a highly networked fashion show promise for gas sensor design. A study carried out by Deb et al. (2007) showed that the sensors based on these films had behavior similar to that of single nanowire devices without much postprocessing. According to Deb et al. (2007), this is possible only when the individual nanowires within the gas-sensing matrix are bonded to each other and form two- and three-dimensional networks of nanowires

## 6.3. SURFACE GEOMETRY

The role of surface geometry in gas-sensing effects, which includes the role of surface steps, terraces, crystallographic defects, and corners, has not been studied in detail, in spite of the fact that the influence of surface faceting in gas adsorption phenomena and catalysis is a well-known phenomenon (Madey et al. 1999; Pederson et al. 2005). The concept of an "active site" is one of the fundamentals in heterogeneous catalysis. This notion reflects experimental evidence that, in general, catalytic surfaces are not uniformly active. Rather, only sites distinguished by a special arrangement of surface atoms (including defects), or by a particular chemical composition, are actually reactive. For example, numerous experiments have shown that in many cases monoatomic steps are highly preferential sites for many surface reactions. To address this issue, Zambelli et al. (1996) studied the dissociative chemisorption of NO on the (0001) surface of ruthenium, which is known to be the most selective catalyst. The surface was exposed to a small dose of NO (0.3 Langmuir, 1 L = $10^{-6}$ torr s) at room temperature, and scanning tunnel microscopy (STM) images were recorded. From the distribution of product atoms observed in the STM micrographs, it was inferred that adsorbed NO molecules diffuse rapidly across surface terraces until they meet a step, where they are observed to dissociate with high probability. Another good example of the structural sensitivity of a gas interaction with a solid-state surface is the behavior of hydrogen at the surface of $SnO_2$. Kohl (1990, 2001) showed that $H_2$ molecules are not activated on smooth $SnO_2$ surfaces of single crystals. The activation means excitation of a bond and, as a possible consequence, ionization, dissociation, or formation of radicals. The same effect was observed for $WO_3$ and $V_2O_5$. However, on the rough surface of sintered $SnO_2$, activation and reaction to $H_2O$ occurs at 470 K. In contrast to $H_2$, as shown by Kohl (1990, 2001), the methane can be activated on a smooth $SnO_2$ surface, forming a $CH_3$ radical.

The important role of surface steps in diffusion processes was established for perovskites as well. Chen et al. (2002) found that the in-diffusion of surface oxygen vacancies takes place mostly at the step edges. Wang et al. (2007a), using a Monte Carlo simulation of the deposition process and a mean-field theory model of the $SrTiO_3$ surface structure, established that, during deposition by laser molecular beam epitaxy, the concentration of oxygen vacancies close to step edges is greater than that on flat terraces and remains stable. It was suggested that oxygen vacancies diffusing at the surface tend to accumulate near step edges due to their slow in-diffusion rate there, and that this in-diffusion dominates the oxidation of the as-deposited film. The step edges act as a route for oxygen vacancies at the surface to move into the film. Wang et al. (2007a) concluded that the in-diffusion rate of oxygen vacancies is limited by step edges. No matter how large the surface diffusion rate is, the oxidation process is dominated by the in-diffusion of oxygen vacancies near step edges.

These observations illustrate that the adsorption properties of many gases are very sensitive to the microscopic surface geometry of metal oxides. Therefore, by expanding our understanding of these processes along with the development of necessary technologies, this phenomenon could be used to improve the selectivity of solid-state gas sensors. For example, it is possible that the increased sensitivity to nitrogen oxide observed after high-temperature annealing and subsequent mechanical milling of $SnO_2$ (Dieguez et al. 2000; Tan et al. 2000; Abe et al. 2005) is a consequence of the appearance of microsteps formed as a result of the mechanical crushing.

In the context of the role of surface morphology in sensor performance, the results of Batzill et al. (2006) deserve special attention. Batzill et al. (2006) analyzed the interaction of water with a $SnO_2$ (101) surface and found that the water interacts weakly with the reduced surface; the small amount of water at the $SnO_2$ surface for $T > 130$ K may be connected with water adsorption at surface defect sites such as step edges (Batzill et al. 2005). This implies that surface geometry may control water influence on gas response as well.

The effect of noble-metal surface clustering, which has not been studied in terms of gas sensing, may be directly connected with surface geometry as well. There are statements (Yamazoe et al. 1991; Kabbabi et al. 1994; Meier et al. 2004; Castañeda 2007) indicating that the dispersion state of surface catalyst particles, which can be characterized by particle size and population, can favorably affect gas-sensing characteristics. It has been shown that improved sensitivity can be achieved if the aggregate distribution of noble metals such as Pd and Pt in the film is characterized by very small particle size coupled with a high number density (Matko et al. 2002). However, it was impossible to find in the literature any correlation between the size of surface clusters and gas sensitivity.

Recent research, however, has shown that the process of surface clustering is structurally sensitive (Chen et al. 2002), and therefore the consequences of surface modification may depend on the surface structure of the metal oxides. The most important consequences pertain to such structural parameters as surface geometry, i.e., the presence of steps, terraces, and facets, the size of planes, the degree of surface reduction, and other parameters (El-Azab et al. 2002; Min et al. 2006).

Recent work (Valden et al. 1998; El-Azab et al. 2002; Lauritsen et al. 2006) has provided direct experimental confirmation that, following thermal treatments, noble-metal clusters accumulate at the step edges of metal oxides (see Figure 2.28). Due to this behavior of noble metals, an increase of terrace area should be accompanied by an increase in the size of clusters, which can be accumulated at the step edges of this terrace. This means that for the same degree of surface coverage by noble metals, the number of clusters will be smaller, while the distance between the clusters will be bigger on the surface with the bigger terrace area (see Figure 2.29).

From this it follows that the appearance of extended, atomically flat surface planes facilitates the process of cluster growth. At the same time, the presence of atomically stepped-like surfaces will provide with high probability the conditions for atomic dispersion of noble metals at the surface of metal oxides. Min et al. (2006) established that Au clusters had higher densities on highly reduced $TiO_2$. It was suggested that reduced Ti sites act as active sites for the nucleation and growth of Au clusters.

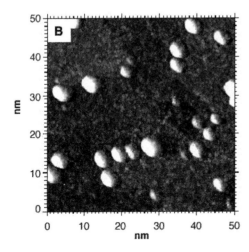

**Figure 2.28.** STM image of Au/$TiO_2$(110) surface after annealing at 850 K for 2 min. The Au coverage was 0.25 monolayer (ML). (Reprinted with permission from Lai et al. 1998. Copyright 1998 Elsevier.)

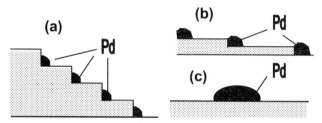

**Figure 2.29.** Schematic illustration of the influence of terrace size on the size of noble metal clusters. (Reprinted with permission from Korotcenkov 2005. Copyright 2005 Elsevier.)

The size of clusters also depends on the nature of the metal deposited. For example, Alfredsson et al. (2004), analyzing the behavior of Pt and Pd atoms on a $ZrO_2$ surface, found that the Pd atoms have higher surface mobility and more easily form larger metal clusters than Pt. The higher mobility of the Pd agrees with other experimental data (Alfredsson et al. 2004). It was established that after annealing at 700 K, the cluster size distribution of Pd on the $TiO_2$ (111) surface moves readily toward larger cluster sizes. Alternatively, the size distribution of Pt is unchanged after annealing, with smaller clusters being stable at higher temperatures. Furthermore, it was found that Pd shows a higher probability of adsorbing during deposition on both terrace and step sites, while Pt is characteristically observed on the steps (see Figure 2.30).

The thickness of deposited noble-metal layers appreciably influences the cluster size as well. As a rule, for maximum sensor response, the concentration of noble-metal additives in the bulk of metal oxides should not exceed 1–2 wt% (Matsushima et al. 1988; Arbiol et al. 2002; Matko et al. 2002). According to Matko et al. (2002), Pt clusters in $SnO_2$ have a mean size smaller than 1.5–2.0 nm. Experimental results indicate that the thickness of the deposited noble-metal layer for films prepared by rheotaxial growth and thermal oxidation (RGTO) (Sberveglieri et al. 1991) should not exceed 2.5 nm to attain the same effect during surface doping. For thin metal oxides this thickness may be even less (Tsud et al. 2001; Veltruska et al. 2001; Korotcenkov et al. 2003a). Research has shown that for

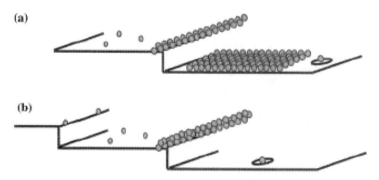

**Figure 2.30.** Possible positions of (a) Pd and (b) Pt clusters on the surface of $ZrO_2$. (Reprinted with permission from Alfredsson et al. 2004. Copyright 2004 Elsevier.)

the formation of an atomically dispersed Pd layer at the $SnO_2$ surface, the Pd thickness should not exceed 1 monolayer. The deposition of several milliliter equivalents of Pd leads to the formation of three-dimensional Pd particles with an average size in the range 3–5 nm (Tsud et al. 2001; Veltruska et al. 2001). Approximately the same size Pd clusters on $SnO_2$ grain surfaces were observed upon bulk doping when the concentration of doping additives was 5 wt% (Arbiol et al. 2002). Similar results were obtained by Alfredsson et al. (2004). Auger spectroscopic studies on Pd and Pt supported on $ZrO_2$ show that both Pd and Pt have layer-by-layer growth at low metal coverage and temperatures. At elevated temperatures, or following annealing, the Pd-clusters are more raftlike (two-dimensional), while the Pt-clusters show three-dimensional islands (Alfredsson et al. 2004).

Based on the catalytic activity of gold (Au) clusters (Haruta 1997; Valden et al. 1998; Wahlstrom et al. 2003; Campbell 2004), one can judge how important it is to form and stabilize noble-metal clusters with a specified size. Gold has long been known to be essentially inert chemically (and therefore catalytically inactive) in its bulk form (Hammer and Norskov 1995). Quite recently, however, it was found that when gold is dispersed in the form of nanoscale clusters and supported on transition-metal oxide surfaces such as $TiO_2$ (Valden et al. 1998; Wahlstrom et al. 2003), Au exhibits very high catalytic activity at low temperature for the partial oxidation of CO, hydrocarbons, hydrogenation of unsaturated hydrocarbons, and reduction of nitrogen oxides. It was shown that the catalytic properties of Au nanoparticles depend specifically on the support, the preparation method, and, critically, on the size of the Au clusters, which are most active when their diameter is smaller than 3.5 nm (Bamwenda et al. 1997; Valden et al. 1998) (see Figure 2.31). Cluster diameters smaller than 3 nm lead to a decrease in reactivity.

It is known that the activity of palladium increases substantially with a decrease in cluster size. For example, Maier et al. (2004) reported that catalytic activity toward electrochemical proton reduction is enhanced by more than two orders of magnitude as the diameter of the palladium particles, measured parallel to the support surface, decreases from 200 to 6 nm. Thus, during fabrication of surface-modified metal-oxide gas sensors, one should learn not to create clusters of specified size and to stabilize their size in the desired range.

It is important to note that each surface catalyst has its own optimal size for providing maximum response for each analyte. Research results (Kohl 1990; Yamazoe et al. 1991; Williams 1999; Cabot et al. 2001; Van de Krol and Tuller 2002; Korotcenkov et al. 2003a) provide good examples of this effect. It is also important to note that the size of noble metal clusters that are optimal for gas sensitivity does not coincide with the optimal size for heterogeneous catalysis (Kabbabi et al. 1994). It therefore follows that the surface noble-metal clusters in gas sensors and heterogeneous catalysis perform different functions.

## 6.4. FILM TEXTURE

Film texture has not been considered so far because in ceramics all grains are oriented arbitrarily. Only since the transition to film technology has film texture started attracting attention. For example, Ning et al. (2007) established that the highly preferred orientation of homogeneous ZnO films decreased the electrical resistance of ZnO ceramics at room temperature. Ning et al. (2007) also found that textured

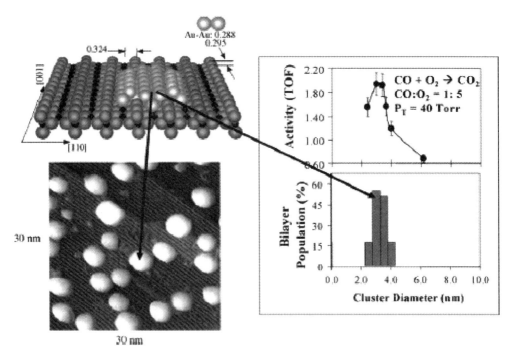

**Figure 2.31.** STM image (bottom left) and schematic (upper left) of Au clusters on $TiO_2$(110) whose population is dominant (lower right) for a $Au/TiO_2$ catalyst most active for CO oxidation (upper right). The activity (at 350 K) is expressed as turnover frequency (TOF) in units of $CO_2$ molecules produced per Au site per second. (Reprinted with permission from Choudhary and Goodman 2005. Copyright 2005 Elsevier.)

ZnO films have more low-angle grain boundaries in comparison with films without texture. Another interesting effect was observed by Korotcenkov et al. (2004a, 2004c): They found that for ozone detection using $In_2O_3$-based sensors, the recovery time increases as the film texture increases (see Figure 2.32). For reducing gases this dependence is observed for both response ($\tau_{res}$) and recovery times ($\tau_{rec}$). Moreover, the change in $\tau_{res}$ and $\tau_{rec}$ in the sensor response to reducing gases was subject to the same regularity of $T_{pyr}$ and film thickness influence as was the change in $\tau_{rec}$ for ozone detection. In other words, the faster response times were obtained from structures formed from arbitrarily oriented crystals. It has been assumed that the influence of the film's texture on gas-sensing characteristics takes place through the change of porosity of the gas-sensing matrix. Korotcenkov et al. (2002, 2004a, 2004b) studied $In_2O_3$ films textured in the (001) direction perpendicular to the substrate. The degree of texture increases when both the sintering temperature and the film thickness increase. One can assume that the indicated parameters of $In_2O_3$ film deposition promote more compact crystallite packing. The decrease of film porosity and gas permeability may be the result of such changes in the $In_2O_3$ film structure.

An attempt to estimate the influence of film texture on the $In_2O_3$ conductivity response to $NO_2$ was made by Steffes et al. (1999). It was found that $In_2O_3$ films with preferential orientation in the (211) direction had higher sensor response to $NO_2$ than the layers with preferential (222) texture. Films studied by Korotcenkov et al. (2002, 2004a, 2004b) had a different texture than the films used by Steffes et al. (1999). Therefore, it was not possibly to compare the results discussed by Korotcenkov et al. (2002, 2004a, 2004b) and Steffes et al. (1999). As a result, we can neither confirm nor refute the above conclusions. However, the existence of such an effect, in addition to the findings presented by Korotcenkov et al. (2005), allows us to conclude that modification of the film texture of polycrystalline films as well as decreasing grain size and film thickness may open a new way to optimize thin-film gas-sensor parameters.

## 6.5. SURFACE STOICHIOMETRY (DISORDERING)

Surface stoichiometry (disordering) of the metal oxide grains may be as important a parameter in gas sensing as film thickness and grain size. This parameter determines the adsorption ability and the surface charge through the number of oxygen vacancies, and as a result, controls the initial surface band bending $(eV_s)$ of the metal oxide and the change of $eV_s$ upon replacement of the surrounding gas.

Considering that the sensing properties of thin-film sensors, such as sensitivity, selectivity, and stability, are strongly related to their microstructures and to the exact stoichiometry of their surfaces, accurate control of these parameters is extremely important for the production of sensors with reproducible behavior (Korotcenkov et al. 2005b; Mašek et al. 2006). For example, a change in the surface oxygen composition of $SnO_2$ is accompanied by a change in the electronic structure of the surface Sn from a valence of IV for the stoichiometric surface to a valence of II for the reduced surface. This change in composition also has a strong influence on the chemical and gas-sensing properties. Batzill et al. (2005) showed for the adsorption of water that water dissociates on the stoichiometric surface but adsorbs weakly on the reduced surface. While this easy reduction of the $SnO_2$ surface can play an important

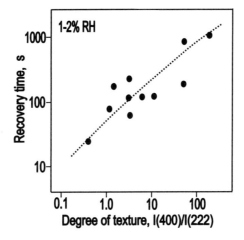

**Figure 2.32.** Influence of $In_2O_3$ film texture on recovery time during ozone detection. Films with thicknesses of 40–300 nm were deposited by spray pyrolysis from 1.0 M $InCl_3$–water solution ($T_{oper}$ = 270°C). (Reprinted with permission from Korotcenkov 2004a. Copyright 2004 Elsevier.)

role in the gas-sensing behavior of this material, it is unlikely to be of general importance for other gas-sensing metal oxides. The particular behavior of the $SnO_2$ surfaces arises from the dual valency of Sn. For other gas-sensing materials, e.g., ZnO, no variation in the lattice-oxygen composition is expected.

Excessive lattice disordering of the surface layer may be accompanied by significant growth of native surface states. Theoretical simulations presented by Brynzari et al. (1999, 2000a, 2001) indicate that an increase of these surface defects may lead to pinning of the surface Fermi-level position, and correspondingly to a drop in response of gas sensors. This means that if we want to maintain effective operation of solid-state gas sensors, the concentration of those states as well as the lattice disordering of the surface layer of metal oxide grains should be minimized. Only in this case will the surface Fermi level not be pinned, because the charge of native surface states becomes comparable to or less than the charge of chemisorbed particles. The indicated surface property creates a condition for modulation of surface band bending of semiconductors with the change of the surrounding atmosphere. The change of surface stoichiometry of metal oxide films may result from changes in deposition and annealing temperature (Brinzari et al. 2000b).

High-resolution transmission electron microscopy (HRTEM) images of $SnO_2$ grains presented by Nayral et al. (2000) (see Figure 2.33) indicate that this effect may occur in metal oxides. In many cases, the particles of $SnO_2$ consist of a well-crystallized core covered with an amorphous layer of tin oxide. Further, such amorphous layers may appear in undoped tin dioxide fabricated by low-temperature technological routes, or in doped material at superfluous concentrations of doping additives.

The same conclusion was reached by Dieguez et al. (2001) on the basis of an analysis of the complete Raman spectrum of nanometric $SnO_2$ particles. Dieguez et al. (2001) found that for grains smaller than 7 nm, two bands appeared in the high-frequency region of spectrum. They proposed that these

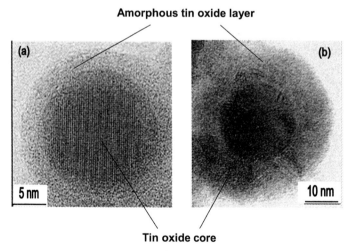

**Figure 2.33.** HREM images of $SnO_2$ nanoparticles showing the crystallized tin oxide core and the amorphous tin oxide shell: (a) $SnO_2$:Pd; (b) undoped $SnO_2$. (Adapted with permission from Nayral et al. 2000. Copyright 2000 Elsevier.)

bands were due to a surface layer of nonstoichiometric $SnO_2$ with a different symmetry than $SnO_2$. The thickness of this layer was calculated to be ~1.1 nm, which is about two to three unit cells.

One can conclude that the presence of such a layer on the surface of small grains is a main reason for lower thermal stability of the metal oxide structure and the enhanced sensitivity to air humidity of sensors fabricated of fine-dispersed material with grain size smaller than 5–7 nm.

The presence of a nonstoichiometric surface layer may also be responsible for a strong interaction between noble metals and metal oxides. The activity of noble-metal catalysts incorporated on solid-state sensors is determined by ether their chemical state, aggregation form, or interaction with metal oxide (Castañeda 2007). The formation of alloys is the main reason for the indicated dependence (Hanys et al. 2006).

Wahlstrom et al. (2004) assumed that the surface stoichiometry of metal oxides, through a charge transfer–induced diffusion mechanism for $O_2$ molecules adsorbed on a metal oxide surface, can also control the surface diffusion of oxygen molecules. Time-resolved scanning tunneling microscopy of the $TiO_2$ (110) surface has shown that the $O_2$ hopping rate depends on the number of surface donors (oxygen vacancies), which determines the density of conduction band electrons. Wahlstrom et al. (2004) assumed that the metal oxides act as reservoirs for oxygen; and the $O_2$ diffusion may be a rate-limiting step in oxidation processes on these metal oxides. Diffusion of oxygen molecules on a metal oxide surface plays a vital role in gas-sensing effects, and therefore these results may have implications for understanding their nature. This mechanism is expected to be an important one for such reducible oxides as $TiO_2$, $Fe_2O_3$, $SnO_2$, and $ZnO$, where shallow donor states provide a rise to a high density of electrons in the conduction band

## 6.6. POROSITY AND ACTIVE SURFACE AREA

From an analysis of the numerous parameters which can affect chemoresistance sensors, one can conclude that in order to achieve maximum gas sensitivity it is necessary either to increase the role of surface conductivity or to increase the contribution of intercrystalline barriers by decreasing either the contact area or the width of the neck (Brynzari et al. 1999, 2000; Barsan and Weimar 2001) (see Figure 2.19).

At present there is no doubt that the simplest method to achieve maximum sensitivity independently on gas-sensing material is to increase the film's porosity (Park et al. 1996; Basu et al. 2001; Ahmad et al. 2003; Hyodo et al. 2003; Lee et al. 2005; Jin et al. 2005; Tesfamichael et al. 2007). In the case of compact layers, the interaction with gases takes place only at the geometric surface (see Figure 2.34).

In porous material the volume of the layer is accessible to the gases, and therefore the active surface is much greater than the geometric area and the sensor response is higher (Barsan et al. 1999). Higher porosity results in a smaller number of contacts with the necks that are not overlapped under interaction with the surrounding gas. Experimental results obtained from a study of gas-sensing characteristics of $SnO_2$ films synthesized by the RGTO technique (Sberveglieri et al. 1991; Sherveglieri 1992b) are in good agreement with this conclusion. The films had high sensitivity in spite of the high level of agglomeration. Further, it was observed that films with minimal contact area between agglomerates showed maximum sensitivity. This was clearly illustrated by the SEM images presented in (Sberveglieri et al. 1991; Sherveglieri 1992b). Increased porosity also decreased the probability of forming so-called

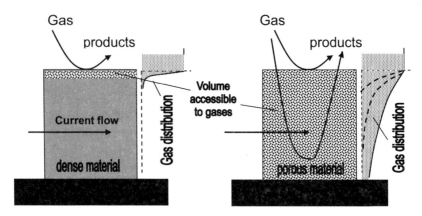

**Figure 2.34.** Schematic representation of the influence of material density on gas penetrability of gas-sensing matrix. (Reprinted with permission from Korotcenkov 2008. Copyright 2008 Elsevier.)

capsulated zones in the volume of the gas-sensing layer. Capsulated zones are isolated from contact with the atmosphere, and therefore their resistance does not depend on the surrounding gas (McAleer et al. 1987). Density, porosity, pore size, and gas permeability can be determined by appropriate measurement techniques (Ishizaki et al. 1998; Nguyen and Do 1999; Wang et al. 2007b). As has been established, all these parameters have a direct correlation with grain size (see Figure 2.35).

The specific surface area parameter, which is rarely used in characterizing gas-sensing materials, deserves more attention. Rumyantseva et al. (2005) have expressed the same opinion. Experiments (Yamazoe et al. 1979; Yamazoe 1991; Li et al. 1999; Ahmad et al. 2003; Rumyantseva et al. 2005) have indicated that the use of material with high specific surface area may lead to the development of highly sensitive metal-oxide gas sensors (see Figure 2.36).

Specific surface area has a more universal meaning than generally used parameters, because it combines such parameters as porosity and grain size. From data on specific surface areas, average grain size may be calculated using the equation (Zhang and Gao 2004; Rumyantseva et al. 2005)

$$d_{SA} = \frac{6}{\rho A} \qquad (2.10)$$

where $d_{SA}$ is the average grain size of spherical particles, $A$ is the surface area of the powder, and $\rho$ is the theoretical density of $SnO_2$. Moreover, in many cases the specific surface area shows better correlation with gas response than does crystallite size (Gas'kov and Rumyantseva 2001; Ahmad et al. 2003).

As a rule, an increase in specific surface area of a gas-sensing material is accompanied by a shift in the sensitivity maximum into the range of lower operating temperatures (see Figure 2.37). One can assume that this effect is caused by the thermally activated nature of gas diffusion inside the gas-sensing matrix (Ruiz et al. 2005).

Specific surface area data, estimated by classical multipoint Brunauer, Emmett, and Teller (BET) adsorption techniques, and grain size, calculated using x-ray diffraction (XRD) data, provide a more

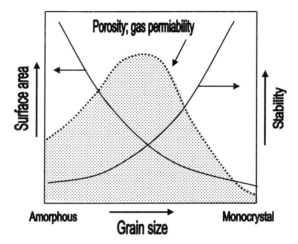

**Figure 2.35.** Schematic illustration of the influence of grain size on parameters of sensing materials.

complete and reliable description of the gas-sensing material. The difference between grain sizes determined by those methods (see Table 2.19) may be used for porosity characterization (Rumyantseva et al. 2005). However, the specific surface area should be determined only after conducting all operations used during the sensor's fabrication.

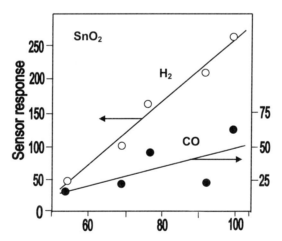

**Figure 2.36.** Relationship between surface areas of $SnO_2$ sensors and their sensitivity to 500 ppm of $H_2$ and CO at 300°C. Sensors were fabricated by pressing and annealing at 450°C calcined $SnO_2$ powders into 14-mm-diameter and 1-mm-thick pellets with two Pt electrodes. $SnO_2$ for these experiments were prepared by a surfactant-templating method. The as-synthesized $SnO_2$ samples were calcined at 450°C. (Data from Li et al. 1999. Reprinted with permission from Korotcenkov 2008. Copyright 2008 Elsevier.)

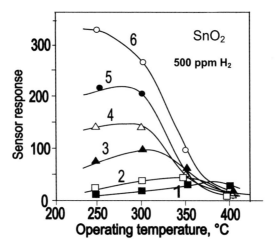

**Figure 2.37.** Influence of surface area of $SnO_2$-based thick-film sensors on response to 500 ppm of $H_2$: 1, 3 m²/g; 2, 54 m²/g; 3, 69 m²/g; 4, 77 m²/g; 5, 92 m²/g; 6, 99 m²/g. Tested sensors were fabricated using technology described in Figure 2.38. (Data from Li et al. 1999. Reprinted with permission from Korotcenkov 2008. Copyright 2008 Elsevier.)

Thus, as the porosity and active surface area of the gas-sensing material increases, the sensor response increases as well (see Figures 2.36 and 2.37). This is consistent with the conclusion of Matko et al. (2002) that the porosity and specific surface area are basic factors that influence the solid/gas interactions and ultimately the material performance in gas detection. This is an important conclusion. However, it is necessary to know that in some cases, dense, nonporous metal oxides may compensate for their lower sensitivity and other shortcomings by having higher temporal and thermal stability (Min and Choi 2004; Kocemba et al. 2001).

Sever studies (Park et al. 1996; De Souza Brito et al. 1998; Korotcenkov et al. 2004a) have shown that gas sensors with higher porosity have faster response. Park et al. (1996) established that the refractive index correlates with film porosity. If the porosity is higher, the refractive index is smaller. It is known that only nonporous metal oxides have refractive indices equal to the tabular data estimated for these materials. Pores in metal oxides can be considered as a component with a refractive index equal to 1.0. Therefore, films with smaller refractive index along with invariability of the main parameters must have a higher degree of porosity and better gas permeability. Figure 2.38 shows that $In_2O_3$ films with low refractive index really have minimal $\tau_{res}$ and $\tau_{rec}$. Korotcenkov et al. (2004b) concluded that for the detection of reducing gases in $In_2O_3$ films, the porosity or the gas permeability is a more important factor than the grain size.

Within the framework of the above discussion, results reported by De Souza Brito et al. (1995) are of interest. The structural properties of $SnO_2$ films synthesized using sol-gel technology were analyzed, and it was found that the pore size increased during sintering within the temperature range from 200 to 700°C as indicated by growth in the grain size. Moreover, it was established that in the low-temperature region ($T_S < 400°C$), the average micropore diameter increased slightly. In the high-temperature region ($T_S > 400°C$), this process was intensified, and the growth of average pore diameter and crystallite size were followed by a drastic reduction of the surface area (see Figure 2.39). The structural evaluation during sintering of compacted $SnO_2$ sol-gel powders was performed using nitrogen adsorption isotherm analysis.

This means that ceramic-based gas sensors with larger grain size due to bigger pore size could have a faster response than sensors fabricated of fine-dispersion metal oxide. As a result, in some cases, annealing at temperatures between 400 and 700°C might result in decreased response times. The linear dependence between average pore size and crystallite size (see Figure 2.39) reveals that the mean coordinate of the pore is constant as the sintering temperature increases from 400 to 700°C (De Souza Brito et al. 1995). Established regularities provide an opportunity for understanding many effects which are discussed in various papers. For example, the growth of porosity of the $SnO_2$ gas-sensing layer with increased grain size is a good explanation for both the increase of $NO_2$ response in films with larger-sized crystallites (Rumyantseva et al. 2005) and the increase of response to CO for $SnO_2$ sensors after high-temperature calcination (Jin et al. 1998).

Note, however, that with an increase in porosity, the influence of film thickness on both the magnitude of sensor response and the response time is appreciably weakened (see Figures 2.40 and 2.41). In

**Table 2.19. Characteristics of $SnO_2$ samples prepared by different routes**

| Route | $T_{an}$, °C | $t$ (XRD), nm | $S_{surf}$, m² g⁻¹ | $t$ (BET), nm | $t_{BET}/t_{EXD}$ | $S$ ($NO_2$) |
|---|---|---|---|---|---|---|
| g | 300 | 4 | 122 | 7 | 1.8 | 15 |
| g | 500 | 9 | 35 | 25 | 2.8 | 60 |
| g | 700 | 22 | 9 | 96 | 4.4 | 110 |
| g | 1000 | 35 | — | — | — | 130 |
| k | 300 | 4 | 175 | 5 | 1.3 | 168 |
| k | 500 | 9 | 65 | 13 | 1.4 | 190 |
| k | 700 | 17 | 28 | 31 | 1.8 | 77 |
| k | 1000 | 26 | — | — | — | — |
| h | 300 | 4 | 135 | 6 | 1.5 | 14 |
| h | 500 | 11 | 26 | 33 | 3.0 | 25 |
| h | 700 | 34 | 12 | 72 | 2.1 | 48 |
| h | 1000 | 35 | — | — | — | — |
| s | 300 | 3 | 180 | 5 | 1.7 | 53 |
| s | 500 | 11 | 69 | 13 | 1.2 | 41 |
| s | 700 | 22 | 14 | 62 | 2.8 | 130 |
| s | 1000 | 36 | — | — | — | — |

$SnO_2$ powders elaborated by wet chemical synthesis using (g) conventional hydrolysis of $SnCl_4$, (k) eryosol technique, and (h, s) specific hydrolysis in various solutions. Films calcined at 500°C for 6 h. Sensor response estimated in dry atmosphere as $S$ ($NO_2$) = $(R - R_O)/R_O$.

*Source:* Data from Rumyantseva et al. 2005. Reprinted with permission from Korotcenkov 2008. Copyright 2008 Elsevier.

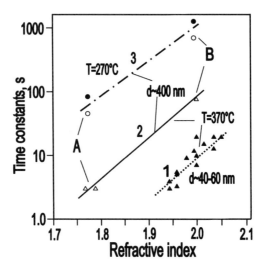

**Figure 2.38.** Time constants of sensor response to CO versus $In_2O_3$ film refraction index: (1, 2) $T_{oper}$ = 370°C; (3) $T_{oper}$ = 270°C; (1) $d \approx 40$–60 nm; (2, 3) $d \approx 400$ nm; (A) $In_2O_3$ films were deposited from 0.2 M $InCl_3$–water solution; (B) from 1.0 M $InCl_3$–water solution. (Reprinted with permission from Korotcenkov 2004a. Copyright 2004 Elsevier.)

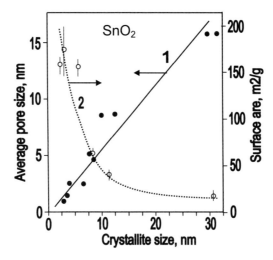

**Figure 2.39.** Average pore size and surface area as a function of mean $SnO_2$ crystallite size. The structural evolution during sintering of compacted $SnO_2$ sol-gel powders was investigated using nitrogen adsorption isotherm analysis. Pressed cylindrical pellets had diameter of 8 mm and height ~2.5 mm. (Data from De Souza Brito et al. 1995. Reprinted with permission from Korotcenkov 2008. Copyright 2008 Elsevier.)

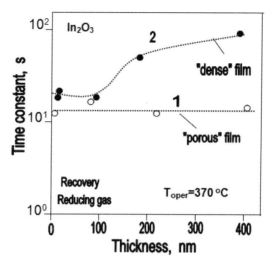

**Figure 2.40.** Dependences of response time during $H_2$ detection ($T_{oper}$ = 370°C) on the thickness of $In_2O_3$ films deposited from (2) high- and (1) low-concentration $InCl_3$-sprayed solutions ($T_{pyr}$ = 475°C): (1), 0.2 M; (2) 1.0 M $InCl_3$ solution. (Reprinted with permission from Korotcenkov 2007a. Copyright 2007 Elsevier.)

particular, Korotcenkov and co-workers (Korotcenkov 2005; Korotcenkov et al. 2001, 2007), studying the influence of the modes of deposition on the structural and gas-sensing properties of $SnO_2$ and $In_2O_3$ films deposited by spray pyrolysis, found that for dense films an increase of film thickness in the range 30–200 nm was accompanied by a decrease in sensitivity to ozone by almost two orders of magnitude; for porous films, the same change of thickness did not affect gas sensitivity.

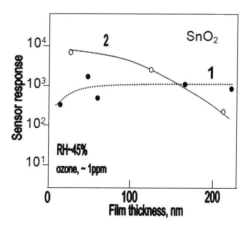

**Figure 2.41.** Influence of film porosity on sensor response dependence from $SnO_2$ film thickness: (1) porous films ($T_{pyr}$ = 460°C); (2) dense film ($T_{pyr}$ = 520°C). (Reprinted with permission from Korotcenkov 2007c. Copyright 2007 Elsevier.)

In addition to the temperature of deposition or synthesis, results presented by Matko et al. (2002) showed that metal oxide doping may also affect the porosity of gas-sensing materials. For example, $SnO_2$:Pt films in the range of 0–0.6% Pt concentration had a maximum porous fraction. However, with a doping concentration higher than 2%, no significant number of pores could be detected.

As shown by Shimuzu et al. (1998), control of porosity provides a means to influence selectivity. Shimuzu et al. (1998) proposed that the limited diffusion of $O_2$ into the interior region of a thick film should lead to a lowered oxygen partial pressure inside the gas-sensing matrix. This then leads to decreased coverage of chemisorbed oxygen on the metal oxide surface. This should result in increased relative sensitivity to $H_2$. The mechanism of this effect is shown in Figure 2.42.

Experiments described by Shimuzu et al. (1998) with thick $SnO_2$ films ($d > 100$ μm) covered by a dense $SiO_2$ layer confirmed this assumption. Thus, porosity control can be one of the methods of influencing gas-sensor selectivity to $H_2$ and other gases for which $H_2$ is one of the products of dissociation. Gas molecules with larger diameters than $H_2$ and $O_2$ will not be able to penetrate into the interior region of a metal oxide film, so the gas sensors will not be sensitive to these gases. Results reported by Hyoda et al. (2002) (see Figure 2.43) confirmed that this effect requires that the pores not exceed a size of 5–10 nm. Only in this case can good selectivity to $H_2$ be obtained.

However, this method can be used only with thick and ceramic gas sensors, where the thickness of the gas-sensing matrix exceeds 10–100 μm. Further, the decrease of pore size will often be accom-

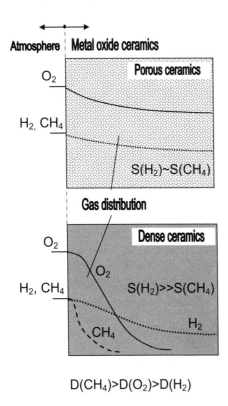

**Figure 2.42.** Schematic illustration of gas distribution in dense and porous gas-sensing matrix for gases with different size ($D$) of molecules. (Reprinted with permission from Korotcenkov 2008. Copyright 2008 Elsevier.)

# DESIRED PROPERTIES FOR SENSING MATERIALS • 139

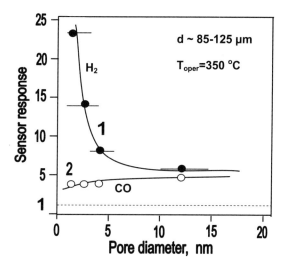

**Figure 2.43.** Influence of pore size on response of ceramic-type $SnO_2$-based gas sensors to $H_2$ and CO. Tested sensors were fabricated using thick-film technology. Paste of the $SnO_2$ powders was calcined at 600°C for 5 h and printed on alumina substrate, on which interdigitated Pt electrodes had been formed. (Data from Hyoda et al. 2002. Reprinted with permission from Korotcenkov 2008. Copyright 2008 Elsevier.)

panied by an increase in response and recovery times, i.e., by a degradation of other important sensor performance parameters. For example, the results presented by Hyoda et al. (2002) indicate that for dense $SnO_2$ ceramics with pore diameters ~3–5 nm, even at $T_{oper} \approx 350$–450°C, the response and recovery times exceed 10 min. Response kinetics were especially effective for CO detection as well as the recovery process for $H_2$. This should be expected, because the diffusion contribution to the total molecular transport through the micropore material depends on the pore size and the length of the path along which the molecule travels. The same correlation has been established by Ahmad et al. (2003). Sensors with minimal active surface area had maximum recovery and response times (see Table 2.20). The crystallite size in all sensors was the same.

Korotcenkov et al. (2004a, 2004b) has shown that porosity effects can be used only for thick films. For example, for $In_2O_3$ films with thickness less than 70–80 nm, no differences in response or recovery times were observed between the "porous" and "dense" films (see Figure 2.40). This means that metal oxide films with thickness $d < 60$–80 nm possess high gas permeability, and such structural parameters as porosity no longer affect the kinetics of gas sensing for thin $In_2O_3$-based gas sensors.

The relationship between material porosity and sensor sensitivity to humidity has not attracted designers' attention. However, it is known that the smaller the pore size, the more probable that water condensation is in it. Numerous capacitive-type humidity sensors have been designed based on this effect. Therefore, one can suppose that during the sensor's storage or break-in period, and when the temperature of the active zone does not exceed room temperature, water condensation in small pores is possible. Long-time contact of water with an oxide's surface could substantially modify the oxide's surface properties, and thus change a sensor's performance parameters.

**Table 2.20. Influence of engineering procedure on structural parameters of SnO$_2$-based sensors and their time constants of response to reducing gas**

| Sensor | $D$ (XRD), nm | $S_{surf}$, m$^2$ g$^{-1}$ | $D$ (BET), nm | Response time (350°C), s | Recovery time (350°C), s |
|---|---|---|---|---|---|
| Sn-CM | 30.15 | 122.44 | 7.0 | 10–20 | 50–60 |
| Sn-DC | 10.06 | 37.3 | 22.97 | 20–30 | 70–80 |
| Sn-HC | 12.33 | 27.7 | 30.94 | 50–55 | 150–160 |
| Sn-HM | 10.77 | 33.4 | 26.22 | 70–75 | 190–200 |

*Source:* Data from Ahmad et al. 2003. Reprinted with permission from Korotcenkov 2007. Copyright 2007 Elsevier.

Material with extremely small pores shows a decreased resistance to poisoning. This is another important problem of gas-sensing materials that is related directly to the need to provide maximum temporal stability of gas-sensor parameters. This problem was discussed earlier for pellistors. Simulations carried out by Gentry and Walsh (1982) for flammable gas-sensing elements showed that the poison resistance of the porous sensors decreased with an increase of the parameter $h_o \sim (1/rD_p)^{1/2}$, where $r$ is the pore radius and $D_p$ is the pore diffusion coefficient.

Accumulation of solid reaction products at or on the sensor surface may also block the active surface sites. With small pores, this process will be accompanied by a sharp decrease of active surface area and a decrease of sensor response. Sulfides and graphite products (Brinzari et al. 2000b) are particularly troublesome for metal oxides

## 6.7. AGGLOMERATION

The agglomeration and aggregation of matter are ubiquitous phenomena that may be expected to occur almost everywhere in nature (Kaye et al. 1989). Their appearance ranges from metallic polycrystals to colloidal aggregates and clusters, as well as the lipid–protein viscoelastic matrices that are often called biomembranes. In most cases the metal oxide films are agglomerated as well (Barsan et al. 1999). One example of a SnO$_2$ agglomerated film is shown in Figure 2.44. However, the role of agglomeration in gas-sensing effects has not been studied in detail yet.

One should note that an agglomeration problem exists for thin-film as well as ceramic and thick-film sensors. It has been established that in many cases synthesized powders agglomerate into big particles. SEM images obtained by Chabanis et al. (2003) indicate that, as a rule, the agglomerates formed during ceramic synthesis are more porous and are larger in size than those formed by the methods of thin-film technology. The agglomerate size for SnO$_2$ prepared using the sol-gel method exceeded 5 μm (Chabanis et al. 2003). For films deposited by spray pyrolysis, the size of agglomerate did not exceed 200–500 nm (Brinzari et al. 2002; Korotcenkov et al. 2002, 2004a, 2005a).

Such a difference in the properties of ceramics and films is natural and follows from the various conditions of synthesis. Agglomerates formed in ceramic-type and thin-film-type metal oxides correspond to two different types of aggregation. In ceramic materials, aggregation is diffusion-controlled, in which the attachment of particles and clusters occurs instantaneously at first contact. In thin films, aggregation is controlled by chemical reaction (Derjaguin 1989; Gadomski et al. 2005). In the latter case, the "first touch" sticking principle does not hold, and the microstructural evolution has to follow certain attachment–detachment sequences of events. Finally, the incorporation takes place in the form of a particle or some cluster of particles with different types of microstructures (Anderson and Lekkerkerker 2002). Chemical reaction–controlled aggregations are compact compared to their diffusive counterparts, which look more irregular, viz., fractal, meaning that their arms or branches expand more visibly. Therefore, in general, in diffusion-controlled reactions the aggregations are not dense (Jullien and Botet 1987; Schmelzer et al. 1999).

The gas-sensing matrix of agglomerated polycrystalline films can shown schematically as an equivalent circuit (see Figure 2.44). We consider the gas-sensing matrix as a three-dimensional network of crystallites (grains) forming a gas-sensing film. Here $R_{(a-a)}$ represents the resistance of interagglomerate contacts, $R_c$ is the resistance of intercrystallite (intergrain) contacts, $R_b$ is the bulk resistance of the crystallites (grains), and $R_{agl}$ is the resistance of the agglomerate.

From this scheme one can see that in a gas-sensing matrix there are four gas-sensitive elements which can affect the sensor response:

1. Intergrain (intercrystallite) contacts
2. Interagglomerate contacts
3. Agglomerates
4. Grains

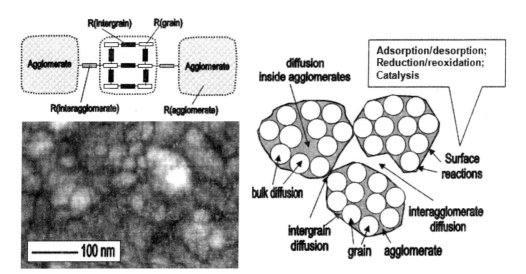

**Figure 2.44.** SEM image of SnO₂ agglomerated film deposited by spray pyrolysis and diagrams illustrating peculiarities of electroconductivity and gas-sensing reactions in agglomerated films. (Reprinted with permission from Korotcenkov 2008. Copyright 2008 Elsevier.)

Here the agglomerate resistance is an integral resistance, representing a three-dimensional grain network. Thus, if there are alterations in the porosity of either agglomerates or the gas-sensing matrix, the role of the various gas-sensing elements in sensor response changes. For example, a decrease of grain size increases the role of grains in gas-sensing effects. A low gas permeability of agglomerates promotes an increase in the influence of interagglomerate contact on sensor response. These observations are in agreement with simulations (Williams and Pratt 2000) affirming that agglomeration of small crystallites into large aggregations is a key phenomenon and can result in great variations of the apparent response characteristics.

Research has shown that both the sensor response and the kinetics depend on the properties of agglomerates. Larger and denser agglomerates exhibit slower response and recovery times. It is important to note that this effect depends on the nature of the detected gas, particularly on the reactivity and diffusion coefficient of this gas in the oxide matrix. A high level of agglomeration also promotes the formation of capsulated zones, i.e., zones with closed porosity. Properties of metal oxides in these zones do not depend on a change in the surrounding gas (McAleer et al. 1987).

Based on an analysis of $SnO_2$ sensors fabricated using the successive ionic layer deposition (SILD) method, Korotcenkov et al. (2003b) concluded that the small size of crystallites (grains) was an essential, but not sufficient, condition to achieve both maximum gas sensitivity and fast response. First, the gas-sensing matrix should be highly porous and the size of the agglomerates must be minimal. Moreover, a paradoxical situation appears for strongly agglomerated structures. In such structures the small size of grains (crystallites) is not an advantage. Agglomerates from smaller grains are more densely packed—i.e., they have smaller gas penetrability. Probably this was the case in the work of Hyoda et al. (2002), where, in spite of minimum grain size (~2 nm), the gas response of $SnO_2$ sensors to $H_2$ was unexpectedly low. Therefore, in accordance with the above concept, such structures can have poorer gas-sensing characteristics than films with larger crystallites. A similar conclusion was drawn by De Souza Brito et al. (1995). They established that films consisting of larger crystallites were more porous than films consisting of small crystallites. Therefore, the former had higher sensor response. In this context, it becomes clear that to perform a comparative analysis of gas-sensing characteristics of metal oxide–based sensors, it is necessary first to consider the size and the porosity of agglomeration, and only after that the crystallite size.

Usually, dense and strongly agglomerated ceramics have a small active surface area (Tan et al. 2000). As was previously shown, this has an important negative impact on the approach for optimization of gas-sensing characteristics. Numerous studies have established that mechanical milling is one effective method for resolving this problem (Koch 1997; Dieguez et al. 2000; Tan et al. 2000, 2004; Abe et al. 2005). Abe et al. (2005) showed that mechanical milling increased the sensitivity to CO more than four times (see Figure 2.45). Dieguez et al. (2000) established that milling allows the possibility to fabricate $SnO_2$ sensors with very good sensitivity to $NO_2$ and a negligible cross-sensitivity to CO, along with high long-term stability.

It is impossible to fully explain this effect by a decrease of grain size, because the grain size was decreased only by 10% as a result of this procedure. At the same time, the influence of mechanical milling on agglomerate size is considerable. The consequences are clearly shown in Figure 2.46 (Tan et al. 2000). Taking into account estimates made from transmission electron microscopy (TEM) micrographs, mechanical milling decreased the agglomerate size more than 20-fold. After mechanical milling, the sensitivity to ethanol increased more than 20-fold as well.

DESIRED PROPERTIES FOR SENSING MATERIALS • 143

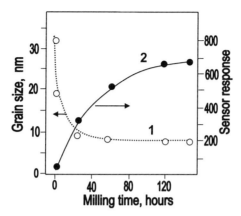

**Figure 2.45.** Influence of milling time on (1) $SnO_2$ grain size and (2) $xSnO_2$-$(1-x)\alpha Fe_2O_3$ sensor response to ethanol (1000 ppm). Sensors were fabricated onto the $Al_2O_3$ substrate with interdigital Au electrodes using thick-film technology. (Data from Tan et al. 2000. Reprinted with permission from Korotcenkov 2008. Copyright 2008 Elsevier.)

However, it is necessary to note that the development of nanocrystalline microstructures during mechanical milling takes place in three stages (Hadjipanayis and Siegel 1994):

- Stage 1. Deformation is localized in shear planes containing a high dislocation density.
- Stage 2. Dislocation annihilation/recombination/rearrangement forms a cell/subgrain structure with nanoscale dimensions; further milling extends this structure throughout the sample.
- Stage 3. The orientation of the grains becomes random.

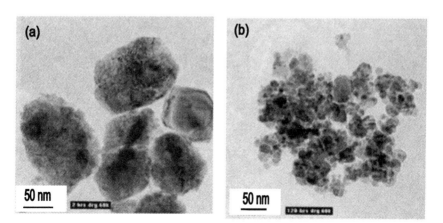

**Figure 2.46.** TEM micrographs for $SnO_2$ powders after (a) 2 h and (b) 129 h milling using high-energy milling in a Fritsh Pulverisette 5 planetary ball milling system. (After Tan et al. 2000. Reprinted with permission from Tan et al. 2000. Copyright 2000 Elsevier.)

Thus, as a result of mechanical milling, additional mechanical tensions appear in the material. The appearance of mechanical tensions and the change in the microstructure may be an important additional factor that influences gas-sensing characteristics. Mechanical tensions should be considered when developing procedures for sensor fabrication. Only thermal treatment at $T \approx 600\text{–}800°C$ seems to relieve these tensions (Cirera et al. 2000). Thus, an additional thermal treatment after mechanical milling may be required.

Based on these facts, one can conclude that thin-film sensors with minimum agglomeration will always have a maximum rate of response. As $T_{pyr}$ increases, the agglomeration of the film decreases. However, at the same time, considerable growth of both grain size and film density takes place, which is undesirable for maximum sensor sensitivity and minimum response time. Therefore, we have to seek a compromise between grain size and the possibility of agglomerate formation when preparing films. Many studies have shown that the easiest approach to this problem is to use ultrathin films with thicknesses of less than 40–60 nm. In such films, neither the agglomerate size nor the film thickness reaches the critical value at which gas penetrability starts to influence the sensor response kinetics.

However, as agglomerates become denser and bigger, the size of the interagglomerate pores increases. Under certain conditions, qualitative changes may take place in the gas-sensing matrix which may affect the metal oxide structure and design parameters for achieving maximum sensitivity. In some cases, mainly for oxidizing gases, the sensor response may increase with growth of film thickness and agglomerate size. The clearest example of such a situation was presented by Rumyantseva et al. (2005). This seems to be a contradiction to what was said before, but the idea becomes clear if one realizes that gas sensitivity may be controlled, not by intercrystallite contacts, but by interagglomerate contacts. The growth of agglomerates increases the size of interagglomerate pores, which facilitates access of active gas into the inner volume of the metal oxide matrix and thus makes it more sensitive. This mechanism was suggested by Korotcenkov et al. (2003) to explain the high sensitivity to ozone of $SnO_2$ films that were synthesized by the SILD method and are characterized by a high level of agglomeration.

## 7. OUTLOOK

In this chapter we have shown that no material satisfies all the desired properties for a sensing material. Further, every new application creates new requirements in terms of properties. advances its own whole requirements to sensors. The literature is therefore replete with examples of materials that have been tested for their applicability to the design of chemical sensors. This process is still ongoing, involving research on entire new families of compounds. Fullerenes, carbon and metal oxide nanotubes, nanowires, nanocomposites, etc., are presently being explored (Baraton et al. 1997; Chao and Shih 1998; Varghese et al. 2001, 2003; Gas'kov et al. 2001; Penza et al. 2001; Baena et al. 2002; Cantalini et al. 2003; Valentini et al. 2003, 2004). We don't know yet where eventually a qualitative leap to a world of nano-sized structures may lead. However, early results are promising.

## 8. ACKNOWLEDGMENTS

The author thanks the Korean BK21 Program for support of his research.

# REFERENCES

Abe S., Choi U.S., Shimanoe K., and Yamazoe N. (2005) Influences of ball-milling time on gas-sensing properties of $Co_3O_4$–$SnO_2$ composites. *Sens. Actuators B* **107**, 516–522.

Adhikari B. and Majumdar S. (2004) Polymers in sensor applications. *Prog. Polym. Sci.* **29**, 699–766.

Ahmad A., Walsh J., and Wheat T.A. (2003) Effect of processing on the properties of tin oxide-based thick-film gas sensors. *Sens. Actuators B* **93**, 538–545.

Alfredsson M., Richard C., and Catlow A. (2004) Predicting the metal growth mode and wetting of noble metals support on c-$ZrO_2$. *Surf. Sci.* **561**, 43–56.

Anderson V.J. and Lekkerkerker H.N.W. (2002) Insights into phase transition kinetics from colloid science. *Nature* **416**, 811–816.

Ando M., Chabicovsky R., and Haruta M. (2001) Optical hydrogen sensitivity of noble metal-tungsten oxide composite films prepared by sputtering deposition. *Sens. Actuators B* **76**, 13–17.

Ando M., Kobayashi T., Iijima S., and Haruta M. (1997) High impact applications, properties and synthesis of exciting new materials. *J. Mater. Chem.* **7**, 1779–1783.

Arbiol J., Peiro F., Cornet A., Morante J.R., Perez-Omil J.A., and Calvino J.J. (2002) Computer image HRTEM simulation of catalytic nanoclusters on semiconductor gas sensor materials supports. *Mater. Sci. Eng. B* **91–92**, 534–536.

Arshak K., Twomey K., and Heffernan D. (2002) Development of a microcontroller-based humidity sensing system. *Sensor Rev.* **22**(2), 150–156.

Badlani M. and Wachs I.E. (2001) Methanol: a "smart" chemical probe molecule. *Catal. Lett.* **75**(3–4), 137.

Badwal S.P.S. (1992) Zirconia-based solid electrolytes: microstructure, stability and ionic conductivity. *Solid State Ionics* **52**, 23–32.

Badwal S.P.S., Ciacchi F.T., and Milosevic D. (2000) Scandia-zirconia electrolytes for intermediate temperature solid oxide fuel cell operation. *Solid State Ionics* **136–137**, 91–99.

Baena J.R., Gallego M., and Valcarcel M. (2002) Fullerenes in the analytical sciences. *Trends Anal. Chem.* **21**(3), 187–198.

Bamwenda G.R., Tsubota S., Nakamura T., and Haruta M. (1997) The influence of the preparation methods on the catalytic activity of platinum and gold supported on $TiO_2$ for CO oxidation. *Catal. Lett.* **44**, 83.

Baraton M.I., Merhari L., Wang J., and Gonsalves K.E. (1997) Dispersion of metal oxide nanoparticles in conjugated polymers: investigation of the $TiO_2$/PPV nanocomposite. In: *MRS Symposium Proceedings, Vol. 501*, Pittsburgh, PA, pp. 59–64.

Barsan N. and Weimar U. (2001) Conduction model of metal oxide gas sensors. *J. Electroceram.* **7**, 143–167.

Barsan N. and Weimar U. (2003) Understanding the fundamental principles of metal oxide based gas sensors: the example of CO sensing with $SnO_2$ sensors in the presence of humidity. *J. Phys. Condensed Matter* **15**, R1–R27.

Barsan N., Heilig A., Kappler J., Weimar U., and Gopel W. (1999b) CO-water interaction with Pd-doped $SnO_2$ gas sensors: simultaneous monitoring of resistance and work function. In: *Proceedings of the 13th European Conference on Solid State Transducers, EUROSENSORS XIII*, September 12–15, 1999, the Hague, The Netherlands, pp. 367–369.

Barsan N., Schweizer-Berberich M., and ·Göpel W. (1999a) Fundamental and practical aspects in the design of nanoscaled $SnO_2$ gas sensors: a status report. *Fresenius J. Anal. Chem.* **365**, 287–304.

Basu S., Saha M., Chatterjee S., Mistry K.Kr., Bandyopadhay S., and Sengupta K. (2001) Porous ceramic sensor for measurement of gas moisture in the ppm range. *Mater. Lett.* **49**, 29–33.

Batzill M. (2006b) Surface science studies of gas sensing materials: $SnO_2$. *Sensors* **6**, 1345–1366.

Batzill M., Bergermayer W., Tanaka I., and Diebold U. (2006a) Tuning the chemical functionality of a gas sensitive material: water adsorption on $SnO_2(101)$. *Surf. Sci.* **600**, L29–L32.

Batzill M., Katsiev K., Burst L.M., Diebold U., Chaka A.M., and Delley B. (2005) Gas-phase-dependent properties of $SnO_2$ (110), (100), and (101) single-crystal surfaces: structure, composition, and electronic properties. *Phys. Rev. B* **72**, 165414.

Bendahan M., Lauque P., Lambert-Mauriat C., Carchano H., and Seguin J.L. (2002) Sputtered thin films of CuBr for ammonia microsensors: morphology, composition and ageing. *Sens. Actuators B* **84**, 6–11.

Bhuiyan A.L. (1984) Some aspects of the thermal stability action of the structure in aliphatic polyamides and polyacrylamides. *Polymer* **25**, 1699–1710.

Blesa M.A., Weisz A.D., Morando P.J., Salfity J.A., Magaz G.E., and Regazzoni A.E. (2000) The interaction of metal oxide surfaces with complexing agents dissolved in water. *Coord. Chem. Rev.* **196**, 31–63.

Bogue R. (2002b) Advanced automotive sensors. *Sens. Rev.* **22**(2), 113–118.

Bogue R.W. (2002a) The role of materials in advanced sensor technology. *Sens. Rev.* **22**(4), 289–299.

Bose A.C., Thangadurai P., and Ramasamy S. (2006) Grain size dependent electrical studies on nanocrystalline $SnO_2$. *Mater. Chem. Phys.* **95**, 72–78.

Brinzari V., Korotcenkov G., and Golovanov V. (2001) Factors influencing the gas sensing characteristics of tin dioxide films deposited by spray pyrolysis: understanding and possibilities for control. *Thin Solid Films* **391**(1/2), 167–175.

Brinzari V., Korotcenkov G., and Matolin V. (2005) Synchrotron radiation photoemission study of indium oxide surface prepared by pyrolysis method. *Appl. Surf. Sci.*, **243**(1–4), 335–344.

Brinzari V., Korotcenkov G., and Schwank J. (1999a) Optimization of thin film gas sensors for environmental monitoring through theoretical modeling. In: S. Buettgenbach (ed.), *Chemical Microsensors and Applications II, Proc. SPIE* **3857**, 186–197.

Brinzari V., Korotcenkov G., Schwank J., Lantto V., Saukko S., and Golovanov V. (2002) Morphological rank of nano-scale tin dioxide films deposited by spray pyrolysis from $SnCl_4.5H_2O$ water solution. *Thin Solid Films* **408**, 51–58.

Brinzari V., Korotcenkov G., Veltruska K., Matolin V., Tsud N., and Schwank J. (2000b) XPS study of gas sensitive $SnO_2$ thin films. In: *Proceeding of Semiconductor International Conference CAS'2000, 10–14 October 2000*, Sinaia, Romania, Vol. **1**, pp. 127–130.

Brinzari V., Korotchenkov G., and Dmitriev S. (1999b) Simulation of thin film gas sensor kinetics. *Sens. Actuators B* **61**(1–3), 143–153.

Brinzari V., Korotchenkov G., and Dmitriev S. (2000a) Theoretical study of semiconductor thin film gas sensitivity: attempt to consistent approach. *J. Electron. Technol.* **33**, 225–235.

Brynn D.H. and Tseung C.C. (1979) The reduction of sulphur dioxide by carbon monoxide on $La_{0.5}Sr_{0.5}CoO_3$ catalyst. *J. Chem. Technol. Biotechnol.* **29**(12), 713.

Cabot A., Dieguez A., Romano-Rodriguez A., Morante J.R., and Barsan N. (2001) Influence of the catalytic introduction procedure on the nano-$SnO_2$ gas sensor performances. Where and how stay the catalytic atoms? *Sens. Actuators B* **79**, 98–106.

Cabot A., Vila A., and Morante J.R. (2002) Analysis of the catalytic activity and electrical characteristics of different modified $SnO_2$ layers for gas sensors. *Sens. Actuators B* **84**, 2–20.

Calatayud M., Markovits A., Menetrey M., Mguig B., and Minot C. (2003) Adsorption on perfect and reduced surfaces of metal oxides. *Catal. Today* **85**, 125–143.

Campbell C.T. (2004) The active site in nanoparticle gold catalysis. *Science* **306**, 234–235.

Cantalini C., Valentini L., Lozzi L., Armentano I., Kenny J.M., and Santucci S. (2003) $NO_2$ gas sensitivity of carbon nanotubes obtained by plasma enhanced chemical vapor deposition. *Sens. Actuators B* **93**, 333–337.

Castañeda L. (2007) Effects of palladium coatings on oxygen sensors of titanium dioxide thin films. *Mater. Sci. Eng. B* **139**, 149–154.

Catlow C.R.A. (ed.) (1997) *Computer Modeling in Inorganic Crystallography*. Academic Press, New York.

Chabanis G., Parkin I., and Williams D.E. (2003) A simple equivalent circuit model to represent microstructure effects on the response of semiconducting oxide-based gas sensors. *Meas. Sci. Technol.* **14**, 76–86.

Chan K.S., Ma J., Jaenicke S., and Chuan G.K. (1994) Catalytic carbon-monoxide oxidation over strontium, cerium and copper-substituted lanthanum manganates and cobaltates. *Appl. Catal. A* **107**, 201.

Chao Y.C. and Shih J.S. (1998) Adsorption study of organic molecules on fullerene with piezoelectric crystal detection system. *Anal. Chim. Acta* **374**, 39–46.

Chen F., Zhao T., Fei Y.Y., Lu H.B., Cheng Z.H., Yang G.Z., and Zhu X.D. (2002) Surface segregation of bulk oxygen on oxidation of epitaxially grown Nb-doped $SrTiO_3$ on $SrTiO_3(001)$. *Appl. Phys. Lett.* **80**, 2889.

Chena C.S., Kruidhof H., Bouwmeestera H.J.M., Verweij H., and Burggraaf A.J. (1996) Oxygen permeation through oxygen ion oxide-noble metal dual phase composites. *Solid State Ionics* **86–88**, 569–572.

Choudhary T.V., and Goodman D.W. (2005) Catalytically active gold: the role of cluster morphology. *Appl. Catal. A* **291**, 32–36.

Cirera A., Cornet A., Morante J.R., Olaizola S.M., Castano E., and Gracia J.R. (2000) Comparative structural study between sputtered and liquid pyrolysis nanocrystalline $SnO_2$. *Mater. Sci. Eng. B* **69–70**, 406–410.

Clifford P.K. (1983) Microcomputational selectivity enhancement of semiconductor gas sensors. In: *Proceeding of International Meeting on Chemical Sensors,* Fukuoka, Japan, 19–22 September 1983, pp. 153–158.

Colladet K., Nicolas M., Goris L., Lutsen L.L., and Vanderzande D. (2004) Low-band gap polymers for photovoltaic applications. *Thin Solid Films* **451–452**, 7–11.

Cox P.A. (1992) *Transition Metal Oxides: An Introduction to Their Electronic Structure and Properties*. Clarendon Press, Oxford.

Dai Z.R., Gole J.L., Stout J.D., and Wang Z.L. (2002) Tin oxide nanowires, nanoribbons, and nanotubes. *Phys. Chem. B* **106**, 1274–1279.

Dai Z.R., Pan Z.W., and Wang Z.L. (2003) Novel nanostructures of functional oxides synthesized by thermal evaporation. *Adv. Funct. Mater.* **13**, 9–24.

Dakin J. and Culshaw B. (eds.) (1988) *Optical Fiber Sensors: Principles and Components, Vol.1*. Artech House, Boston.

De Fresart E., Darville J., and Gilles J.M. (1982) Influence of the surface reconstruction of the work function and surface conductance of (110) $SnO_2$. *Appl. Surf. Sci.* **11/12**, 637–651.

De Souza Brito G.E., Santilli C.V., and Pulcenelli S.H. (1995) Evolution of the fractal structural during sintering of $SnO_2$ compacted sol-gel powders. *Colloids Surf. A* **97**, 217–225.

Dean J.A. (ed.) (1972) *Lange's Handbook of Chemistry*, 13th ed. McGraw-Hill, New York.

Deb B., Desai S., Sumanasekera G.U., and Sunkara M.K. (2007) Gas sensing behaviour of mat-like networked tungsten oxide nanowire thin films. *Nanotechnology* **18**, 285501.

Derjaguin B.V. (1989) *Theory of Stability of Colloids and Thin Films*. Consultants Bureau, New York, London.

Dhandapani B. and Oyama S.T. (1997) Gas phase ozone decomposition catalysis. *Appl. Catal. B* **11**, 129–166.

Dieguez A., Romano-Rodriguez A., Vila A., and Morante J.R. (2001) The complete Raman spectrum of nanometric $SnO_2$ particles. *Appl. Phys.* **90**, 1550–1557.

Dieguez A., Romano-Rodrýguez A., Alay J.L., Morante J.R., Barsan N., Kappler J., Weimar U., and Gopel W. (2000) Parameter optimisation in $SnO_2$ gas sensors for $NO_2$ detection with low cross-sensitivity to CO: sol–gel preparation, film preparation, powder calcination, doping and grinding. *Sens. Actuators B* **65**, 166–168.

Dimitrov V. and Komatsu T. (2002) Classification of simple oxides: a polarizability approach. *J. Solid State Chem.* **163**, 100–112.

Dmitriev S., Lilach Y., Button B., Moskovits M., and Kolmakov A. (2007) Nanoengineered chemiresistors: the interplay between electron transport and chemisorption properties of morphologically encoded $SnO_2$ nanowires. *Nanotechnology* **18**, 055707.

Doll T. and Eisele I. (1998) Gas detection with work function sensors. In: *Proceeding of SPIE Conference on Chemical Microsensors and Applications*, Boston, November 1998, vol. **3539**, pp. 96–105.

Doll T., Lechner J., Eisele I., Schierbaum K.D., and Gopel W. (1996) Ozone detection in the ppb range with work function sensors operating at room temperature. *Sens. Actuators B* **34**, 506–510.

Dufour L.C. and Nowotny J. (eds.) (1988*) Surface and Near Surface Chemistry of Oxide Materials*. Elsevier, Amsterdam.

Eglitis R.I., Heifets E., Kotomin E.A., Majer J., and Borstel G. (2003) First-principles calculations of perovoskite thin film. *Mater. Sci. Semin. Proc.* **5**, 129–134.

Eisele I., Doll T., and Burgmair M. (2001) Low power gas detection with FET sensors. *Sens. Actuators B* **78**, 19–25.

El-Azab A., Gan S., and Liang Y. (2002) Binding and diffusion of Pt nanoclusters on anatase $TiO_2$(001)-(1x4) surface. *Surf. Sci.* **506**, 93–104.

Emiroglu S., Barsan N., Weimar U., and Hoffmann V. (2001) In-situ diffuse reflectance infrared spectroscopy study of CO adsorption on $SnO_2$. *Thin Solid Films* **391**, 176–185.

Epling W.S., Peden C.H.F., Henderson M., and Diebolod U. (1998) Evidence for oxygen adatoms on $TiO_2$ (110) resulting from $O_2$ dissociation at vacancy sites. *Surf. Sci.* **412/413**, 333–343.

Esch H., Huyberechts G., Mertens R., Maes G., Manca J., DeCeuninck W., and De Schepper L. (2000) The stability of Pt heater and temperature sensing elements for silicon integrated tin oxide gas sensors. *Sens. Actuators B* **65**, 190–192.

Fang Q., Chetwynd D.G., and Gardner J.W. (2002) Conducting polymer films by UV-photo processing. *Sens. Actuators A* **99**, 74–77.

Fleischer M. and Meixner H. (1997) Fast gas sensors based on metal oxides, which are stable at high temperatures. *Sens. Actuators B* **43**, 1–10.

Fleischer M. and Meixner H. (1998) Selectivity in high-temperature operated semiconductor gas-sensors. *Sens. Actuators B* **52**, 179–187.

Freund H.J., Kuhlenbeck H., and Staemmler V. (1996) Oxide surfaces. *Rep. Prog. Phys.* **59**, 283–347.

Freund H.J., Kuhlenbeck H., Libuda J., Rupprechter G., Baumer M., and Hamann H. (2001) Bridging the pressure and materials gaps between catalysis and surface science: clean and modified oxide surfaces. *Topics Catal.* **15** (2–4), 201–209.

Freund J. and Umbach E. (1993) (eds.) *Adsorption on Ordered Surfaces of Ionic Solids and Thin Films*. Springer Series in Surface Sciences, vol. **33**. Springer-Verlag, Heidelberg.

Gadomski A., Rub J.M., Luczka J., and Ausloos M. (2005) On temperature-and space-dimension dependent matter agglomerations in a mature growing stage. *Chem. Phys.* **310**, 153–161.

Galdikas A., Kaciulis S., Mattogno G., Mironas A., Senuliene D., and Setkus A. (1998) Stability and oxidation of the sandwich type gas sensors based on thin metal films. *Sens. Actuators B* **48**, 376–382.

Galdikas A., Mironas A., Senuliene D., and Setkus A. (1996) CO gas induced resistance switching in $SnO_2$:ultra-thin-Pt sandwich structure. *Sens. Actuators B* **32**, 87–92.

Gardner J.M. and Bartlett P.N. (1999) *Electronic Noses: Principles and Applications*. Oxford University Press, Oxford.

Gardner J.M. and Bartlett P.N. (eds.) (1992) *Sensors and Sensory Systems for an Electronic Nose*. Kluwer, Dordrecht, The Netherlands.

Gas'kov A.M. and Rumyantseva M.N. (2001) Nature of gas sensitivity in nanocrystalline metal oxides. *Russ. J. Appl. Chem.* **74**(3), 440–444.

Geistlinger H. (1993) Electron theory of thin film gas sensors. *Sens. Actuators B* **17**, 47–60.

Geistlinger H. (1994) Accumulation layer model for $Ga_2O_3$ thin-film gas sensors based on the Volkenstein theory of catalysis. *Sens. Actuators B* **18–19**, 125–131.

Gellinger P.J. and Bouwmeester H.J.M. (2000) Solid state aspects of oxidation catalysis. *Catal. Today* **58**, 1–53.

Gellinger P.J. and Bouwmeester H.J.M. (eds.) (1997) *The CRC Handbook of Solid State Electrochemistry*. CRC Press, Boca Raton, FL.

Gentry S.J. and Walsh P.T. (1982) The theory of poisoning of catalytic flammable gas-sensing elements. In: Mosely P.T. and Tofied B.C. (eds.), *Solid State Gas Sensors*. Adam Hilger. Bristol and Philadelphia, pp. 32–50.

Gillet M., Aguir K., Bendahan M., and Mennini P. (2005) Grain size effect in sputtered tungsten trioxide thin films on the sensitivity to ozone. *Thin Solid Films* **484**, 358–363.

Golovanov V., Maki-Jaskari M.A, Rantala T.T., Korotcenkov G., Brinzari V., Cornet A., and Morante J. (2005a) Experimental and theoretical studies of the indium oxide-based gas sensors deposited by spray pyrolysis. *Sens. Actuators* **106**, 563–571.

Golovanov V., Pekna T., Kiv A., Litovchenko V., Korotcenkov G., Brinzari V., Cornet A., and Morante J. (2005b) The influence of structural factors on sensitivity of $SnO_2$-based gas sensors to CO in humid atmosphere. *Ukr. Phys. J.* **50**, 374–380.

Gopel W. (1996) Ultimate limits in the miniaturization of chemical sensors. *Sens. Actuators A* **56**, 83–102.

Gopel W. and Reinhard G. (1998) Metal oxide sensors: new devices through tailoring interfaces on the atomic scale. In: *Sensor Update, Part 3*. Wiley-VCH, Weinheim, 47–121.

Gopel W., Hesse J., and Zemel J.N. (eds.) (1989–1995) *Sensors, A Comprehensive Survey*, Vols. 1–9. Wiley-VCH, Weinheim.

Gordon M.J., Gaur S., Kelkar S., and Baldwin R.M. (1996) Low temperature incineration of mixed wastes using bulk metal oxide catalysts. *Catal. Today* **28**, 305–317.

Gurlo A., Ivanovskaya M., Barsan N., Schweizer-Berberich M., Weimar U., Gopel W., and Dieguez A. (1997) Grain size control in nanocrystalline $In_2O_3$ semiconductor sensors. *Sens. Actuators B* **44**, 327–333.

Hadjipanayis G.C. and Siegel R.W. (eds.) (1994) *Nanophase Materials*. Kluwer, Dordrecht, The Netherlands.

Hagen J.E. and Kvaal K. (1998) Electronic nose and artificial neural network. *Meat Sci.* **49**(1), S273–S286.

Hammer B. and Norskov J.K. (1995) Why gold is the noblest of all the metals. *Nature* **376**, 238–240.

Hamnett A. and Goodenough J.B. (1984) Binary transition-metal oxides. In: Madelung O. (ed.), *Semiconductors: Physics of Nontetrahedrally Bonded Binary Compounds III*, Vol. 17, part g, Landolt-Börnstein New Series, Group III, Springer-Verlag, Berlin.

Han P.G., Wong H., and Poon M.C. (2001) Sensitivity and stability of porous polycrystalline silicon gas sensor. *Colloids Surf. A* **179**, 171–175.

Hanys P., Janecek P., Matolin V., Korotcenkov G., and Nehasil V. (2006) XPS and TPD study of $Rh/SnO_2$ system-reversible process of substrate oxidation and reduction. *Surf. Sci.* **600**, 4233–4238.

Harsanyi G. (1994) *Polymer Films in Sensor Applications*. Technomic, Lancaster.

Harsanyi G. (2000) Polymer films in sensor applications: a review of present uses and future possibilities. *Sensor Rev.* **20**(2), 98–105.

Haruta M. (1997) Size- and support dependency in the catalysis of gold. *Catal. Today* **36**, 153.

Heeg J., Kramer C., Wolter M., Michaelis S., Plieth W., and Fisher W.J. (2001) Polythiophene-$O_3$ surface reactions studied by XPS. *Appl. Surf. Sci.* **180**, 36–41.

Henrich V.E. and Cox P.A. (1994) *The Surface Science of Metal Oxides*. Cambridge University Press, Cambridge.

Heo Y.W., Norton D.P., Tien L.C., Kwon Y., Kang B.S., Ren F., Pearton S.J., and LaRoche J.R. (2004) ZnO nanowire growth and devices. *Mater. Sci. Eng. R* **47**, 1–47.

Holzinger M., Maier J., and Sitte W. (1996) Fast $CO_2$-selective potentiometric sensor with open reference electrode. *Solid State Ionics* **86–88**, 1055–1062.

Horiuchi T., Hidaka H., Fukui T., Kubo Y., Horio M., Suzuki K., and Mori T. (1998) Effect of added basic metal oxides on $CO_2$ adsorption on alumina at elevated temperatures. *Appl. Catal. A* **167**, 195–202.

Houser E.J., Mlsna T.E., Nguyen V.K., Chung R., Mowery E.L., and McGill R.A. (2001) Rational materials design of sorbent coatings for explosives: applications with chemical sensors. *Talanta* **54**, 469–485.

Huang X.J. and Choi Y.K. (2007) Chemical sensors based on nanostructured materials. *Sens. Actuators B* **122**, 659–671.

Hyoda T., Nishida N., Shimizu Y., and Egashira M. (2002) Preparation and gas sensing properties of thermally stable mesoporous $SnO_2$. *Sens. Actuators B* **83**, 209–215.

Hyodo T., Abe S., Shimuzu Y., and Egashira M. (2003) Gas sensing properties of ordered mesoporous $SnO_2$ and effects of coatings thereof. *Sens. Actuators B* **93**, 590–600.

Ihokura K. and Watson J. (1994) *The Stannic Oxide Gas Sensor, Principle and Applications*. CRC Press, Boca Raton, FL.

Ishihara T. and Matsubara S. (1998) Capacitive type gas sensors. *J. Electroceram.* **2**, 215–228.

Ishizaki K., Komarneni S., and Nanko M. (1998) *Porous Materials: Process Technology and Applications*. Kluwer, Dordrecht, The Netherlands, p. 202.

Ivanovskaya M., Kotsikau D., Faglia G., and Nelli P. (2003) Influence of chemical composition and structural factors of $Fe_2O_3/In_2O_3$ sensors on their selectivity and sensitivity to ethanol. *Sens. Actuators B* **96**, 498–503.

Jamnik J., Kamp B., Merkle R., and Maier J. (2002) Space charge influenced oxygen incorporation in oxides: in how far does it contribute to the drift of Taguchi sensors? *Solid State Ionics* **150**, 157–166.

Jianping L., Yue W., Xiaoguang, Qing M., Li W., and Jinghong H. (2000) $H_2S$ sensing properties of the $SnO_2$-based thin films. *Sens. Actuators B* **65**, 111–113.

Jin C.J., Yamazaki T., Shirai Y., Yoshizawa T., Kikuta T., Nakatani N., and Takeda H. (2005) Dependence of $NO_2$ gas sensitivity of $WO_3$ sputtered films on film density. *Thin Solid Films* **474**, 255–260.

Jin Z., Zhou H.J., Jin Z.L., Savinell R.F., and Lin C.C. (1998) Application of nano-crystalline porous tin oxide film for CO sensing. *Sens. Actuators B* **52**, 188–194.

Johnson C.L., Schwank J.W., and Wise K.D. (1994) Integrated ultra-thin-film gas sensors. *Sens. Actuators B* **20**, 55–62.

Johnson D.A. (1982) *Some Thermodynamic Aspects of Inorganic Chemistry*. Cambridge University Press, Cambridge.

Jolivet J.P. (2000) *Metal Oxide Chemistry and Synthesis: From Solution to Solid State*. Wiley, Chichester, UK.

Jullien R. and Botet R. (1987) *Aggregation and Fractal Aggregates*. World Scientific, Singapore.

Kabbabi A., Gloaguen F., Andolfatto F., and Durand R. (1994) Particle size effect for oxygen reduction and methanol oxidation on Pt/C inside a proton exchange membrane. *J. Electroanal. Chem.* **373**, 251–254.

Kanazawa E., Sakai G., Shimanoe K., Kanmura Y., Teraoka Y., Miura N., and Yamazoe N. (2001) Metal oxide semiconductor $N_2O$ sensors for medical use. *Sens. Actuators B* **77**, 72–77.

Kanda K. and Maekawa T. (2005) Development of a $WO_3$ thick-film-based sensors for the detection of VOC. *Sens. Actuators B* **108**, 97–101.

Kaur M., Gupta SA.K., Betty C.A., Saxena V., Katti V.R., Gadkari S.C., and Yakhmi J.V. (2005) Detection of reducing gases by $SnO_2$ thin films: an impedance spectroscopy study. *Sens. Actuators B* **107**, 360–365.

Kawabe T., Tabata K., Suzuki E., Ichikawa Y., and Nagasawa Y. (2001) Morphological effects of $SnO_2$ thin film on the selective oxidation of methane. *Catal. Today* **71**, 21–29.

Kaye B.H. (1989) *A Random Walk Through Fractal Dimensions*. VCH, Weinheim.

Kindery W.D., Bowen H.K., and Uhlmann D.R. (1976) *Introduction to Ceramics*. 2nd ed., Wiley, New York.

Kocemba I., Szafran S., Rynkowski J., and Paryjczak T. (2001) The properties of strongly pressed tin oxide-based gas sensors. *Sens. Actuators B* **79**, 28–32.

Koch C.C. (1997) Synthesis of nanostructured materials by mechanical milling: problems and opportunities. *NanoStruct. Mater.* **9**, 13–22.

Kohl D. (1990) The role of noble metals in the chemistry of solid-state gas sensors. *Sens. Actuators B* **1**, 158–165.

Kohl D. (2001) Function and application of gas sensors. *J. Phys. D* **34**, R125–R149.

Kong X.H., Sun X.M., and Li Y.D. (2003) Synthesis of ZnO nanobelts by carbothermal reduction and their photoluminescence properties. *Chem. Lett.* **32**, 546–547.

Korotcenkov G. (2005) Gas response control through structural and chemical modification of metal oxides: State of the art and approaches. *Sens. Actuators B* **107**, 209–232.

Korotcenkov G. (2007a) Metal oxides for solid state gas sensors. What determines our choice? *Mater. Sci. Eng. B* **139**, 1–23.

Korotcenkov G. (2007b) Practical aspects in design of one-electrode semiconductor gas sensors: Status report. *Sens. Actuators B* **121**, 664–678.

Korotcenkov G. (2008) The role of morphology and crystallographic structure of metal oxides in response of conductometric-type gas sensors. *Mater. Sci. Eng. Rev.* **61**, 1–39.

Korotcenkov G. and Stetter J.R. (2008) Comparative study of $SnO_2$- and $In_2O_3$-based ozone sensors. *ECS Trans.* **6**, 29–41.

Korotchenkov G., Brynzari V., and Dmitriev S. (1998) Kinetics characteristics of $SnO_2$ thin film gas sensors for environmental monitoring. In: Buettgenbach S. (ed.), *Chemical Microsensors and Applications. Proc. SPIE* **3539**, 196–204.

Korotcenkov G., Brinzari V., DiBattista M., Schwank J., and Vasiliev A. (2001) Peculiarities of $SnO_2$ thin film deposition by spray pyrolysis for gas sensor application. *Sens. Actuators B* **77**, 244–252.

Korotcenkov G., Cerneavschi A., Brinzari V., Cornet A., Morante J., Cabot A., and Arbiol J. (2002) Crystallographic characterization of $In_2O_3$ films deposited by spray pyrolysis. *Sens. Actuators B* **84**, 37–42.

Korotcenkov G., Brinzari V., Boris Y., Ivanov M., Schawnk J., and Morante J. (2003a) Surface Pd doping influence on gas sensing characteristics of $SnO_2$ thin films deposited by spray pyrolysis. *Thin Solid Films* **436**, 119–126.

Korotcenkov G., Macsanov V., Tolstoy V., Brinzari V., Schwank J., and Faglia G. (2003b) Structural and gas response characterization of nano-size $SnO_2$ films deposited by SILD method. *Sens. Actuators B* **96**, 602–609.

Korotcenkov G., Brinzari V., Cerneavschi A., Ivanov M., Golovanov V., Cornet A., Morante J., Cabot A., and Arbiol J. (2004a) The influence of film structure on $In_2O_3$ gas response. *Thin Solid Films* **460**, 315–323.

Korotcenkov G., Brinzari V., Cerneavschi A., Ivanov M., Cornet A. Morante J., Cabot A., and Arbiol J. (2004b) $In_2O_3$ films deposited by spray pyrolysis: Gas response to reducing (CO, $H_2$) gases. *Sens. Actuators B* **98**(2–3), 236–243.

Korotcenkov G., Cerneavschi A., Brinzari V., Vasiliev A., Cornet A. Morante J., Cabot A., and Arbiol J. (2004c) $In_2O_3$ films deposited by spray pyrolysis as a material for ozone gas sensors. *Sens. Actuators B* **99**, 304–310.

Korotcenkov G., Boris I., Brinzari V., Luchkovsky Yu., Karlotsky G., Golovanov V., Cornet A., Rossinyol E., Rodriguez J., and Cirera A. (2004d) Gas sensing characteristics of one-electrode gas sensors on the base of doped $In_2O_3$ ceramics. *Sens. Actuators B* **103**, 13–22.

Korotcenkov G., Brinzari V., Golovanov V., and Blinov Y. (2004e) Kinetics of gas response to reducing gases of $SnO_2$ films deposited by spray pyrolysis. *Sens. Actuators B* **98**, 41–45.

Korotcenkov G., Cornet A., Rossinyol E., Arbiol J., Brinzari V., and Blinov Y. (2005a) Faceting characterization of $SnO_2$ nanocrystals deposited by spray pyrolysis from $SnCl_4$-$5H_2O$ water solution. *Thin Solid Films* **471**, 310–319.

Korotcenkov G., Brinzari V., Ivanov M., Cerneavschi A., Rodriguez J., Cirera A., Cornet A., and Morante J. (2005b) Structural stability of $In_2O_3$ films deposited by spray pyrolysis during thermal annealing. *Thin Solid Films* **479**, 38–51.

Korotcenkov G., Golovanov V., Cornet A., Brinzari V., Morante J., and Ivanov M. (2005c) Distinguishing feature of metal oxide films' structural engineering for gas sensor application. *J. Phys.: Confer. Ser.* (IOP) **15**, 256–261.

Korotcenkov G., Blinov I., and Stetter J.R. (2007a) Kinetics of indium oxide-based thin film gas sensor response: The role of "redox" and adsorption/desorption processes in gas sensing effects. *Thin Solid Films* **515**, 3987–3996.

Korotcenkov G., Brinzari V., Stetter J.R., Blinov I., and Blaja V. (2007b) The nature of processes controlling the kinetics of indium oxide-based thin film gas sensor response. *Sens. Actuators B* **128**, 51–63.

Korotcenkov G., Blinov I., Ivanov M., and Stetter J.R. (2007c) Ozone sensors on the base of $SnO_2$ films deposited by spray pyrolysis. *Sens. Actuators B* **120**, 679–686.

Kosima I., Adachi H., and Yasumori I. (1983) Electronic structures of the $LaBO_3$ (B = Co, Fe, Al) perovskite oxides related to their catalysis. *Surf. Sci.* **130**, 50.

Kotsikau D., Ivanovskaya M., and Orlik D. (2004) Gas-sensitive properties of thin and thick film sensosrs on $Fe_2O_3$-$SnO_2$ nanocomposites. *Sens. Actuators B* **101**, 227–231.

Koudelka-Hep M. (ed.) (2001) Proceedings of the 8th International Meeting on Chemical Sensors, Part 1. 2–5 July 2000, Basel, Switzerland. *Sens. Actuators B* **76**(1–3), 1–673.

Kreuer K.D. (1997) On the development of proton conducting materials for technological aaplications. *Solid State Ionics* **97**, 1–15.

Krilov O.V. and Kisilev V.F. (1981) *Adsorption and Catalysis on the Transition Metals and Their Oxides*. Chemistry Press, Moscow.

Kulkarni D. and Wachs I.E. (2002) Isopropanol oxidation by pure metal oxide catalysts: number of active surface sites and turnover frequencies. *Appl. Catal. A* **237**, 121–137.

Kulwicki B.M. (1991) Humidity sensors. *J. Am. Ceram. Soc.* **74**, 697–708.

Kumar D. and Sharma R.C. (1998) Advances in conductive polymers. *Eur. Polym. J.* **34**(8), 1053–1060.

Kupriyanov L.Y. (ed.) (1996) *Semiconductor Sensors in Physico-Chemical Studies*. Elsevier, Amsterdam.

Labeau M., Schmatz U., Delabouglise G., Roman J., Vallet-Regi M., and Gaskov A. (1995) Capacitance effects and gaseous adsorption on pure and doped polycrystalline tin oxide. *Sens. Actuators B* **26–27**, 49.

Labidi A., Jacolin C., Bendahan M., Abdelghani A., Guerin J., Aguir K., and Maaref M. (2005) Impedance spectroscopy on $WO_3$ gas sensor. *Sens. Actuators B* **106**, 713–718.

Lai X., St. Clair T.P., and Goodman D.W. (1998) Scanning tunneling microscopy studies of metal clusters supported on TiO2(110): morphology and electronic structure. *Prog. Surf. Sci.* **59**, 25–52.

Lamoreaux R.H., Hildenbrand D.L., and Brewer L. (1987) High-temperature vaporization behavior of oxides: II. Oxides of Be, Mg, Ca, Cr, Ba, B, Al, Ga, In, Ti, Si, Ge, Sn, Pb, Zn, Cd, and Hg. *J. Phys. Chem. Perf. Data* **16**(3), 419–443.

Lampe U., Fleischer M., Reitmeier N., Meixner H., McMonagle J.B., and Marsh A. (1996) New materials for metal oxide sensors. In: *Sensors Update, Part 2*. Wiley-VCH, Weinheim.

Lantto V., Rantalla T.S., and Rantalla T.T. (2000) Experimental and theoretical studies on the receptor and transducer function of $SnO_2$ gas sensors. *Electron. Technol.* **33**(1/2), 22–30.

Lauritsen J.V., Vang R.T., and Besenbacher F. (2006) From atom-resolved scanning tunneling microscopy (STM) studies to the design of new catalysts. *Catal. Today* **111**, 34–43.

Leblanc E., Perier-Camby L., Thomas G., Gibert R., Primet M., and Gelin P. (2000) $NO_x$ adsorption onto dehydroxylated or hydroxylated tin dioxide surface. Application to $SnO_2$-based sensors. *Sens. Actuators B* **62**, 67–72.

Lee G.G. and Kang S.-J.K. (2005) Formation of large pores and their effect on electrical properties of $SnO_2$ gas sensors. *Sens. Actuators B* **107**, 392–396.

Lee J.H. and Park S.J. (1990) Temperature dependence of electrical conductivity in polycrystalline tin oxide. *J. Am. Ceram. Soc.* **73**(9), 2771–2774.

Li C., Zhang D., Han S., Liu X., Tang T., and Zhou C. (2003) Diameter-controlled growth of single-crystalline $In_2O_3$ nanowires and their electronic properties. *Adv. Mater.* **15**, 143–146.

Li G.J., Zhang X.H., and Kawi S. (1999) Relationships between sensitivity, catalytic activity and surface areas of $SnO_2$ gas sensors. *Sen. Actuators B* **60**, 64–70.

Liang C., Meng G., Lei Y., Phillip F., and Zhang L. (2001) Catalytic growth of semiconducting $In_2O_3$ nanofibers. *Adv. Mater.* **13**, 1330–1333.

Litzelman S.J., Rothschild A., and Tuller H.L. (2005) The electrical properties and stability of $SrTi_{0.65}Fe_{0.35}O_{3-\delta}$ thin films for automotive oxygen sensor applications. *Sens. Actuators B* **108**, 231–237.

Lu J.G., Chang P., and Fan Z. (2006) Quasi-one-dimensional metal oxide materials—synthesis, properties and applications. *Mater. Sci. Eng. R.* **52**, 49–91.

Lucas E., Decker S., Khaleel A., Seitz A., Fultz S., Ponce A., Li W., Carnes C., and Klabunde K. (2001) Nanocrystalline metal oxides as unique chemical reagent/sorbents. *Chem. Eur. J.* **7**, 2505–2510.

Lundstrem I. (1996) Approaches and mechanisms to solid state based sensing. *Sens. Actuators B* **35–36**, 11–19.

Madey T.E., Pelhos K., Wu Q., Barnes R., Ermanoski I., Chen W., Kolodziej J.J., and Rowe J.E. (2002) Nanoscale surface chemistry. *Proc Natl. Acad. Sci. USA* **99**, 6503–6508.

Madou M.J. and Morrison S.R. (1987) *Chemical Sensing with Solid State Devices*. Academic Press, Harcourt Brace Jovanovich, Boston, New York.

Maier J., Holzinger M., and Sitte W. (1994) Fast potentiometric $CO_2$ sensors with open reference electrodes. *Solid State Ionics* **74**, 5–9.

Maier J., Schiotz J., Liu P., Norskov J.K., and Stimming U. (2004) Nano-scale effects in electrochemistry. *Chem. Phys. Lett.* **390**, 440–444.

Majoo S., Gland J.L., Wise K.D., and Schwank J.W. (1996) A silicon micromachined conductometric gas sensor with a maskless Pt sensing film deposited by selected-area CVD, *Sens. Actuators B* **35–36**, 312–319.

Maki-Jaskar M. and Rantalla T.T. (2001) Band structure and optical parameters of the $SnO_2$ (110) surface. *Phys. Rev. B* **64**, 075407-1-7.

Maki-Jaskar M. and Rantalla T.T. (2002) Theoretical study of oxygen-deficient $SnO_2$ (110) surfaces. *Phys. Rev. B* **65**, 245428-1-8.

Mašek K., Libra J., Skála T., Cabala M., Matolín V., Cháb V., and Prince K.C. (2006) SRPES investigation of tungsten oxide in different oxidation states. *Surf. Sci.* **600**, 1624–1627.

Masel R.I. (1996) Introduction to surface reactions. In: *Principles of Adsorption and Reaction on Solid State Surfaces*. Wiley, New York, pp. 438–481.

Matko I., Gaidi M., Chenevier B., Charai A., Saikaly W., and Labeau M. (2002) Pt doping of $SnO_2$ thin films. *J. Electrochem. Soc.* **149**, H153–H158.

Matsuguchi M., Kuroiwa T., Miyagishi T., Suzuki S., Ogura T., and Sakai Y. (1998) Stability and reliability of capacitive-type relative humidity sensors using crosslinked polyimide films. *Sens. Actuators B* **52**, 53–57.

Matsushima S., Teraoka Y., Miura N., and Yamazoe N. (1988) Electronic interaction between metal additives and tin dioxide in tin dioxide-based gas sensors. *Jap. J. Appl. Phys.* **27**, 1798–1802.

McAleer J.F., Moseley P.T., Norris J.O., and Williams D.E. (1987) Tin dioxide gas sensors: I. Aspects of the surface chemistry revealed by electrical conductance variations. *J. Chem. Soc. Faraday Trans.* I, **83**, 1323–1346.

Meier J., Schiotzb J., Liub P., Norskov J.K., and Stimminga U. (2004) Nano-scale effects in electrochemistry. *Chem. Phys. Lett.* **390**, 440–444.

Meixner H. and Lampe U. (1996) Metal oxide sensors. *Sens. Actuators B* **33**, 198–202.

Menesklou W., Schreiner H.J., Moos R., Hardtl K.H., and Ivers-Tiffee E. (2000) $Sr(Ti,Fe)O_3$: material for a temperature independent resistive oxide sensors. In: Wun-Fogle M., Uchino K., Ito Y., and Gotthardt R. (eds.), *Materials for Smart Systems III*, MRS Proceedings Vol. **604**. Pittsburgh, PA, pp. 305-310.

Michel H.J., Leiste H., Schierbaum K.D., and Halbritter J. (1998) Adsorbates and their effects on gas sensing properties of sputtered $SnO_2$ films. *Appl. Surf. Sci.* **126**, 57–64.

Min B.K. and Choi S.D. (2004) $SnO_2$ thin film gas sensor fabricated by ion beam deposition. *Sens. Actuators B* **98**, 239–246.

Min B.K., Wallace W.T., and Goodman D.W. (2006) Support effects on the nucleation, growth and morphology of gold nano-clusters. *Surf. Sci.* **600**, L7–L11.

Minot C. (2001) Theoretical approaches of the reactivity at MgO(100) and $TiO_2$(110) surfaces. In: Chaer Nascimento M.A. (ed.), *Progress in Theoretical Chemistry and Physics, Vol. 7*,. Kluwer, Dordrecht, The Netherlands, pp. 241–249.

Minot C., Fahmi A., and Ahdjoudj J. (1995) Periodic HF calculations of the adsorption of small molecules on $TiO_2$. In: Farrugia, L.J. (ed.), *The Synergy Between Dynamics and Reactivity at Clusters and Surfaces*. Kluwer, Drymen, Scotland, pp. 257–270.

Monkman G. (2000) Monomolecular Langmuir-Blodgett films—tomorrow's sensors? *Sensor Rev.* **20**(2), 127–131.

Morrison S.R. (1982) Chemisorption on nonmetalic surfaces. In: Anderson J.R., and Boudart M. (eds.), *Catalysid Science and Technology*. Springer-Verlag, Berlin.

Morrison S.R. (1987) Mechanism of semiconductor gas sensor operation. *Sens. Actuators* **11**, 283–287.

Moseley P.T. (1997) Solid state gas sensors. *Meas. Sci. Technol.* **8**, 223–237.

Moseley P.T., Norris J.O.W. and Williams D.E. (eds.) (1991) *Techniques and Mechanisms in Gas Sensing*. Adam Hilger, Bristol, UK.

Mrowec S. (1978) On the defect structure in nonstoichiometric metal oxides. *Ceram. Int.* **4**(2), 47–58.

Murch G.E. and Nowick A.S. (eds.) (1984) *Diffusion in Crystalline Solids*. Academic Press, New York.

Najafi N., Wise K.D., and Schwank J.W. (1994) A micromachined ultra-thin-film gas detector. *IEEE Trans. Electron. Devices* **41**, 1770–1777.

Nayral C., Viala E., Colliere V., Fau P., Senocq F., Maisonnat A., and Chaudret B. (2000) Synthesis and use of a novel $SnO_2$ nanomaterial for gas sensing. *Appl. Surf. Sci.* **164**, 219–226.

Nguyen C. and Do D.D. (1999) Adsorption of supercritical gases in porous media: determination of micropore size distribution. *J. Phys. Chem. B* **103**, 6900–6908.

Ning J.L., Jiang D.M., Kim K.H., and Shim K.B. (2007) Influence of texture on electrical properties of ZnO ceramics prepared by extrusion and spark plasma sintering. *Ceram. Int.* **33**, 107–114.

Nitta T. (1981) Ceramic humidity sensor. *Ind. Eng. Chem. Prod. Res. Dev.* **20**, 669–674.

Nitta T., Terada Z., and Hayakawa S. (1980) Humidity-sensitive electrical conduction of $MgCr_2O_4$-$TiO_2$ porous ceramics. *J. Am. Ceram.Soc.* **63**, 295–300.

Noguera C. (1995) *Physics and Chemistry at Oxide Surfaces*. Cambridge University Press, Cambridge.

Nowick A.S. (1991) Defects in ceramic oxides. *MRS Bull.* **16**(11), 38–41.

Obando L.A. and Booksh K.S. (1999) Tunning dynamic range and sensing of white-light multimode, fiber-optic surface plasmon resonance sensors. *Anal. Chem.* **71**, 5116–5122.

Oelerich W., Klassen T., and Bormann R. (2001) Metal oxides as catalysts for improved hydrogen sorption in nanocrystalline Mg-based materials. *J. Alloys Compounds* **315**, 237–242.

Ogawa H., Nishikawa M., and Abe A. (1982) Hall measurement studies and an electrical conduction model of tin oxide ultrafine particle films. *J. Appl. Phys.* **53**, 4448–4455.

Pan C.A. and Ma T.P. (1980) Work function of $In_2O_3$ film as determined from internal photoemission. *Appl. Phys. Lett.* **37**, 714–716.

Park S.S. and Mackenzie J.D. (1996) Thickness and microstructure effect on alcohol sensing of tin oxide thin films. *Thin Solid Films* **274**, 154–159.

Pasierb P., Komornicki S., Kozinski S., Gajerski R., and Rekas M. (2004) Long-term stability of potentiometric $CO_2$ sensors based on Nasicon as a solid electrolyte. *Sens. Actuators B* **101**, 47–56

Pauling L. (1929) The principles determining the structure of complex ionic crystals. *J. Am. Chem. Soc.* **51**(4), 1010–1026.

Pederson F.A., Greely J., and Norskov J.K. (2005) Understanding the effects of steps, strain, poisons, and alloying metane activation on Ni surfaces. *Catal. Lett.* **105**, 9–13.

Penza M., Antolini F., and Vittori-Antisari M. (2004) Carbon nanotubes as SAW chemical sensors materials. *Sens. Actuators B* **100**, 47–59.

Post M.L., Tunney J.J., Yang D., Du X., and Singleton D.L. (1999), Material chemistry of perovskite compounds as chemical sensors. *Sens. Actuators B* **59**, 190–194.

Rabek J.F. (1995) *Polymer Photodegradation: Mechanism and Experimental Methods*. Chapman & Hall, London.

Rao C.N.R., Kulkarni G.U., Thomas P.J., and Edwards P.P. (2002) Size-dependent chemistry: Properties of nanocrystals. *Chem. Eur. J.* **8**, 28–35.

Razumovskii S.D. and Zaikov G.Y. (1982) Effect of ozone on saturated polymers. *Polymer Sci. U.S.S.R.* **24**(10), 2805–2325.

Rothschild A. and Komem Y. (2004) The effect of grain size on the sensitivity of nanocrystalline metal oxide gas sensors. *J. Appl. Phys.* **95**, 6374–6380.

Rothschild A., Litzelman S.J., Tuller H.L., Menesklou W., Schneider T., and Ivers-Tiffee E. (2005) Temperature-independent resistive oxygen sensors based on $SrTi_{1-x}Fe_xO_{3-\delta}$ solid solutions. *Sens. Actuators B* **108**, 223–230.

Ruiz A.M., Cornet A., Shimanoe K., Morante J.R., and Yamazoe N. (2005) Effects of various metal additives on the gas sensing performances of $TiO_2$ nanocrystals obtained from hydrothermal treatments. *Sens. Actuators B* **108**(1–2), 34–40.

Rumyantseva M.N., Gaskov A.M., Rosman N., Pagnier T., and Morante J.R. (2005) Raman surface vibration modes in nanocrystalline $SnO_2$: correlation with gas sensor performances. *Chem. Mater.* **17**, 893–901.

Ryabtsev S.V., Shaposhnick A.V., Lukin A.N., and Domashevskaya E.P. (1999) Application of semiconductor gas sensors for medical diagnostics. *Sens. Actuators B* **59**, 26–29.

Sadaoka Y. (1992) Organic semiconductor gas sensors. In: Sberveglieri G. (ed.), *Gas Sensors*. Kluwer, Dordrecht, The Netherlands, pp. 187–218.

Samsonov G.V. (1973) *The Oxide Handbook*. IFI/Plenum, New York.

Sandler S.R. and Karo W. (1974) *Polymer Synthesis*. Academic Press, New York.

Satterfield C.N. (1991) *Hreterogeneous Catalysis in Industrial Practice*. McGraw-Hill New York.

Sberveglieri G. (1992b) Classical and novel techniques for the preparation of $SnO_2$ thin-film gas sensors. *Sens. Actuators B* **6**, 239–247.

Sberveglieri G. (1995) Recent developmenta insemiconducting thin film gas sensors. *Sens. Actuators B* **23**, 103–109.

Sberveglieri G. (ed.) (1992a) *Gas Sensors—Principles Operation and Development*. Kluwer, Dordrecht, The Netherlands.

Sberveglieri G., Gropelli S., Nelli P., and Camanzi A. (1991) A new technique for the preparation of highly sensitive hydrogen sensors based on $SnO_2(Bi_2O_3)$ thin films. *Sens. Actuators B* **5**, 253–255.

Schierbaum K.D., Weimar U., and Gopel W. (1992) Comparison of ceramic, thick-film and thin-film chemical sensors based upon $SnO_2$. *Sens. Actuators B* **7**, 709–716.

Schierbaum K.D., Weimar U., Gopel W., and Kowalkowski R. (1991) Conductance, work function and catalytic activity of $SnO_2$-based gas sensors. *Sens. Actuators B* **3**, 205–214.

Schmelzer J., Ropke G., and Mahnke R. (1999) *Aggregation Phenomena in Complex Systems*. Wiley-VCH, Weinheim.

Schneider W.F., Hass K., Miletic M., and Gland J.I. (2002) Dramatic cooperative effects in adsorption of $NO_x$ on MgO(001). *Phys. Chem. B* **106**(30), 7405–7413.

Semancik S., Cavicchi R.E., Kreider K.G., Suehle J.S., and Chaparala P. (1996) Selected-area deposition of multiple active films for conductometric microsensors arrays. *Sens. Actuator B* **34**, 209–212.

Seo M.G., Kwang B.H., Chai Y.S., Song K.D., and Lee D.D. (2000) $CO_2$ gas sensor using lithium ionic conductor with inside heater. *Sens.Actuators B* **65**, 346–348.

Shaw M.P. (1985) *Handbook on Semiconductors*. North Holland, Amsterdam, pp. 51–60.

Shek C.H., Lai J.K.L., and Lin G.M. (1999) Investigation of interface defects in nanocrystalline $SnO_2$ by positron annihilation. *J. Phys. Chem. Solids* **601**, 189–193.

Shimizu Y. and Egashira M. (1999) Basic aspects and challenges of semiconductor gas sensors. *MRS Bull.* **24**, 18–24.

Shimuzu Y., Maekawa T., Nakamura T., and Egashira M. (1998) Effects of gas diffusivity and reactivity on gas sensing properties of thick film $SnO_2$-based sensors. *Sens. Actuators B* **46**, 163–168.

Skotheim T.A., Elsenbaumer R.L., and Reynolds J.R. (eds.) (1998) *Handbook of Conducting Polymers*. Marcel Dekker, New York.

Skouras E.D., Burganos V.N., and Payatakes A.C. (1999) Simulation of gas diffusion and sorption in nanoceramic semiconductors. *J. Chem. Phys.* **110**(18), 9244–9253.

Sol C. and Tilley J.D. (2001) Ultraviolet laser irradiation induced chemical reactions of some metal oxides. *Mater. Chem.* **11**, 815–820.

Somorjai G. (1981) *Chemistry in Two Dimensions Surfaces*. Cornell University Press. Ithaca, NY.

Sorensen O.T. (ed.) (1981) *Non-stoichiometric Oxides*. Academic Press, New York.

Spivey J.J. (1987) Complete catalytic oxidation of volatile organics. *Ind. Eng. Chem. Res.* **26**, 2165–2180.

Steffes H., Imawan C., Solzbacher F., and Obermeier E. (1999) Reactively RF-sputtered $In_2O_3$ thin films for the detection of $NO_2$. In: *Proceedings of European Conference Eurosensors XIII*, The Hague, The Netherlands, 12–15 September 1999, pp. 871–874.

Suzuki T. and Yamazaki T. (1990) Effect of annealing on the gas sensitivity of thin oxide ultra-thin films. *J. Mater. Sci. Lett.* **9**, 750–751.

Szuber J. and Gopel W. (2000) Photoemission studies of the electronic properties of the space charge layer of $SnO_2$ (110) surface. *Electron. Technol.* **33**(1/2), 216–281.

Talazac L., Brunet J., Battut V., Blanc J.P., Pauly A., Germain J.P., Pellier S., and Soulier C. (2001) Air quality evaluation by monolithic InP-based resistive sensors. *Sens. Actuators B* **76**, 258–264.

Tamaki J., Shimanoe K., Yamada Y., Yamamoto Y. Miura N., and Yamazoe N. (1998) Dilute hydrogen sulfide sensing properties of $CuO$-$SnO_2$ thin film prepared by low-pressure evaporation method. *Sens. Actuators B* **49**, 121–125.

Tan O.K., Cao W., Hu Y., and Zhu W. (2004) Nanostructured oxides by high-energy ball milling technique: application as gas sensing materials. *Solid State Ionics* **172**, 309–316.

Tan O.K., Zhu W., Yan Q., and Kong L.B. (2000) Size effect and gas sensing characteristics of nanocrystalline $xSnO_{(1-x)}$ $\alpha$-$Fe_2O_3$ ethanol sensors. *Sens. Actuators B* **65**, 361–365.

Tanner R.E., Liang Y., and Altman E.I. (2002) Structure and chemical reactivity of adsorbed carboxylic acids on anatase $TiO_2$ (001). *Surf. Sci.* **506**, 251–271.

Tesfamichael T., Motta N., Bostrom T., and Bell J.M. (2007) Development of porous metal oxide thin films by co-evaporation. *Appl. Surf. Sci.* **253**, 4853–4859.

Tess M.E. and Cox J.A. (1999) Chemical and biochemical sensors based on advances in materials chemistry. *J. Pharmaceut. Biomed. Anal.* **19**, 55–68.

Thompson M. and Stone D.C. (1997) *Surface-Launched Acoustic Wave Sensors: Chemical Sensing and Thin-Film Characterization*. Wiley, New York.

Trasatti S. (1987) Oxide/aqueous solution interfaces. Interplay of surface chemistry and electrocatalysis. *Mater. Chem. Phys.* **16**, 157–174.

Trasatti S. (ed.) (1980 and 1981) *Electrodes of Conductive Metallic Oxides. Part A and B*. Elsevier, Amsterdam.

Traversa E., Gnappi G., Montenero A., and Gusmanoa G. (1996) Ceramic thin films by sol-gel processing as novel materials for integrated humidity sensors. *Sens. Actuators B* **31**, 59–70.

Tsiulyanu D., Marian S., Liess H.D., and Eisele I. (2004) Effect of annealing and temperature on the $NO_2$ sensing properties of tellurium based films. *Sens. Actuators B* **100**, 380–386.

Tsud N., Johanek V., Stara I., Veltruska K., and Matolin V. (2001) XPS, ISS and TPD study of Pd-Sn interaction on Pd-$SnO_2$ systems. *Thin Solid Films* **391**, 204–208.

Ustaze S., Guillemont L., Verucchi R., and Esaulov V.A. (1998) Electron capture on surfaces with electronegative adsorbates and surface poisoning. *Surf. Sci.* **397**, 361–473.

Valden M., Lai X., and Goodman D.W. (1998) Onset of catalytic activity of gold clusters on titania with the appearance of nonmetallic properties. *Science* **281**, 1647.

Valentini L., Cantalini C., Armentano I., Kenny J.M., Lozzi L., and Santucci S. (2004) Highly sensitive and selective sensors based on carbon nanotubes thin films for molecular detection. *Diamond Rel. Mater.* **13**, 1301–1305

Valentini L., Cantalini C., Lozzi L., Armentano I., Kenny J.M., and Santucci S. (2003) Reversible oxidation effects on carbon nanotubes thin films for gas sensing applications. *Mater. Sci. Eng. C* **23**, 523–529.

Van de Krol R. and Tuller H.L. (2002) Electroceramics—the role of interfaces. *Solid State Ionics* **150**, 167–179.

Varghese O.K., Gong D., Paulose M., Grimes C.A., and Dickey E.C. (2003) Crystallization and high-temperature structural stability of titanium oxide nanotube arrays. *J. Mater. Res.* **18**(1), 156–165.

Varghese O.K., Kichambre P.D., Gong D., Ong K.G., Dickey E.C., and Grimes C.A. (2001) Gas sensing characteristics of multi-wall carbon nanotubes. *Sens. Actuators B* **81**, 32–41.

Vasiliev R.B., Dorofeev S.G., Rumyantseva M.N., Ryabova L.I., and Gaskov A.M. (2006) Impedance spectroscopy of the ultradisperse $SnO_2$ ceramic with variable grain size. *Phys. Techn. Semicond.* **40**, 108–111 (in Russian).

Veltruska K., Tsud N., Brinzari V., Korotcenkov G., and Matolin V. (2001) CO adsorption on Pd clusters deposited on pyrolytically prepared $SnO_2$ studied by XPS. *J. Vacuum* **61**, 129–134.

Voorhoeve R.J., Remeika J.P., Freeland P.E., and Matthias B.T. (1972) Rare earth oxides of manganese and cobalt rival platinum for the treatment of carbon monoxide in auto exhaust. *Science* **177**(46), 353.

Wachs I.E. (1996) Raman and IR studies of surface metal oxide species on oxide supported metal oxide catalysts. *Catal. Today* **27**, 437–455.

Wahlstrom E., Lopez N., Schaub R., Thostrup P., Ronnau A., Africh C., Lagsgaard E., Norskov J.K., and Besenbacher F. (2003) Bonding of gold nanoclusters to oxygen vacancies on rutile $TiO_2(110)$. *Phys. Rev. Lett.* **90**, 026101.

Wahlstrom E., Vestergaard E.K., Schaub R., Ronnau A., Vestergaard M., Lagsgaard E., Stensgaard I., and Besenbacher F. (2004) Electron transfer–induced dynamics of oxygen molecules on the $TiO_2$ (110) surface. *Science* **303**, 511–513.

Wallace R.M. (2004) Challenges for the characterization and integration of high-$K$ dielectrics. *Appl. Surf. Sci.* **231–232**, 543–551.

Walton D.J. (1990) Electrically conducting polymers. *Mater. Design* **11**(3), 142–152.

Wang C.C., Akbar S.A., and Madou M.J. (1998) Ceramic based resistive sensors. *J. Electrocer.* **2**(4), 273–282.

Wang J., Gan M., and Shi J. (2007b) Detection and characterization of penetrating pores in porous materials. *Mater. Character.* **58**, 8–12.

Wang X., Fei Y., Lu H., Jin K.J., Zhu X.D., Chen Z., and Yang G. (2007a) In-diffusion of oxygen vacancies near step edges dominates the oxidation of perovskite films. *J. Phys.: Cond. Matter* **19**, 026206.

Wang X., Yee S.S., and Carey W.P. (1995) Transition between neck-controlled and grain-boundary-controlled sensitivity of metal-oxide gas sensors. *Sens. Actuators B* **25**, 454–457.

Weast R.C., Astle M., and Beyer W. (eds.) (1988) *CRC Handbook of Chemistry and Physics*, 69th ed. CRC Press, Boca Raton, FL.

Werle P., Slemr F., Maurer K., Kormann R., Mucke R., and Janker B. (2002) Near-and mid-infrared laser-optical sensors for gas analysis. *Optics Lasers Eng.* **37**, 101–1114.

Williams D. (1999) Semiconducting oxides as gas-sensitive resistors. *Sens. Actuators B* **57**, 1–16.

Williams D.E. and Pratt K.F.E. (1997) Self-diagnostic gas-sensitive resistors in sour gas applications. *Sens. Actuators B* **45**, 147–153.

Williams D.E. and Pratt K.F.E. (1998) Classification of reactive sites on the surface of polycrystalline tin dioxide. *J. Chem. Soc. Faraday Trans.* **94**, 3493–3500.

Williams D.E. and Pratt K.F.E. (2000) Microstructure effects on the response of gas-sensitive resistors based on semiconducting oxides. *Sens. Actuators B* **70**, 214–221.

Williams G. and Coles G.S.V. (1998b) Gas sensing properties of nanocrystalline metal oxide powders produced by a laser evaporation technique. *J. Mater. Chem.* **8**, 1657–1664.

Wilson D.M., Dunman K., Roppel T., and Kalim R. (2000) Rank extraction in tin-oxide sensor arrays. *Sens. Actuators B* **62**, 199–210.

Woormann H., Kohl D., and Heiland G. (1979) Work function and band bending on clean cleaved zinc oxide surfaces. *Surf. Sci.* **80**, 261–264.

Xu C., Tamaki J., Miura N., and Yamazoe N. (1990) Relationship between gas sensitivity and microstructure of porous $SnO_2$. *J. Electrochem. Soc. Jpn.* **58**, 1143–1148.

Xu C., Tamaki J., Miura N., and Yamazoe N. (1991) Grain size effects on gas sensitivity of porous $SnO_2$-based elements. *Sens. Actuators B* **3**, 147–155.

Yakovlev Y.P., Baranov A.N., Imenkov A.N., and Mikhailova M.P. (1991) Optoelectronic LED-photodiode pairs for moisture and gas sensors in spectral range 1.8–4.8 μm. In: Wolfbeis S. (ed.), *Chemical and Medical Sensors. Proc. SPIE* **1510**, 170–177.

Yamazoe N. (1991) New approaches to improvement semiconducting gas sensors. *Sens. Actuators B* **5**, 7–18.

Yamazoe N. and Miura N. (1992a) Some basic aspects of semiconductor gas sensors. In: Yamauchi S. (ed.) *Chemical Sensors Technology*, vol. **4**. Kodansha, Tokyo, pp. 20–41.

Yamazoe N. and Miura N. (1992b) New approaches in the devices of gas sensors. In: Sberveglieri G. (ed.) *Gas Sensors*. Kluwer, Dordrecht, The Netherlands.

Yamazoe N., Fuchigami J., Kishikawa M., and Seiyama T. (1979) Interactions of tin oxide surface with $O_2$, $H_2O$ and $H_2$. *Surf. Sci.* **86**, 335–344.

Yamazoe N., Kurokawa Y., and Seiyama T. (1983) Effects of additives on semiconductor gas sensors. *Sens. Actuators* **4**, 283–289.

Yamazoe N., Matsushima S., Maekawa T., Tamaki J., and Miura N. (1991) Control of Pd dispersion in $SnO_2$-based sensors. *Catal. Sci. Technol.* **1**, 201–205.

Yanagida, H. (1990) Intelligent ceramics. *Ferroelectrics* **102**, 251–257.

Zambelli T., Trost J., Wintterlin J.G., and Ertl G. (1996) Diffusion and atomic hopping of N atoms on Ru(0001) studied by scanning tunneling microscopy. *Phys. Rev. Lett.* **76**, 795–798.

Zapola P., Jaffe J.B., and Anthony C. Hess A.C. (1999) *Ab-initio* study of hydrogen adsorption on the ZnO (101:0) surface. *Surf. Sci.* **422**, 1–7.

Zemel J. (1988) Theoretical description of gas film interaction on $SnO_x$. *Thin Solid Films* **163**, 89–195.

Zhang J. and Gao L. (2004) Synthesis of antimony-doped tin oxide (ATO) nanoparticles by the nitrate–citrate combustion method. *Mater. Res. Bull.* **39**, 2249–2255.

Zhang S., Yongqing D.S., and Du F.H. (2003) Recent advances of superhard nanocomposite coatings: a review. *Surf. Coat. Technol.* **167**, 113–119.

Zhang Y.C., Tagawa H., Asakura S., Mizusaki J., and Narita H. (1997) Solid-state $CO_2$ sensor with $Li_2CO_3$–$Li_3PO_4$–$LiAlO_2$ electrolyte and $LiCoO_2$–$Co_3O_4$ as a solid reference electrode. *J. Electrochem. Soc.* **144**(12), 4345–4350.

Ziolek M., Kujawa J., Saur O., Aboulayt A., and Lavalley J.C. (1996) Influence of sulfur dioxide adsorption on the surface properties of metal oxides. *J. Mol. Catal. A: Chem.* **112**, 125–132.

CHAPTER 3

# COMBINATORIAL CONCEPTS FOR DEVELOPMENT OF SENSING MATERIALS

R. A. Potyrailo

## 1. INTRODUCTION

Rational design of sensing materials based on prior knowledge is a very attractive approach because it may avoid time-consuming synthesis and testing of numerous materials candidates (Honeybourne 2000; Shtoyko et al. 2004; Njagi et al. 2007). However, to be quantitatively successful, rational design (Newnham 1988; Akporiaye 1998; Ulmer II et al. 1998; Lavigne and Anslyn 2001; Suman et al. 2003; Hatchett and Josowicz 2008) requires detailed knowledge regarding relation of intrinsic properties of sensing materials to a set of their performance properties. This knowledge is typically obtained from extensive experimental and simulation data. However, with the increase of structural and functional complexity of materials, the ability to rationally define the precise requirements that result in a desired set of performance properties becomes increasingly limited (Schultz 2003). Thus, in addition to rational design, a variety of sensing materials, ranging from dyes and ionophores to biopolymers, organic and hybrid polymers, and nanomaterials, have been discovered using detailed experimental observations or simply by chance (McKusick et al. 1958; Pedersen 1967; Bühlmann et al. 1998; Svetlicic et al. 1998; Walt et al. 1998; Steinle et al. 2000; Martin et al. 2001; Hu et al. 2004; Potyrailo and Sivavec 2004). Such an approach in development of sensing materials reflects a more general situation in materials design that is "still too dependent on serendipity" and with only limited capability for rational materials design (Eberhart and Clougherty 2004).

Conventionally, detailed experimentation with sensing materials candidates for screening and optimization consumes a tremendous amount of time and project cost. Thus, developing sensing materials is a recognized challenge because it requires extensive experimentation to achieve not only the best short-term performance but also long-term stability, manufacturability, and other practical requirements. Numerous practical challenges in rational sensing material design provide tremendous prospects for the application of combinatorial methodologies for the development of sensing materials.

This chapter demonstrates the broad applicability of combinatorial technologies in discovery and optimization of new sensing materials. We discuss general principles of combinatorial materials screening, followed by a discussion of the opportunities facilitated by combinatorial technologies for discovery and optimization of new sensing materials. We further critically analyze results of materials development using discrete and gradient materials arrays and provide examples from a wide variety of sensors based on various energy-transduction principles that involve radiant, mechanical, and electrical types of energy.

## 2. GENERAL PRINCIPLES OF COMBINATORIAL MATERIALS SCREENING

Combinatorial materials screening is a process that couples the capability for parallel production of large arrays of diverse materials together with different high-throughput measurement techniques for various intrinsic and performance properties, followed by navigation through the collected data to identify "lead" materials (Jandeleit et al. 1999; Takeuchi et al. 2002; Xiang and Takeuchi 2003; Potyrailo and Amis 2003; Koinuma and Takeuchi 2004; Potyrailo et al. 2004a; Potyrailo and Takeuchi 2005b; Potyrailo and Maier 2006). The terms *combinatorial materials screening* and *high-throughput experimentation* are typically applied interchangeably to all types of automated parallel and rapid sequential evaluation processes for materials and process parameters that include truly combinatorial permutations or selected subsets.

Individual aspects of accelerated materials development have been known for decades. These include combinatorial and factorial experimental designs (Birina and Boitsov 1974), parallel synthesis of materials on a single substrate (Kennedy et al. 1965; Hoffmann 2001), screening of materials for performance properties (Hoogenboom et al. 2003), and computer data processing (Anderson and Moser 1958, Eash and Gohlke 1962). However, in 1970, Hanak (1970) suggested an integrated materials-development workflow. Its key aspects included (1) complete compositional mapping of a multicomponent system in one experiment; (2) simple, rapid, nondestructive, all-inclusive chemical analysis; (3) testing of properties by a scanning device; and (4) computer data processing. In 1995, Xiang, Schultz, and co-workers (1995) initiated applications of combinatorial methodologies in materials science. Since then, combinatorial tools have been employed to discover and optimize a wide variety of materials (see Table 3.1).

A typical combinatorial materials development cycle is outlined in Figure 3.1 (Potyrailo and Mirsky 2008, 2009). Compared to the initial idea of Hanak (1970), the modern workflow has several new and important aspects, such as design/planning of experiments, data mining, and scale-up. In combinatorial screening of materials, concepts originally conceived as highly automated have been recently refined to have more human input, with only an appropriate level of automation. For the throughput of 50–100 materials formulations per day, it is acceptable to perform certain aspects of the process manually (Potyrailo et al. 2002, 2003a). To address numerous materials-specific properties, a variety of high-throughput

**Table 3.1. Examples of materials developed using combinatorial screening techniques**

| MATERIALS | REFERENCE |
|---|---|
| Superconductor materials | Xiang et al. 1995 |
| Ferroelectric materials | Chang et al. 1998 |
| Magnetoresistive materials | Briceño et al. 1995 |
| Luminescent materials | Danielson et al. 1998 |
| Agricultural materials | Wong and Robertson 1999 |
| Structural materials | Zhao 2001 |
| Hydrogen storage materials | Olk 2005 |
| Organic light-emitting materials | Zou et al. 2001 |
| Ferromagnetic shape-memory alloys | Takeuchi et al. 2003 |
| Thermoelastic shape-memory alloys | Cui et al. 2006 |
| Heterogeneous catalysts | Holzwarth et al. 1998 |
| Homogeneous catalysts | Cooper et al. 1998 |
| Polymerization catalysts | Lemmon et al. 2001 |
| Electrochemical catalysts | Reddington et al. 1998 |
| Electrocatalysts for hydrogen evolution | Greeley et al. 2006 |
| Polymers | Brocchini et al. 1997 |
| Zeolites | Lai et al. 2001 |
| Metal alloys | Ramirez and Saha 2004 |
| Materials for methanol fuel cells | Jiang et al. 2005 |
| Materials for solid oxide fuel cells | Lemmon et al. 2004 |
| Materials for solar cells | Hänsel et al. 2002 |
| Automotive coatings | Chisholm et al. 2002 |
| Waterborne coatings | Wicks and Bach 2002 |
| Vapor-barrier coatings | Grunlan et al. 2003 |
| Marine coatings | Stafslien et al. 2006 |
| Fouling-release coatings | Ekin and Webster 2007 |

characterization tools are required. Characterization tools are used for rapid and automated assessment of single or multiple properties of the large number of samples fabricated together as a combinatorial array or "library" (MacLean et al. 2000; Potyrailo and Amis 2003; Potyrailo and Takeuchi 2005a).

In addition to the parallel synthesis and high-throughput characterization instrumentation, which differ significantly from conventional equipment, the data management approaches also differ from conventional data evaluation (Potyrailo and Maier 2006). In a well-developed combinatorial workflow, design and synthesis protocols for materials libraries are computer-assisted, materials synthesis and library preparation are carried out with computer-controlled manipulators, and property screening and materials characterization are also software-controlled. Further, materials synthesis data as well as property and characterization data are collected into a materials database. This database contains information on starting components, their descriptors, process conditions, materials testing algorithms, and performance properties of libraries of sensing materials. Data in such a database are not just stored, they

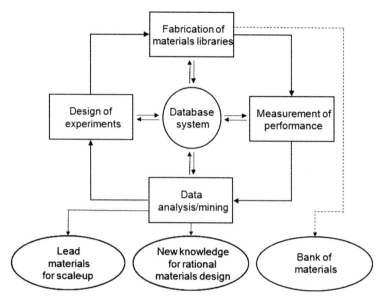

**Figure 3.1.** Typical process for combinatorial materials development.

are also processed with the proper statistical analysis, visualization, modeling, and data-mining tools. Combinatorial synthesis of materials provides a good possibility for formation of banks of combinatorial materials (Potyrailo and Mirsky 2009). Such banks can be used, for example, for further investigation of materials of interest for new applications or as reference materials.

## 3. OPPORTUNITIES FOR SENSING MATERIALS

The development process for a sensor system with a new sensing material can be described using technology readiness levels (TRLs) as shown in Figure 3.2. The concept of TRLs is an accepted way to assess technology maturity (2005). These TRLs provide a scale from TRL 1 (least mature) to TRL 9 (most mature) that describes the maturity of a technology with respect to a particular use. Sensor development includes several phases, including discovery with initial observations, feasibility experimentation, and laboratory-scale detailed evaluation (TRLs 1–4), followed by validation of components and a whole-system prototype in the field (TRLs 5–6), and then testing of the system prototype in the operational environment (TRL 7) and tests and end-use operation of the actual system (TRLs 8–9).

At the initial concept stage, performance of the sensing material is matched with the appropriate transducer for signal generation. The stage of laboratory-scale evaluation is very labor-intensive because it involves detailed testing of sensor performance. Several key aspects of this evaluation include optimization of the sensing material composition and morphology, its deposition method, detailed evaluation of response accuracy, stability, precision, selectivity, shelf-life, long term stability of the response, and key noise parameters (e.g., material instability because of temperature, potential poisons). Thus, as illustrated in Figure 3.2, combinatorial methodologies for the development of sensing materials have broad opportunities in TRLs 1–5.

## 4. DESIGNS OF COMBINATORIAL LIBRARIES OF SENSING MATERIALS

The broad goals of combinatorial development of sensing materials are to discover and optimize performance parameters and to optimize fabrication parameters of sensing materials. The key performance and fabrication parameters of sensing materials are outlined in Figure 3.3. Factors affecting performance of sensing material films are also summarized in Figure 3.3 and can be categorized as those originating from the sample, the sample/film interface, the bulk of the film, and the film/substrate interface. Depending on the real-world application, the qualities of the sensing materials are often

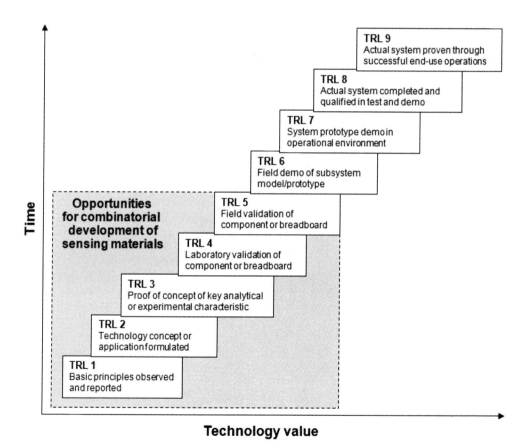

**Figure 3.2.** Opportunities for combinatorial development of sensing materials across technology readiness levels.

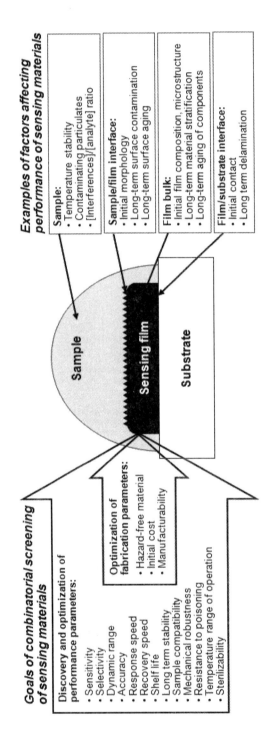

**Figure 3.3.** Broad goals of combinatorial development of sensing materials and examples of factors affecting materials performance.

weighted differently. For example, response speed with millisecond time resolution is critical in gas sensors for intensive care, while a much slower response speed is sufficient in home blood glucose biosensors (Pickup and Alcock 1991; Newman and Turner 2005). Specific requirements for medical *in vivo* sensors and bioprocess sensors include sample compatibility (Meyerhoff 1993; Clark and Furey 2006). Resistance to gamma radiation during sterilization, drift-free performance, and cost are the most critical specific requirements for sensors in disposable bioprocess components (Clark and Furey 2006).

Combinatorial experimentation is performed by arranging materials candidates as discrete and gradient sensing materials arrays. A wide variety of array fabrication methods have been reported, as summarized in Table 3.2. The specific type of library layout depends on the required density of space to be explored, available library-fabrication capabilities, and capabilities of high-throughput characterization techniques. Upon array fabrication, the array is exposed to an environment of interest and steady-state or dynamic measurements are acquired to assess materials performance. Serial scanning analysis (e.g., by optical or impedance spectroscopy) is often performed to provide more detailed information about materials properties than parallel analysis (e.g., imaging) does. When monitoring a dynamic process (e.g., response/recovery time, aging) of sensing materials arranged in an array with a scanning system, the maximum number of elements in the sensor library that can be measured with the required temporal resolution may be limited by the data-acquisition ability of the scanning system (Potyrailo and Hassib 2005). In addition to measurements of materials performance parameters, it is important to characterize intrinsic materials properties (Göpel,1998).

To demonstrate the broad applicability of combinatorial technologies in discovery and optimization of sensing materials, in the following sections we critically analyze results of materials development using discrete and gradient materials arrays and provide examples from a wide variety of sensors based on various energy-transduction principles that involve radiant, mechanical, and electrical types of energy.

## 5. DISCOVERY AND OPTIMIZATION OF SENSING MATERIALS USING DISCRETE ARRAYS

### *5.1. RADIANT ENERGY TRANSDUCTION SENSORS*

Sensors based on radiant energy transduction can be categorized on the basis of the five parameters that completely describe a light wave: amplitude, wavelength, phase, polarization state, and time-dependent waveform. The majority of the development of sensing materials for these types of sensors relies on colorimetric and fluorescent materials properties. At present, organic fluorophores dominate sensing applications because of the diversity of their functionality and well-understood synthesis methods, but new semiconducting nanocrystal labels have several advantages (photostability, relatively narrow emission spectra, and broad excitation spectra (Alivisatos 2004; Medintz et al. 2005)) over organic fluorophores. Thus, finding a solution to complement the existing organic fluorescent reagents with more photostable yet chemically or biologically responsive nanocrystals is very attractive. It is known that a variety of photoluminescent materials are sensitive to the local environment (Ko and Meyer 1999). In particular, polished or etched bulk CdSe semiconductor crystals (Lisensky et al. 1988; Seker et al. 2000 )

**Table 3.2. Examples of fabrication methods of discrete and gradient materials arrays**

| Types of arrays of sensing material | Fabrication methods | References |
|---|---|---|
| **Discrete arrays** | Ink jet printing | Lemmo et al. 1997; Calvert 2001; de Gans and Schubert 2004 |
| | Robotic liquid dispensing | Apostolidis et al. 2004; Hassib and Potyrailo 2004 |
| | Robotic slurry dispensing | Scheidtmann et al. 2005 |
| | Microarraying | Schena 2003 |
| | Automated dip-coating | Potyrailo et al. 2004d |
| | Electropolymerization | Mirsky and Kulikov 2003; Mirsky et al. 2004 |
| | Chemical vapor deposition | Taylor and Semancik 2002 |
| | Pulsed-laser deposition | Aronova et al. 2003 |
| | Spin coating | Cawse et al. 2003; Amis 2004 |
| | Screen printing | Potyrailo et al. 2007 |
| | Electrospinning | Yoon et al. 2009 |
| **Gradient arrays** | *In situ* photopolymerization | Dickinson et al. 1997 |
| | Microextrusion | Potyrailo et al. 2003d, 2005b; Potyrailo and Wroczynski 2005 |
| | Solvent casting | Potyrailo et al. 2003c, 2004; Potyrailo and Hassib 2005 |
| | Colloidal self-assembly | Potyrailo et al. 2008b |
| | Surface-grafted orthogonal polymerization | Bhat et al. 2004 |
| | Ink jet printing | Turcu et al. 2005 |
| | Temperature-gradient chemical vapor deposition | Taylor and Semancik 2002 |
| | Thickness-gradient chemical vapor deposition | Sysoev et al. 2004 |
| | 2-D thickness-gradient evaporation of two metals | Klingvall et al. 2005 |
| | Gradient surface coverage and gradient particle size | Baker et al. 1996 |

and nanocrystals (Nazzal et al. 2003; Vassiltsova et al. 2007) were shown to be sensitive to environmental changes. To better understand the environmental sensitivity of semiconductor nanocrystals upon their incorporation into polymer films, mixtures of multisize CdSe nanocrystals were incorporated into numerous rationally selected polymeric matrices (see Table 3.3) to produce thin films. These films were further screened for their photoluminescence (PL) response to vapors of different polarity upon excitation with a 407-nm laser (Potyrailo and Leach 2006; Leach and Potyrailo 2006).

It was discovered that CdSe nanocrystals of different sizes (2.8 and 5.6 nm in diameter) and passivated with tri-$n$-octylphosphine oxide had dramatically different PL response patterns upon exposure to methanol and toluene after incorporation into polymeric matrices (see Figure 3.4A). As an example, Figure 3.4B shows response patterns of gas-dependent PL of the two-size CdSe nanocrystals in poly(methyl methacrylate) (PMMA, polymer 2 in Table 3.3) sensor film. The difference in the response patterns of the nanocrystals was attributed to the combined effects of the dielectric medium surrounding the nanocrystals, their size, and their surface oxidation state. The sensing films were tested for 16 h under continuous laser excitation and exhibited high stability of PL intensity (Potyrailo and Leach 2006).

To evaluate polymer matrices quantitatively, K-nearest-neighbor (KNN) cluster analysis was employed as a data-mining tool. Cluster analysis is often used in assessing the diversity of materials according to their compositional or performance properties and in developing structure–property relationship models (Otto 1999; Potyrailo 2001). In KNN analysis, links are made between nearest neighbors of adjoining clusters. A measure that accounts for the different scales of variables and their correlations is the Mahalanobis distance (Otto 1999). Results of cluster analysis of PL response patterns upon exposure to methanol and toluene after incorporation into polymeric matrices are demonstrated in the dendrogram in Figure 3.4C. The dendrogram was constructed by performing principal-component analysis on the data from Figure 3.4A and then using Mahalanobis distance on three principal components (PCs). From this dendrogram, it is clear that polymers 6 and 7 (see Table 3.3) were the most similar in their vapor response the the CdSe nanocrystals studied, as demonstrated by a very small distance to K-nearest neighbors between them. Polycaprolactone (polymer 4 in Table 3.3) was the most different from the rest of the polymers, as indicated by the largest diversity distance to the K-nearest neighbor. Such data-mining tools provide a means to evaluate polymer matrices quantitatively. Coupled with quantitative structure–property relationship simulation tools that incorporate molecular descriptors, new knowledge generated from high-throughput experiments may provide additional insights for rational design of gas sensors

### Table 3.3. Polymer matrices for incorporation of different-size CdSe nanocrystals

| POLYMER NO. | POLYMER TYPE | RATIONALE FOR SELECTION AS SENSOR MATRIX |
| --- | --- | --- |
| 1 | Poly(trimethyl-silyl) propyne | Polymer with largest known solubility of oxygen, candidate for efficient oxidation of CdSe nanocrystals |
| 2 | Poly(methyl methacrylate) | Polymer for solvatochromic dyes |
| 3 | Silicone block polyimide | Polymer with very high partition coefficient for sorbing organic vapors |
| 4 | Polycaprolactone | Polymer for solvatochromic dyes |
| 5 | Polycarbonate | Polymer with high $T_g$ for sorbing of organic vapors |
| 6 | Polyisobutylene | Polymer with low $T_g$ for sorbing of organic vapors |
| 7 | Poly(dimethylaminoethyl) methacrylate | Polymer for surface passivation of semiconductor nanocrystals |
| 8 | Polyvinylpyrrolidone | Polymer for sorption of polar vapors |
| 9 | Styrene-butadiene ABA block copolymer | Polymer for sorbing of nonpolar vapors |

*Source:* Data from Leach and Potyrailo 2006.

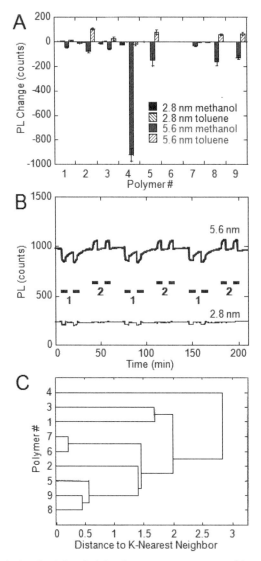

**Figure 3.4.** Diversity of steady-state photoluminescence response of two-size (2.8- and 5.6-nm) mixtures of CdSe nanocrystals to polar (methanol) and nonpolar (toluene) vapors. (A) Magnitude of PL change in nine polymer matrices listed in Table 3.3. (B) Gas-dependent PL of two-size CdSe nanocrystals sensor film (polymer #2) with emission of 2.8-nm nanocrystals at 511 nm and emission of 5.6-nm nanocrystals at 617 nm. (C) Results of KNN cluster analysis of PL response patterns upon exposure to methanol and toluene after incorporation into nine polymer matrices. Numbers 1 and 2 in (B) are replicate exposures of sensor film to methanol (6% vol.) and toluene (1.5% vol.), respectively. [(A) Reprinted with permission from Leach and Potyrailo 2006. Copyright 2006 Materials Research Society. (B) Reprinted with permission from Potyrailo and Leach 2006. Copyright 2006 American Institute of Physics.]

based on incorporated semiconductor nanocrystals. In the future, such work promises to complement existing solvatochromic organic dye sensors with more photostable and reliable sensor materials.

Optimizing sensor materials that have been formulated is a cumbersome process because theoretical predictions are often limited by practical issues, such as poor solubility and compatibility of formulation components (Mills et al. 1998; Bedlek-Anslow et al. 2000; Wang et al. 2003; Apostolidis et al. 2004). These practical issues represent significant knowledge gaps that prevent a more efficient rational design of sensor materials. Thus, combinatorial methodologies have been demonstrated for the development of multicomponent sensor materials for gaseous (Potyrailo and Brennan 2004; Amis 2004; Apostolidis et al. 2004) and ionic (Potyrailo 2004; Hassib and Potyrailo 2004, 2005a; Chojnacki et al. 2004) species. Because polymer matrices (Hartmann and Trettnak 1996; Draxler et al. 1995; Mohr and Wolfbeis 1996; Conway et al. 1997; Dickinson et al. 1997; Walt et al. 1998; Kolytcheva et al. 1999; Amao 2003) and plasticizers (Preininger et al. 1996; Kolytcheva et al. 1999; Legin et al. 2002; Peper et al. 2003; Penco et al. 2004) are known to affect the response of sensors for gases and liquids, automated screening was applied to determine which polymers and plasticizers were best to use in constructing oxygen-sensing materials based on Ru(4,7-diphenylphenanthroline) fluorophore. Structures of polymers 1–9 and plasticizers 10–13 are presented in Figures 3.5 and 3.6, respectively. Following the initial study screening polymer matrices 1–9 (Figure 3.7A), focused libraries were constructed with plasticizers 10–13 to tune sensor sensitivity (Figure 3.7B). While in general the sensitivity of the sensor coatings increased with the plasticizer concentration, because of an increase in the permeability of oxygen in the polymer matrix, it was unexpectedly found that plasticizers 12 and 13 showed an initial decrease of sensitivity at low concentrations. By combining manual and automated steps in the preparation of discrete sensor film arrays, it was possible to reduce the time needed to screen sensor materials by at least 1000 (Apostolidis et al. 2004).

Applying polymers with an intrinsic conductivity also permits development of chemical and biological sensors (Leclerc 1999; McQuade et al. 2000; Janata and Josowicz 2002; Dai et al. 2002; Bobacka et al. 2003). A variety of conjugated organic monomers readily undergo polymerization and form linear polymers. For example, acetylene, *p*-phenylenevinylene, *p*-phenylene, pyrrole, thiophene, furane, and aniline form conducting polymers that are widely employed in sensors (Bidan 1992; Albert et al. 2000; McQuade et al. 2000; Gomez-Romero 2001). However, as prepared, conducting polymers lack selectivity and often are unstable. Thus, such polymers are chemically modified to reduce these undesirable effects. Modification methods include side-group substitution of heterocycles, doping of polymers, charge compensation upon polymer oxidation by incorporation of functionalized counterions, formation of organic–inorganic hybrids, incorporation of various biomaterials (e.g., enzymes, antibodies, nucleic acids, cells), and others (Bidan 1992; Gill and Ballesteros 1998; Gill 2001). Variations in polymerization conditions (oxidation potential, oxidant, temperature, solvent, electrolyte concentration, monomer concentration, etc.) can be also employed to produce diverse polymeric materials from the same monomer, because polymerization conditions affect sensor-related polymer properties (morphology, molecular weight, connectivity of monomers, conductivity, band gap, etc.) (McQuade et al. 2000; Barbero et al. 2003).

Recently, a combinatorial approach for the colorimetric differentiation of organic solvents was developed (Yoon et al. 2009). A polydiacetylene (PDA)–embedded electrospun fiber mat, prepared with aminobutyric acid–derived diacetylene monomer (PCDA-ABA), displayed colorimetric stability when

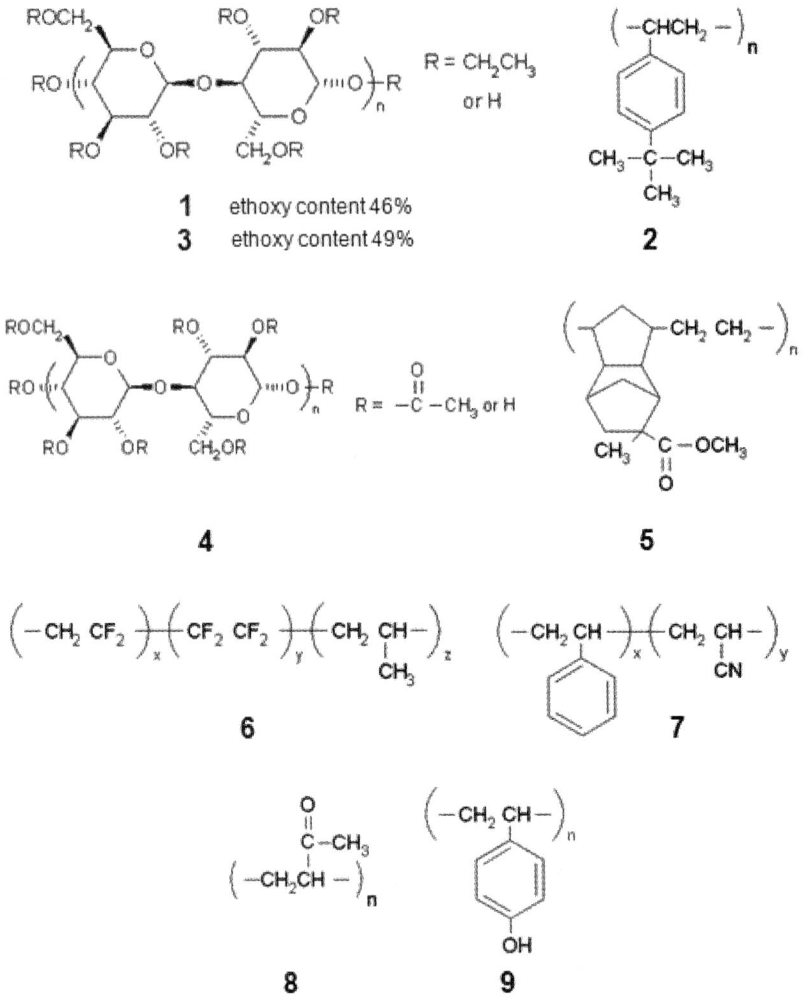

**Figure 3.5.** Structures of polymers 1–9 employed for construction of oxygen-sensing materials based on Ru(4,7-diphenylphenanthroline) fluorophore.

exposed to common organic solvents. In contrast, a fiber mat prepared with the aniline-derived diacetylene (PCDA-AN) exhibited a solvent-sensitive color transition. Arrays of PDA-embedded microfibers were constructed by electrospinning poly(ethylene oxide) solutions containing various ratios of two diacetylene monomers. Unique color patterns were developed when the conjugated polymer-embedded electrospun fiber arrays were exposed to common organic solvents in a manner which enabled direct colorimetric differentiation of the solvents. Results of these experiments are presented in Figure 3.8. Scanning electron microscopy (SEM) images of electrospun fiber mats encapsulated with DA monomers prepared from pure PCDA-ABA, pure PCDA-AN, and a 1:1 molar mixture of PCDA-ABA and PCDA-AN are presented in Figure 3.8A. No significant morphological differences were observed among these electrospun fiber mats and polymer fibers with an average diameter of ~1 μm. Color patterns of the combinatorial arrays of fiber mats derived from different combinations of DA monomers (see Figure 3.8B) demonstrated the significance of the combinatorial approach for sensor development. This methodology enables the generation of a compositionally diverse array of sensors starting with only two DA monomers for visual differentiation of organic solvents.

In addition to sensing materials for radiant sensors, plasmonic nanostructures have also attracted significant interest for chemical and biological sensing because of their potential for a dramatic improvement of sensor performance with a concurrent simplification of instrument design. Significant improvements have been accomplished in developing nanoplasmonic structures for surface-enhanced Raman spectroscopic (SERS) sensors and for localized surface plasmon resonance (LSPR) sensors.

Combinatorial approaches have been applied to optimize SERS substrates based on nanoparticles. These approaches include 2-D variation of nanoparticle density to discover the existence of an optimal surface coverage for the most effective SERS enhancement (Baker et al. 1996) and combinatorial exploration of the roughness effects of metallic substrates (Kahl et al. 1998). It was proposed (Baker et al. 1996) that the SERS enhancement of optimized periodic structures can be much larger than that of simple island films, and the power of solution-based combinatorial approaches was demonstrated for synthesis of surfaces exhibiting nanometer-scale variation in mixed-metal composition and architecture. The SERS response with variable surface coverage of colloidal gold and the amount of silver coating on Au nanoparticles (Ag staining) was studied in detail. A gradient in particle coverage was produced along the direction of immersion of a glass slide by fixed-rate immersion of the (3-mercaptopropyl)trimethoxysilane-coated glass slide into an aqueous solution of 12-nm-diameter colloidal Au particles. The slide was further rotated by 90° and fixed-rate immersion was performed into an Ag ion–containing solution for Ag staining of Au. This second immersion step produced a gradient in particle size over the new immersion direction. The resulting surface exhibited a continuous variation in nanometer-

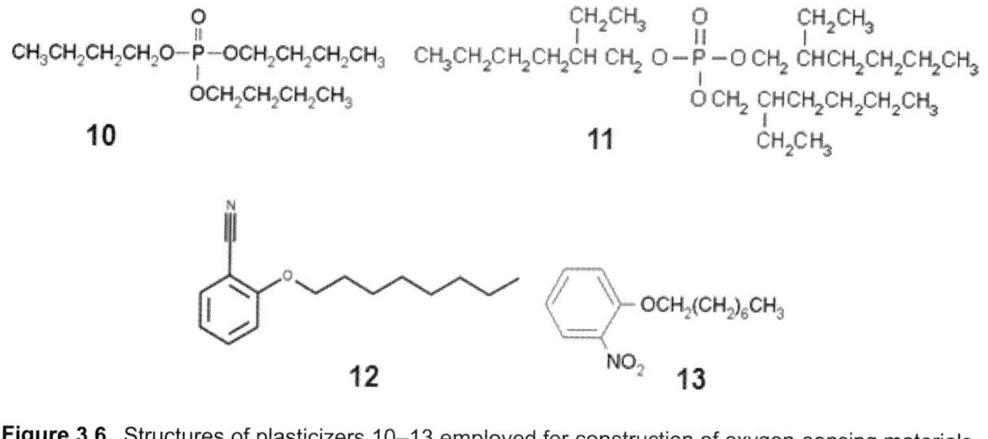

**Figure 3.6.** Structures of plasticizers 10–13 employed for construction of oxygen-sensing materials based on Ru(4,7-diphenylphenanthroline) fluorophore.

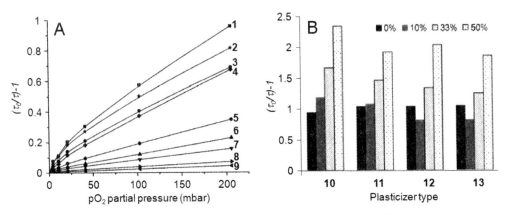

Figure 3.7. Results of combinatorial screening of steady-state responses of formulated optical gas sensor materials. (A) Stern-Volmer plots of oxygen quenching of Ru(4,7-diphenylphenanthroline)–based fluorophore in polymers 1–9 as changes in the fluorescence decay time. (B) Effect of type and concentration of plasticizers 10–13 on the sensitivity of fluorescent oxygen-sensing materials in polymer 1. (Reprinted with permission from Apostolidis et al. 2004. Copyright 2004 American Chemical Society.)

scale morphology as defined by particle coverage and particle sizes. The SERS signal for adsorbed $p$-nitrosodimethylaniline was measured over a 2 × 2 cm sample and exhibited continuous gradients in Au coverage and Ag cladding thickness as shown on a spatial map of background-corrected SERS intensity for the phenyl-nitroso stretch at 1168 cm$^{-1}$. Detailed investigation revealed a region that was over $10^3$-fold more SERS-enhancing than the least active sites (see Figure 3.9). The nanometer-scale morphology at positions of interest was determined by atomic force microscopy, and the results showed significant changes in SERS enhancement factor with only small alterations in surface morphology.

To provide more fundamental insight into design of nanoplasmonic sensors, several computational methods have been applied (Christensen and Fowers 1996, Schatz 2007, Montgomery et al. 2008, Zhao et al. 2008). For simulations of LSPR sensors, computational data on refractive index sensitivity of LSPR structures agrees qualitatively with experiments (Stewart et al. 2006; McMahon et al. 2007; Zheng et al. 2008; Lee et al. 2008a; Chen et al. 2009; Potyrailo et al. 2009a). However, numerous practical factors affect the refractive index sensitivity of LSPR structures, which are difficult to assess quantitatively. For example, the discrepancy between experimental and theoretical results has been attributed to surface contamination during fabrication or during experiments (Zheng et al. 2008), to small but very sensitive deviations between the experimental array structures and the theoretical idealized configurations (McMahon et al. 2007), to the effects of directly transmitted background (Chen et al. 2009), and to contribution from the edges of small-dimension nanohole arrays (Genet and Ebbesen 2007; Yang et al. 2008). Overall, examples of important factors can be grouped as (1) possible edge effects of relatively small-size arrays; (2) surface roughness effects; (3) surface chemistry (contamination) effects; (4) imperfections in the fabrication of individual features; (4) array defects; and (5) general effects from optical setup alignment (Potyrailo et al. 2009a). Thus, in addition to simulations, often different LSPR structures are explored experimentally as combinatorial arrays.

The dependence of the visible-light transmission efficiency on the shape and size of nanoapertures in Au films was explored in quantitative detail experimentally by Matteo et al. (2004). The goal of these experiments was to find the shape and size of a subwavelength aperture with the largest visible-light transmission efficiency (see Figure 3.10). The aperture shapes studied included C apertures, same-area squares, same-area rectangles, and square apertures of the same near-field spot size (see Figure 3.10A). These studies were inspired by computational results and microwave experiments showing that C-shaped apertures can overcome large light attenuation due to their cross-sectional shape. However, utilizing these apertures in the visible regime presented design challenges. To make these subwavelength apertures at optical frequencies, focused ion beam milling (FIB) was employed (see Figure 3.10B). The apertures were formed in 200/5-nm (Au/Cr) film with the critical dimension $d$ for all the apertures ranging from 40 to 55 nm in 5-nm steps. The transmission properties of these individual apertures were measured using white light for comparison of the performance of C apertures to square and rectangular apertures. Figures 3.10C and 3.10D show transmitted colors and intensities of studied arrays for horizontal and vertical polarizations, respectively. Both the C apertures and the rectangular apertures were strongly resonant at the horizontal polarization and brighter than the square apertures. The C apertures changed from red to a blue-green in appearance as their size decreased, while the other apertures

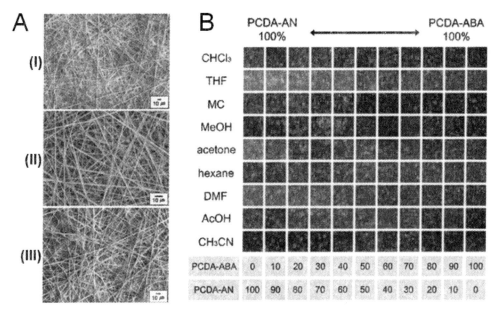

Figure 3.8. Combinatorial approach for colorimetric differentiation of organic solvents based on conjugated polymer-embedded electrospun fibers. (A) SEM images of electrospun fiber mats embedded with (I) PCDA-ABA, (II) PCDA-AN, and (III) 1:1 molar ratio of PCDA-ABA and PCDA-AN after UV irradiation. (B) Photographs of the polymerized PDA-embedded electrospun fiber mats after exposure to organic solvents at 25°C for 30 s. (Color photographs can be seen in the original paper.) (Reprinted with permission from Yoon et al. 2009. Copyright 2009 Wiley-VCH.)

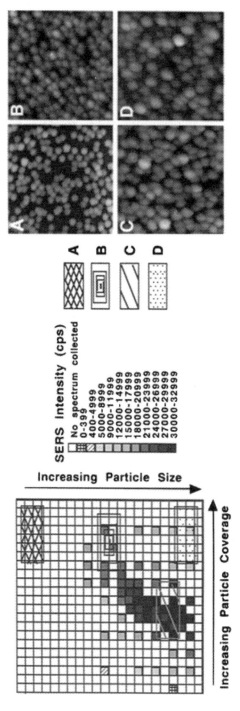

**Figure 3.9.** Combinatorial discovery of the most active SERS region from a 2-D gradient of variable surface coverage of 12-nm-diameter Au nanoparticles and variable thickness of Ag shell. (Left) Background-subtracted SERS intensity map for the 1168-cm$^{-1}$ phenyl-nitroso stretch of *p*-nitrosodimethylaniline. Each shaded box represents one SERS spectrum collected in a 1- × 1-mm$^2$ area. (Right) AFM images of nanoparticles (500 × 500 nm$^2$) from regions A–D. (Color photographs can be seen in the original paper.) (Reprinted with permission from Baker et al. 1996. Copyright 1996 American Chemical Society.)

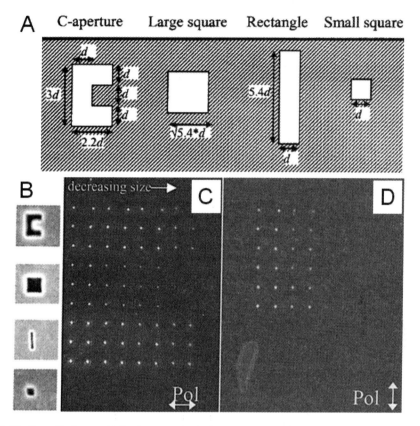

**Figure 3.10.** Quantitative analysis of arrays of subwavelength LSPR apertures of different geometries. (A) Design details and (B) SEM images of C aperture, same-area square, same-area rectangle, and square aperture aimed to produce the same near-field spot size. (C) and (D) Photographs taken through microscope eyepiece of the entire sample under horizontal and vertical polarization, respectively. Each group of three rows was fabricated with the same shape but with decreasing $d$ size, from 55 nm to 40 nm in 5-nm steps. (Color photographs can be seen in the original paper.) (Reprinted with permission from Matteo et al. 2004. Copyright 2004 American Institute of Physics.)

exhibited a much smaller size-dependent resonance shift, with the rectangular apertures strongly transmitting red light and the same-area square apertures having a dim blue color. Strikingly, under vertical polarization, the size-dependent color of the C apertures was lost, with all sizes having a dim blue color, while the rectangular apertures shorted out and the squares remained unchanged. The small square apertures, designed to produce the same spot size as the C aperture, did not even transmit enough light to exceed the noise level of the detector. Throughput enhancements of the C-shaped apertures over square apertures of the same area ranging from 13 times for the $d = 40$-nm apertures to 22 times for the $d = 55$-nm apertures were observed. The reason for the improved performance of the larger apertures was the rounding and tapering effects of smaller structures, which is inherent in FIB milling.

Round apertures (round nanoholes) have been demonstrated in numerous LSPR sensors (Anker et al. 2008; Stewart et al. 2008; Potyrailo et al. 2009a) because of the relative simplicity of their fabrication. To demonstrate multiplexed, high-spatial-resolution, and real-time analysis of biorecognition events, small-dimension (≤20 µm$^2$) ordered arrays of subwavelength holes were made with FIB milling (Yang et al. 2008). Nanohole arrays were perforated on a super-smooth Au film (220 nm thickness, roughness rms < 2.7 Å) deposited onto a fluoropolymer (fluorinated ethylene propylene copolymer, FEP) substrate. The smooth Au surface provided a superb environment for fabricating nanometer features and uniform immobilization of biomolecules. The fluorinated FEP polymer chosen as the substrate was chemically inert, transparent in the visible region, and had a refractive index of 1.341 at $\lambda$ = 590 nm, close to that of biological solutions. With the matched refractive indices on both sides of the Au film, the intensity of extraordinary transmission through the nanoholes was shown to be strongly enhanced (Martín-Moreno et al. 2001; Krishnan et al. 2001; Yang et al. 2008). The refractive index matching between FEP and biological solutions contributed to ~20% improvement in sensing performance. For quantitative assessment of sensing performance, a series of nanohole patterns with different nanohole diameters, periodicities, and numbers was fabricated. Figure 3.11A shows a SEM micrograph of the top view of a typical nanohole array. Figure 3.11B illustrates an intensity image of a pattern of thirty 9 × 9 nanohole arrays spaced by 11 µm. In the fabricated arrays, the hole diameter was changed from 130 to 180 nm and the hole periodicity/diameter ratio was varied from 1.67 to 3.00. By analyzing the transmission spectral features from series of nanohole arrays, it was determined that the hole diameters only slightly affected the transmission maxima, and the periodicity was the dominant factor.

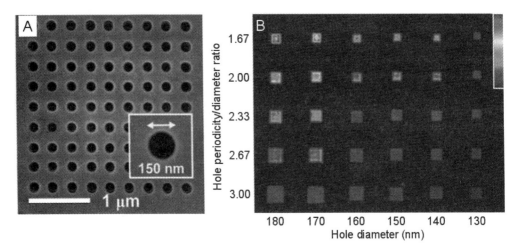

Figure 3.11. Fabricated series of 30-nanohole patterns with different nanohole diameters and hole periodicity/diameter ratios. (A) SEM micrograph of the top view of a typical 9 × 9 nanohole array. (B) Intensity image at 540 nm of a pattern of thirty 9 × 9 nanohole arrays spaced by 11 µm. In the fabricated arrays, the hole diameter was changed from 130 to 180 nm and the hole periodicity/diameter ratio was varied from 1.67 to 3.00. (Color photographs can be seen in the original paper.) (Reprinted with permission from Yang et al. 2008. Copyright 2008 American Chemical Society.)

Application of nanoslits in LSPR sensors promises significant further improvement in sensor performance (Lee et al. 2007; Potyrailo et al. 2008a; Lee et al. 2008b; Wu et al. 2009). Detection sensitivities of gap plasmons in gold nanoslit arrays were studied and compared with surface plasmons on the outside surface (Lee et al. 2008b). The nanoslit arrays (each 150 μm × 150 μm) were fabricated in a 130-nm-thick Au film with various slit widths ranging from 20 to 100 nm and different periods between the nanoslits ranging from 400 to 900 nm (see Figure 3.12). With transverse-magnetic (TM) incident waves, the nanoslit arrays showed two distinguishable transmission peaks corresponding to the resonances of gap plasmons and surface plasmons. The surface sensitivities for both modes were compared by coating thin $SiO_2$ film and different biomolecules on the nanoslit arrays. Experimental results demonstrated that gap plasmons were more sensitive than conventional surface plasmons, with an increase of detection sensitivity with a decrease of slit width. Experiments with BSA biomolecules (molecular weight 66 kDa) demonstrated that the detection sensitivity of SPR mode was not significantly improved when the slit width was changed, whereas the cavity mode was very sensitive when slit width was smaller than 50 nm. Interestingly, a slight decrease in sensitivity of the cavity mode was observed for the smallest slits tested, with widths of 20–30 nm.

## 5.2. MECHANICAL ENERGY TRANSDUCTION SENSORS

Sensors based on mechanical energy transduction can be categorized on the basis of transducer functionality and include cantilevers and acoustic-wave devices. The mass loading and/or changes in the viscoelastic properties of the sensing materials lead to the transducer response.

A 2-D multiplexed cantilever array platform was developed for an elegant combinatorial screening of vapor responses of alkane thiols with different functional end groups (Lim et al. 2006; Raorane et al. 2008). The cantilever sensor array chip (size 2.5 × 2.5 cm) had ~720 cantilevers and was fabricated using surface and bulk micromachining techniques. The optical readout has been developed for parallel analysis of deflections from individual cantilevers. Figure 3.13A illustrates a general view of the 2-D cantilever array system. To evaluate the performance of this 2-D sensor array for screening of sensing materials, nonpolar and polar vapors such as toluene and water vapor were selected as analytes. The screening system was tested with three candidate alkane thiol materials with different functional end groups as sensing films: mercaptoundecanoic acid $SH-(CH_2)_{10}-COOH$ (MUA), mercaptoundecanol $SH-(CH_2)_{11}-OH$ (MUO), and dodecanethiol $SH-(CH_2)_{11}-CH_3$ (DOT). Each type of sensing films had different chemical and physical properties because the –COOH group is acidic in nature and can dissociate to give a –COO– group, the –OH group does not dissociate easily but can form hydrogen bonds with polar molecules, and the –$CH_3$ group is inert to polar molecules, so the only interactions it can have result from van der Waals and hydrophobic effects. Results of these experiments are presented in Figure 3.13B.

Since toluene is an organic solvent, it is likely to act via van der Waals interaction with the thiol film on the gold side. Thus, the van der Waals intermolecular interaction is generally attractive: It brings thiol molecules close to toluene molecules. This in turn brings the thiol molecules closer to each other, inducing shrinkage in the Au layer, and this results in upward deflection as shown in Figure 3.13B. In the case of DOT, because the –$CH_3$ group is nonpolar, it has maximum contact area with toluene and thus exhibits van der Waals interaction. This tendency is reduce as the end groups become more polar

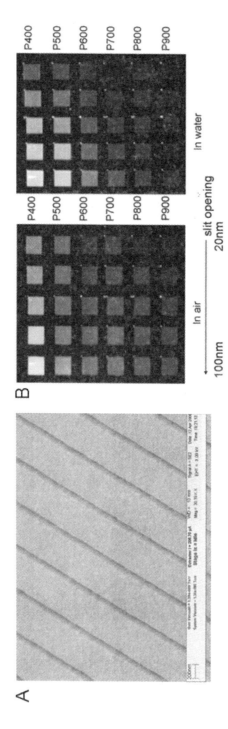

**Figure 3.12.** Experiments with arrays of nanoslits in a gold film with variable slit width and periodicity. (A) Example of a fabricated nanoslit array (period 600 nm, slit width 80 nm). (B) Transmission optical images of various nanoslit arrays in air and in water with variable array periods (400–900 nm) and slit widths (20–100 nm). Area of each nanoslit array was 150 μm × 150 μm. The colors of the nanoslit arrays (shown in shades of gray) were due to extraordinary transmissions of surface plasmon resonances. (Color photographs can be seen in the original paper.) (Reprinted with permission from Lee et al. 2008b. Copyright 2008 Elsevier.)

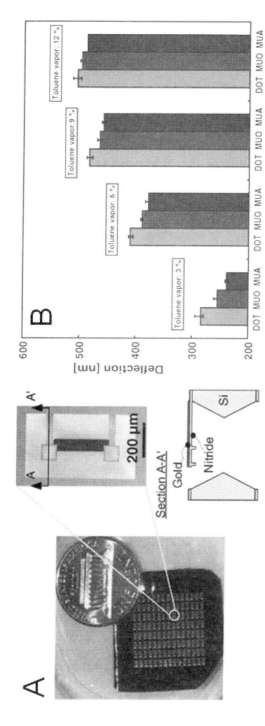

**Figure 3.13.** Combinatorial vapor-response screening of alkane thiols with different functional end groups using a 2-D multiplexed cantilever array system. (A) General view of the fabricated cantilever array chip. (B) Steady-state deflection values of cantilevers upon exposure to toluene vapor at four concentration levels (3, 6, 9, and 12% by mass). (Reprinted with permission from Lim et al. 2006. Copyright 2006 Elsevier.)

in nature. Hence, the –OH group of MUO undergoes more van der Waals interaction than the –COOH group of MUA. As a result, the induced stress in the Au layer is maximum for DOT, medium for MUO, and least for MUA. The water vapor experiment was performed with relative humidity (RH) levels ranging from 8.8 to 61.8% RH. An upward deflection was recorded for all thiols, indicating that the Au film was under compression. The response ranking of three thiols to water vapor was opposite compared to the response to toluene. The largest response was of cantilevers coated with MUA, followed by those coated with MUO, and the smallest response was of cantilevers coated with DOT. This multiplexed cantilever sensor platform was further proposed as a search tool for sensing materials with improved selectivity (Lim et al. 2006).

Polymeric materials are widely used for sensing because they provide the ability for room-temperature sensor operation, rapid response and recovery times, and long-term stability over several years (Hierlemann et al. 1995; Potyrailo and Sivavec 2004). In gas sensing with polymeric materials, polymer–analyte interaction mechanisms include dispersion, dipole induction, dipole orientation, and hydrogen bonding (Grate et al. 1997; Grate 2000). These mechanisms facilitate partial selectivity of response of different polymers to diverse vapors. An additional molecular selectivity in response is added by applying molecular imprinting of target vapor molecules into polymers and formulating polymers with molecular receptors. Several models have been developed to calculate polymer responses (Grate and Abraham 1991; Abraham 1993; Maranas 1996; Wise et al. 2003; Belmares et al. 2004), but the most widely employed model is based on linear solvation energy relationships (LSER) (Grate and Abraham 1991; Abraham 1993). This LSER method has been applied as a guide to select a combination of available polymers to construct an acoustic-wave sensor array based on thickness shear mode (TSM) resonators for determination of organic solvent vapors in the headspace above groundwater (Potyrailo et al. 2004b). Field testing of the sensor system (Potyrailo et al. 1999) demonstrated that its detection limit with available polymers was too high (several parts per million) to meet the requirements for detection of groundwater contaminants. However, a new polymer for sensing has been found (silicone block polyimide 14, Figure 3.14) that has a partition coefficient of >200,000 to part-per-billion concentrations of trichloroethylene (TCE) and that provides at least 100 times more sensitive response for detection of chlorinated organic solvent vapors than other known polymers (Sivavec and Potyrailo 2002; Potyrailo and Sivavec 2004).

For development of materials for more selective part-per-billion detection of chlorinated solvent vapors in the presence of interferencing substances, six families of polymeric materials were fabricated based on polymer 14. Performance of these polymeric materials was evaluated with respect to the differences in partition coefficients to the analytes perchloroethylene (PCE), trichloroethylene (TCE), and *cis*-dichloroethylene (*cis*-DCE) and interfering carbon tetrachloride, toluene, and chloroform. For quantitative screening of sensing materials candidates, a 24-channel TSM sensor system was built that matched a 6 × 4 microtiter wellplate format (Figures 3.15A and 3.15B). The sensor array was positioned in a gas flow-through cell and kept in an environmental chamber. Comprehensive materials screening was performed at three levels (Potyrailo et al. 2003b, 2004c). In the primary (discovery) screening, materials were exposed to a single analyte concentration. In the secondary (focused) screening, the best materials subset was exposed to analytes and interfering compounds. Finally, in the tertiary screening, the remaining materials were tested under conditions mimicking long-term application. While all the screenings were valuable, the tertiary screening provided the most intriguing data because aging of base

COMBINATORIAL CONCEPTS FOR DEVELOPMENT OF SENSING MATERIALS • 181

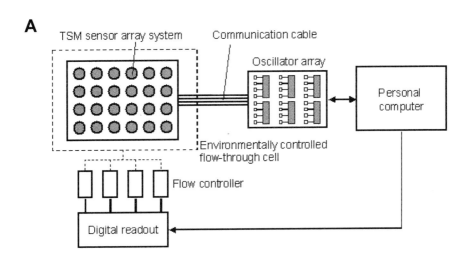

**Figure 3.14.** Structure of vapor-sensing polymer 14.

**Figure 3.15.** Approach for high-throughput evaluation of sensing materials for field applications. (A) Setup schematic of a 24-channel TSM sensor array for gas-sorption evaluation of sorbing polymeric films. (B) Photo of 24 sensor crystals (including two reference sealed crystals) in a gas flow cell. [(A) Reprinted with permission from Potyrailo et al. 2004c. Copyright 2004 American Institute of Physics.]

polymers and copolymers is difficult or impossible to model (Ulmer II et al. 1998). From the tertiary screening, the decrease in materials response to the nonpolar analyte vapors and the increase in response to a polar interference vapor were quantified.

For detailed evaluation of diversity of the fabricated materials, principal-components analysis (PCA) tools (Beebe et al. 1998) were applied as shown in Figure 3.16. PCA is a multivariate data analysis tool that projects the data set onto a subspace of lower dimensionality with collinearity removed. PCA achieves this objective by explaining the variance of the data matrix in terms of the weighted sums of the original variables with no significant loss of information. These weighted sums of the original variables are called principal components (PCs). The capability of discriminating among six vapors using eight types of polymers was evaluated using a scores plot (see Figure 3.16A). It demonstrated that these six vapors are well separated in the PCA space when these eight types of polymers are used for determinations. To understand what materials induce the most diversity in the response, a loadings plot was constructed (see Figure 3.16B). The bigger the distance between the films of the different types, the better was the differences between these films. The loadings plot also demonstrates the reproducibility of the response of replicate films of the same materials. Such information served as an additional input into materials selection for the tertiary screening. However, material selection on the basis of PCA alone does not guarantee optimal discrimination of particular vapors in the test set, because PCA measures variance, not discrimination (Grate 2000). Thus, cluster analysis tools—e.g., those shown in Figure 3.4C—can be also applied.

This 24-channel TSM sensor array system was further applied for high-throughput screening of solvent resistance of a family of polycarbonate copolymers prepared from the reaction of bisphenol A (BPA), hydroquinone (HQ), and resorcinol (RS) with the goal of using these copolymers as solvent-resistant supports for deposition of solvent-containing sensing formulations (Potyrailo et al. 2005a). The mass increase of the crystal was determined during periodic exposure of the TSM crystals to polymer/

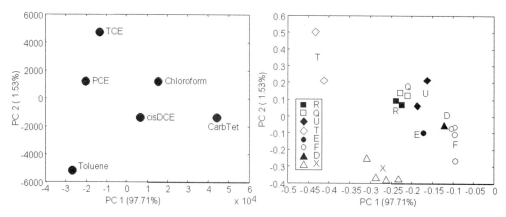

**Figure 3.16.** Application of PCA tools for determination of differences in the response pattern of the sensor materials toward analytes (PCE, TCE, and *cis*-DCE) and interferences (carbon tetrachloride, toluene, and chloroform): (left) scores and (right) loadings plots of the first two principal components. (Adapted with permission from Potyrailo et al. 2004c. Copyright 2004 American Institute of Physics.)

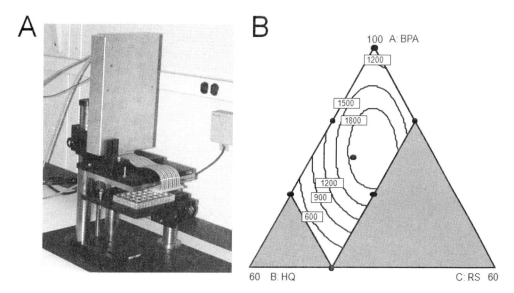

**Figure 3.17.** Application of the 24-channel TSM sensor array system for mapping of solvent resistance of polycarbonate copolymers. (A) General view of the screening system with a 6 × 4 microtiter wellplate positioned below the sensor array. (B) Example of property/composition mapping of solvent-resistance of polycarbonate copolymers in tetrahydrofuran. Numbers in the contour lines are normalized sensor frequency shift values (hertz per milligram of polymer in a well of the microtiter wellplate). [(A) Reprinted with permission from Potyrailo et al. 2004d. Copyright 2004 American Chemical Society. (B) Reprinted with permission from Potyrailo et al., 2006a. Copyright 2006 American Chemical Society.]

solvent combinations (Figure 3.17A) (Potyrailo et al. 2004d) and was found to be proportional to the amount of polymer dissolved and deposited onto the sensor from a polymer solution. The high mass sensitivity of the resonant TSM sensors (10 ng), use of only a minute volume of solvent (2 mL), and parallel operation (matching a layout of available 24 microtiter wellplates) made this system a good fit with available polymer combinatorial synthesis equipment. These parallel determinations of polymer–solvent interactions also eliminated errors associated with serial determinations. The data were further mined to construct detailed solvent-resistance maps of polycarbonate copolymers and to determine quantitative structure–property relationships (see Figure 3.17B) (Potyrailo et al. 2006a). The application of this sensor-based polymer screening system provided a lot of stimulating data, which would be difficult to obtain using a conventional one-sample-at-a-time approach.

## 5.3. ELECTRICAL ENERGY TRANSDUCTION SENSORS

Sensors based on electrical energy transduction are applicable for combinatorial screening of sensing materials when these materials undergo electrically detectable changes, for example, changes in resistance or conductance during polymerization reactions and exposure to species of interest, changes in resistance due to swelling of polymers, interactions of metal-oxide semiconducting surfaces with oxidizing or reducing species, etc. Typical devices for these applications include electrochemical and electronic transducers (Zemel 1990; Suzuki 2000; Wang 2002; Hagleitner et al. 2002, 2003).

The simplicity of microfabrication of electrode arrays and their subsequent application as transducer surfaces makes sensors based on electrical energy transduction among the most employed tools in combinatorial materials screening. The possibility of regulating polymerization on solid conductive surfaces by application of corresponding electrochemical potentials suggested a realization of this process in the form of multiple polymerization regions on multiple electrodes of an electronic sensor system (Mirsky and Kulikov 2003; Kulikov and Mirsky 2004). Arranging such polymerization electrodes in an array eliminated the need for dispensing systems and allowed an electrically addressable immobilization. This approach has been demonstrated on electropolymerization of aniline that was performed independently on different electrodes of the array (Mirsky and Kulikov 2003; Kulikov and Mirsky 2004). Thin-layer polymerization of defined mixtures of monomers was performed directly on the 96 interdigital addressed electrodes of an electrode array on an area of less than 20 mm × 20 mm (see Figure 3.18A). The electrodes had an interdigital configuration designed for four-point measurements and were fabricated by lithography on an oxidized silicon wafer. Computer-controlled addition of analyte species provided automated investigation of the influence of different substances on the synthesized polymers. This system has been applied for combinatorial electrochemical copolymerization of mixtures of the

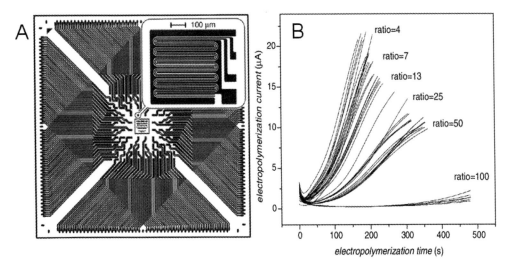

**Figure 3.18.** Application of a microfabricated electrode sensor array for multiple electropolymerizations and characterization of resulting conducting polymers as sensor materials. (A) Layout of the interdigital addressed electrode array. Inset, detailed structure of the single electrode for four-point measurements. (B) Current kinetics during combinatorial electropolymerization at constant potential for different ratios of nonconductive additive to conductive monomer in the polymerization mixture. (Adapted with permission from Kulikov and Mirsky 2004. Copyright 2004 Institute of Physics.)

nonconductive monomer aminobenzoic acid and the conductive monomer aniline at various ratios to form diverse polymers. Six groups of polymers were formed at the same conditions on 12 electrodes per group with ratios of monomers ranging from 100:1 to 4:1 to study the effects of the nonconductive aniline derivative (see Figure 3.18B). Electrical characterization of the polymers demonstrated that incorporation of aminobenzoic acid as a nonconductive additive in the polymer structure disturbed the conductive polymer chains and led to a strong decrease of polymer conductance, corresponding to predicted behavior. Exposure to HCl gas showed that polymer conductance increased with HCl concentration.

Numerous copolymers have been screened using this electropolymerization system. Example screening results for different binary copolymers are presented in Figure 3.19 (Potyrailo and Mirsky 2008). Introduction of nonconductive monomers into the polymer decreased the polymer conductance and therefore decreased the difference between conductive and insulating polymer states. This caused a decrease of absolute sensitivity (Figure 3.19A). Normalization to the polymer conductance without

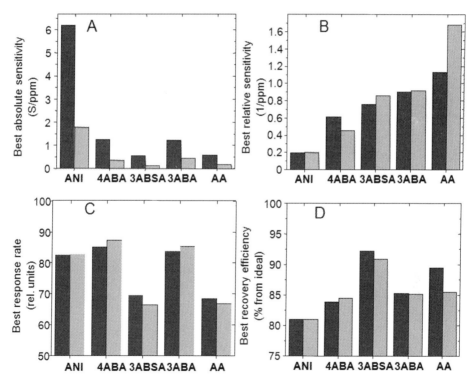

**Figure 3.19.** Selected results of screening of sensing materials for their response to HCl gas: (A) best absolute sensitivity; (B) best relative sensitivity; (C) best response rate; (D) best recovery efficiency, performed by heating. Sensor materials: ANI indicates polyaniline; 4ABA, 3ABSA, 3ABA, AA indicate polymers synthesized from aniline and 4-aminobenzoic acid, 3-aminobenzenesulfonic acid, 3-aminobenzoic acid, and anthranilic acid, respectively. Gray and black bars are the results obtained by 2- and 4-point techniques, respectively. (Reprinted with permission from Potyrailo and Mirsky 2008. Copyright 2008 American Chemical Society.)

analyte exposure compensated for this effect and demonstrated that the polymer synthesized from the mixture of aminobenzoic acid and aniline possessed the highest relative sensitivity (Figure 3.19B). This effect may be explained by the strong dependence of polymer conductance on the defect number in the polymer chains. In comparison with pure polyaniline, this copolymer had better recovery efficiency but slower response time (Figures 3.19C and 3.19D). The high-throughput screening system was capable of reliably ranking sensing materials and required only ~20 min of manual interaction with the system and ~14 h of computer-controlled combinatorial screening, compared to ~2 weeks of laboratory work using traditional electrochemical polymer synthesis and materials evaluation (Mirsky and Kulikov 2003).

An interesting approach has been recently reported (Setasuwon et al. 2008) to enhance the diversity of responses of sensors with conventional polymer films (see Figure 3.20). Eight sensors with unique characteristics were made by combinatorial stacking of three layers of two types of carbon black–formulated composites such as carbon black with ethylene-co-vinyl acetate (composite E) and carbon black with poly(vinyl alcohol) (composite A). Eight sensors were constructed on interdigitated electrodes with 200-μm spacing on 10- × 15-mm glass substrates by spin-coating three consecutive layers of composites A and E. As shown in Figure 3.20A, eight sensors resulted from all combinations of three layers of two different composites, EEE, EEA, EAE, EAA, AAE, AEA, AEE, and AAA, where the first letter indicates the first layer applied onto the electrodes. These sensors were subjected to 15 solvent vapors with a broad range of dielectric constants from 2 to 80. The test set of solvents consisted of water, dimethylformamide (DMF), methanol, ethanol, acetone, isopropanol, 1-butanol, 1,2-dichlorobenzene, 1-octanol, tetrahydrofuran (THF), ethyl acetate, chlorobenzene, chloroform, toluene, and hexane. Results of the response patterns of all sensors to 15 tested solvents are presented in Figure 3.20B. The responses of the sensors were processed for calculation of resolution factor, the pairwise parameter representing the ability to resolve the responses of one analyte vapor from the others. If the detector responses are assumed to have a normal distribution, the resolution factor values of 1.0, 2.0, and 3.0 indicate 76%, 92%, and 98% confidence, respectively, of correctly identifying one analyte from the other of a specific pair. It was found that by stacking layers of two types of carbon black–formulated composites, it was possible to improve the selectivity of vapor measurements using resistive sensors. The resolution factor was affected by both the dielectric constant and the boiling point of the solvents tested.

Semiconducting metal oxides are another type of sensing materials that benefit from combinatorial screening technologies. Semiconducting metal oxides are typically used as gas-sensing materials that change their electrical resistance upon exposure to oxidizing or reducing gases. Over the years, significant technological advances have been made that resulted in practical and commercially available sensors, and now new materials are being developed that further improve the performance of these sensors. To enhance response selectivity and stability, an accepted approach is to formulate multicomponent materials that contain additives in metal oxides. Introducing additives into base metal oxides can change a variety of materials properties, including concentration of charge carriers, energy spectra of surface states, energy of adsorption and desorption, surface potential and intercrystallite barriers, phase composition, sizes of crystallites, catalytic activity of the base oxide, stabilization of a particular valence state, formation of active phases, stabilization of the catalyst against reduction, the electron exchange rate, etc. Dopants can be added at the preparation stage (bulk dopants) that will affect the morphology, electronic properties of the base material, and catalytic activity. However, the fundamental effects of volume dopants

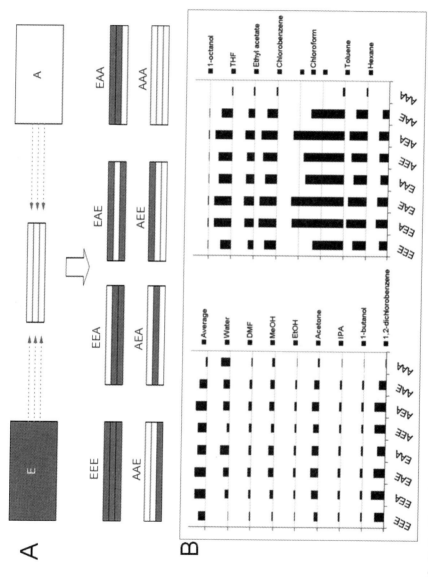

**Figure 3.20.** Method for enhancement of the diversity of responses of sensors with conventional polymer films. (A) Combinatorial stacking of three layers of two types of carbon black–formulated composites, such as carbon black with ethylene-co-vinyl acetate (composite E) and carbon black with poly(vinyl alcohol) (composite A), onto interdigitated electrodes. Eight resulting sensor configurations are EEE, EEA, EAE, EAA, AEE, AEA, AAE, and AAA. (B) Results of the response patterns of eight sensors, EEE, EEA, EAE, EAA, AEE, AEA, AAE, and AAA, to 15 tested solvents. (Reprinted with permission from Setasuwon et al. 2008. Copyright 2008 American Chemical Society.)

on base materials are not yet predictable (Siemons et al. 2007). Addition of dopants to the preformed base material (surface dopants) can lead to different dispersion and segregation effects, depending on the mutual solubility (Franke et al. 2006) and influence on the overall oxidation state of the metal oxide surface (Korotcenkov 2005; Franke et al. 2006; Barsan et al. 2007; Siemons et al. 2007).

To improve the productivity of materials evaluation by using combinatorial screening, a 36-element sensor array was employed to evaluate various surface-dispersed catalytic additives on $SnO_2$ films (Semancik 2002, 2003). Catalysts were deposited by evaporation to nominal thicknesses of 3 nm, and then the microhotplates were heated to effect the formation of a discontinuous layer of catalyst particles on the $SnO_2$ surfaces. The layout of the fabricated 36-element library is shown in Figure 3.21A. The response characteristics of $SnO_2$ with different surface-dispersed catalytic additives are presented in Figure 3.21B. These radar plots show sensitivity results to benzene, hydrogen, methanol, and ethanol for operation at three temperatures.

To expand the capabilities of screening systems, it is attractive to characterize not only the conductance of the sensing materials with DC measurements but also their complex impedance spectra (Barsoukov and Macdonald 2005). The use of complex impedance spectroscopy provides the capability to test both ion- and electron-conducting materials and to study electrical properties of sensing materials that are determined by the material microstructure, such as grain boundary conductance, interfacial polarization, and polarization of the electrodes (Simon et al. 2002, 2005). A 64-multielectrode array has been designed and built for high-throughput impedance spectroscopy ($10–10^7$ Hz) of sensing materials (see Figure 3.22A) (Simon et al. 2002). In this system, an array of interdigital capacitors was screen-printed onto a high-temperature-resistant $Al_2O_3$ substrate. To ensure high quality of the determinations, parasitic effects caused by the leads and contacts were compensated by a software-aided calibration (Simon et al. 2002). After the system validation with doped $In_2O_3$ and automation of the data evaluation (Simon et al. 2005), the system was implemented for screening of a variety of additives and matrices with the long-term goal of developing materials with improved selectivity and long-term stability. Sensing films were applied using robotic liquid-phase deposition based on optimized sol-gel synthesis procedures. Surface doping was achieved by the addition of appropriate salt solutions followed by library calcination. Screening results at 350°C of thick films of $In_2O_3$ base oxide surface doped with various metals are presented as bar diagrams in Figure 3.22B (Sanders and Simon 2007). It was found that some doping elements lead to changes in both the conductivity in air as well as the gas-sensing properties toward oxidizing ($NO_2$, NO) and reducing ($H_2$, CO, propene) gases. Correlations between the sensing and the electrical properties in reference atmosphere indicated that the effect of the doping elements was due to an influence on the oxidation state of the metal oxide surface rather that to an interaction with the test gas. This accelerated approach for generating reliable systematic data was then coupled to data-mining statistical techniques that resulted in the development of (1) a model associating the sensing properties and the oxidation state of the surface layer of the metal oxide based on oxygen spillover from doping element particles to the metal oxide surface and (2) an analytical relation for the temperature-dependent conductivity in air and nitrogen that described the oxidation state of the metal oxide surface taking into account sorption of oxygen (Sanders and Simon 2007).

This high-throughput complex impedance screening system was further employed for reliable screening of a wide variety of less explored material formulations. Polyol-mediated synthesis has been known as an attractive method for preparation of nanoscale metal oxide nanoparticles (Feldmann 2001).

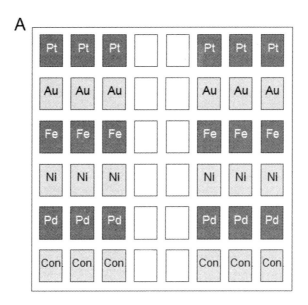
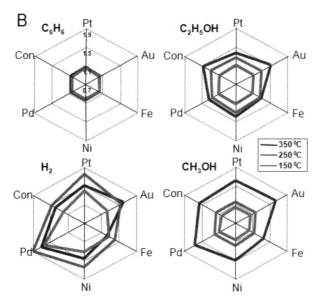

**Figure 3.21.** Combinatorial study of effects of surface dispersion of metals into CVD-deposited $SnO_2$ films. (A) Layout of a 36-element library for study of the sensing characteristics of $SnO_2$ films with 3 nm of surface-dispersed Pt, Au, Fe, Ni, or Pd (Con. = control). Each sample was made with six replicates. (B) Radar plots of sensitivity results to benzene, hydrogen, methanol, and ethanol for operation at 150, 250, and 350°C. (Color photographs can be seen in the original paper.) (Reprinted with permission from Semancik 2002.)

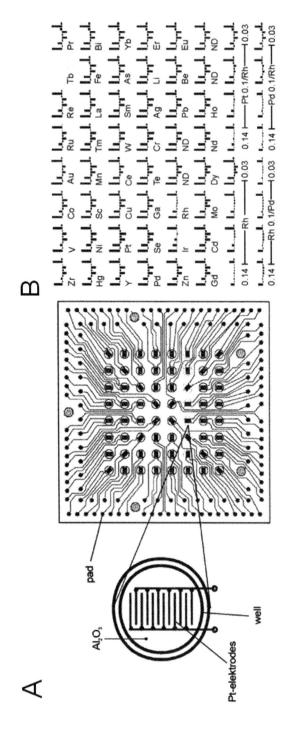

**Figure 3.22.** Screening of sensor metal oxide materials using complex impedance spectroscopy and a multielectrode 64-sensor array. (A) Layout of 64-sensor array. (B) Relative gas sensitivities at 350°C of the $In_2O_3$ base oxide materials library surface-doped with multiple salt solutions, concentration 0.1 atom % if not denoted otherwise; ND = undoped. Sequence of test gases and their concentrations (with air in between) was $H_2$ (25 ppm), CO (50 ppm), NO (5 ppm), $NO_2$ (5 ppm), propene (25 ppm). [(A) Reprinted with permission from Simon et al. 2002. Copyright 2002 American Chemical Society. (B) Reprinted with permission from Sanders and Simon 2007. Copyright 2007 American Chemical Society.]

It requires only low annealing temperatures and provides the opportunity to tune the composition of the materials by mixing the initial components on the molecular level (Siemons and Simon 2006, 2007). To explore previously unknown combinations of $p$-type semiconducting nanocrystalline $CoTiO_3$ with different volume dopants as sensing materials, the polyol-mediated synthesis method was used to synthesize nanometer-sized $CoTiO_3$ followed by volume doping with Gd, Ho, K, La, Li, Na, Pb, Sb, and Sm (all at 2 at%). The SEM-estimated primary particle size of the volume-doped $CoTiO_3$ materials was in the range from 30 to 140 nm, with the smallest particle size for $CoTiO_3$:La and the largest for $CoTiO_3$:K.

The significant amount of data collected during experiments with numerous sensing materials candidates facilitated successful efforts to develop data-mining techniques (Sieg et al. 2006, 2007) and a database system (Frantzen et al. 2005). The data-mining tools, based on hierarchical clustering maps (see Figure 3.23), have been applied to identify several promising materials candidates such as $In_{99.5}Co_{0.5}O_x$, $W_{99}Co_{0.5}Y_{0.5}O_x$, $W_{98.3}Ta_{0.2}Y_1Mg_{0.5}O_x$, $W_{99.5}Ta_{0.5}O_x$, and $W_{99.5}Rh_{0.5}O_x$ with different gas-selectivity patterns (Frenzer et al. 2006).

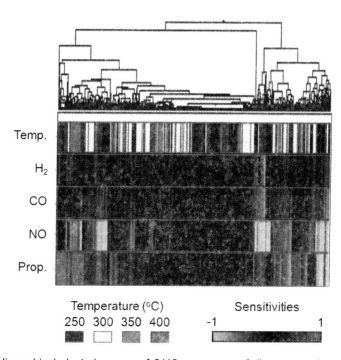

**Figure 3.23.** Hierarchical clustering map of 2112 responses of diverse sensing materials to $H_2$, CO, NO, and propene (Prop.) at four temperatures established from high-throughput constant-current measurements and processed with Spotfire data-mining software (clustering algorithm was "complete linkage" of the Euclidean distances). (Color photographs can be seen in the original paper.) (Reprinted with permission from Frenzer et al. 2006. Copyright 2006 Molecular Diversity Preservation International.)

## 6. OPTIMIZATION OF SENSING MATERIALS USING GRADIENT ARRAYS

Sensor material optimization can be performed using gradient sensor materials. Spatial gradients in sensing materials can be generated by varying the nature and concentration of starting components, processing conditions, thickness, and some other parameters. Once a gradient sensor array is formed, it is important to estimate the possibilities to adequately measure the variation of properties along the gradient. These can be intrinsic (thickness, chemical composition, morphology, etc.) or performance (response magnitude, selectivity, stability, immunity to poisoning, etc.) properties.

### 6.1. VARIABLE CONCENTRATION OF REAGENTS

Optimization of concentrations of formulation components can require significant effort because of the nonlinear relationship between additive concentration and sensor response (Papkovsky et al. 1995; Dickinson et al. 1997; Collaudin and Blum 1997; Mills 1998; Levitsky et al. 2001; Eaton 2002; Florescu and Katerkamp 2004; Basu et al. 2005). Concentration-gradient sensor material libraries have been employed for detailed optimization of sensor materials. One-, two-, and three-component composition gradients were made by flow-coating individual liquid formulations onto a flat substrate and allowing them to merge under diffusion control while still containing solvents (Potyrailo and Hassib 2005). This method combines the fabrication of gradients of materials composition with recording the materials' response before and after analyte exposure and taking the ratio or difference of responses. These gradient films were applied for optimization of sensor materials formulations for analysis of ionic and gaseous species (Potyrailo and Hassib 2005). A very low reagent concentration in the film is expected to produce only a small signal change. A small signal change is also expected when the reagent concentration is too high. Thus, the optimal reagent concentration depends on the analyte concentration and the activity of the immobilized reagent.

Concentration optimization of a colorimetric reagent was performed in a polymer film for detection of trace concentrations of chlorine in water. A concentration gradient of a near-infrared cyanine dye was formed in a poly(2-hydroxyethyl methacrylate) hydrogel sensing film. The optical absorption profile $A_0(x)$ was obtained before analyte exposure to map the reagent concentration gradient in the film. A subsequent scanning across the gradient after analyte exposure (1 ppm of chlorine) resulted in determination of the optical response profile $A_E(x)$. The difference in responses, $\Delta A(x) = A_0(x) - A_E(x)$, revealed the spatial location of the optimal concentration of the reagent that produced the largest signal change (see Figure 3.24A). Sensing films with the optimized concentration of the cyanine dye for chlorine determinations in industrial water were further screen-printed as a part of sensing arrays (Potyrailo et al. 2007) onto conventional optical disks. The quantitative readout of changes in film absorbance was performed in a conventional optical disk drive in a recently developed lab-on-a-disk system (Potyrailo et al. 2006b, 2007).

This concentration-optimization method was also applied to optimize sensor material formulations for analysis of gaseous species. Figure 3.24B shows optimization of concentration of Pt octaethylporphyrin in a polystyrene film for detection of oxygen by fluorescence quenching. These data demonstrate

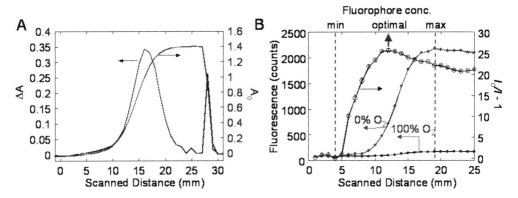

**Figure 3.24.** Optimization of formulated sensing materials using sensing films with gradient reagent concentration along the film length. (A) Concentration optimization of a colorimetric chlorine-responsive reagent in a formulated polymeric poly(2-hydroxyethyl methacrylate) hydrogel sensing film for detection of ions in water. Exposure, 1 ppm of chlorine. (B) Concentration optimization of an oxygen-responsive Pt octaethylporphyrin fluorophore in polystyrene sensing film for detection of oxygen in air.

the simplicity, yet tremendous value, of such determinations for rapid assessment of sensor film formulations. One can see whether the optimal concentration has been reached or exceeded, depending on the nonlinearity and decrease of the sensor response at the highest tested additive concentration. Unlike traditional concentration optimization approaches (Papkovsky et al. 1995; Basu et al. 2005), this new method provides a more dense evaluation mesh and opens opportunities for time-affordable optimization of concentration of multiple formulation components with tertiary and higher gradients.

## 6.2. VARIABLE THICKNESS OF SENSING FILMS

The effect of thickness of sensing films on the stability of the response in water to ionic species has been also evaluated using gradient-thickness sensing films (Potyrailo and Hassib 2005). Sensor reagent stability in a polymer matrix upon water exposure is one of the key requirements. For deposition of gradient sensor regions, several sensor coatings were flow-coated onto a 2.5-mm-thick polycarbonate sheet. Typical coating dimensions were 1–1.5 cm wide and 10–15 cm long. To produce thickness gradients, the coatings were positioned vertically until solvent evaporation in air at room temperature. The coating thickness was further evaluated using optical absorbance or profilometry. An example of a gradient-sensor coating array is shown in Figure 3.25A. The gradient thickness of sensing films was determined from the absorbance of the film-incorporated bromothymol blue reagent. When these arrays were further exposed to a pH 10 buffer (Figure 3.25B), an "activation" period was observed before leaching of the reagent from the polymer matrix was detected from the absorbance decrease. This activation period was roughly proportional to the film thickness. However, the leaching rate was independent of the film thickness, as indicated by the similar slopes of the response curves at 3–9.5 h exposure time.

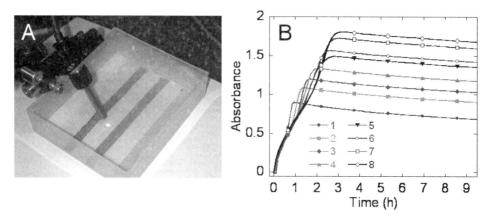

**Figure 3.25.** Application of gradient-thickness sensor film arrays for evaluation of reagent leaching kinetics. (A) Three gradient-thickness sensor film arrays with different loadings of an analyte-sensitive indicator. (B) Reagent leaching kinetics at pH 10. (Reprinted with permission from Potyrailo and Hassib 2005. Copyright 2005 American Institute of Physics.)

## 6.3. VARIABLE 2-D COMPOSITION

Sensors based on the change in the work function of the catalytic metal gate (Pd, Pt, Rh, Ir) due to chemical reactions on the metal surface (Lundström et al. 1975, 1993) are attractive for detection of various gases (hydrogen sulfide, ethylene, ethanol, various amines, and others). The chemical reaction mechanisms in these sensors depend on the specific gas molecules. Optimization of materials for these sensors involves several degrees of freedom (Eriksson et al. 2006). To simplify screening of the desired material compositions and to reduce the common problem of batch-to-batch differences of hundreds of individually made sensors for materials development, the scanning light pulse technique (SLPT) has been developed (Lundström et al. 1991, 2007; Löfdahl et al. 2000). In the SLPT, a focused light beam is scanned over a large-area, semitransparent, catalytic metal–insulator–semiconductor structure, and the photocurrent generated in the semiconductor depletion region is measured to create a 2-D response pattern of the sensing film (a.k.a. "chemical image").

These chemical images were used to optimize properties such as chemical sensitivity, selectivity, and stability (Klingvall et al. 2005). When combined with surface-characterization methods, this information led to increased knowledge of gas response phenomena. It was suggested that a 2-D gradient made from two types of metal films as a double-layer structure should provide new capabilities for sensor materials optimization, capabilities that are not available with thickness gradients of single metal films (Klingvall et al. 2003). To make a 2-D gradient, the first metal film was evaporated on the insulator with the linear thickness variation in one dimension by moving a shutter with a constant speed in front of the substrate during evaporation. On top of the first gradient-thickness film, a second metal film was evaporated with a linear thickness variation perpendicular to that of the first film. As validation of the 2-D array deposition, the responses of devices with 1-D thickness gradients of Pd, Pt, and Ir films to several gases have been studied with SLPT and demonstrate results similar to those of corresponding discrete components (Klingvall et al. 2005).

The 2-D gradients have been used for studies and optimization of two-metal structures (Klingvall et al. 2003, 2005) and for determination of the effects of insulator surface properties on the magnitude of sensing response (Eriksson et al. 2005). Two-dimensional gradients of Pd/Rh film compositions were also studied to identify materials compositions for the most stable performance (Klingvall et al. 2005). The Pd/Rh film compositions were tested for their response stability to 1000 ppm of hydrogen upon aging for 24 h at 400°C while exposed to 250 ppm of hydrogen. This accelerated aging experiment of the 2-D gradient film surface demonstrated the existence of two most stable local regions. One region was a "valley" of stable response, shown as a dark color in Figure 3.26. The other region was a thicker part of the two-component film, with a ~20-nm-thick Rh film and a ~23-nm-thick Pd film. This new knowledge inspired new questions about the position stability of the "valley" and the possibility of improving sensor stability by using an initial annealing process.

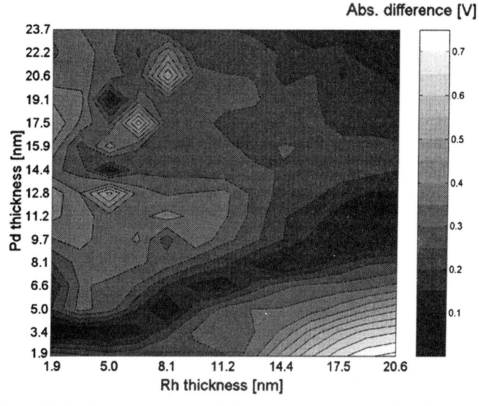

**Figure 3.26.** Results of accelerated aging of 2-D combinatorial library of Rh/Pd film. Chemical response image to 1000 ppm of hydrogen as the differential response after and before the accelerated aging; the most stable regions have the darkest color. (Reprinted with permission from Klingvall et al. 2005. Copyright 2005 The Institute of Electrical and Electronics Engineers, Inc.)

## 6.4. VARIABLE OPERATION TEMPERATURE AND DIFFUSION-LAYER THICKNESS

At present, in conductometric sensors, semiconducting metal oxides are used as gas-sensing materials that change their electrical resistance upon exposure to oxidizing or reducing gases. Over the years, significant technological advances have been made that have resulted in practical and commercially available sensors, and now new materials are being developed that further improve the performance of these sensors. Realizing the opportunities that arise with the temperature dependence of the sensor response, temperature gradient–based sensors that utilize a single metal-oxide thin film segmented by electrodes have been developed (Goschnick et al. 1998, 2005; Arnold et al. 2002a, 2002b; Koronczi et al. 2002; Schneider et al. 2004; Sysoev et al. 2004). In addition to the spatial temperature-gradient heater, one design for the sensor chip also had a $SiO_2$ or $Al_2O_3$ membrane with a gradient thickness from 2 to 50 nm (see Figure 3.27A). These ceramic membranes provided an additional response selectivity (Goschnick 2001) through thickness-dependent gas transport.

To fabricate such a temperature- and membrane-gradient sensor, a gas-sensitive $SnO_2$:Pt film (Pt content of 0.8 at%) was deposited onto a thermally oxidized Si wafer by RF magnetron sputtering using a shadow mask. Next, Pt strip electrodes and two meander-shaped thermoresistors were sputtered on the same side of the substrate as the $SnO_2$ film, under a shadow mask for structuring the films. The arrangement of the electrodes subdivided the monolithic $SnO_2$ film into 38 sensor segments on an area of $4 \times 8$ mm². Finally, Pt heaters were deposited onto the back of the substrate to operate the chip with a 50°C temperature gradient from 310 to 360°C (Sysoev et al. 2004). The application of a temperature gradient increased the gas discrimination power of the sensor by 35%. The sensor with a $SiO_2$ gradient-thickness membrane was employed for detection of gaseous precursors of smoldering fires induced by overheated cable insulation (see Figure 3.27B) (Arnold et al. 2002b).

## 7. EMERGING WIRELESS TECHNOLOGIES FOR COMBINATORIAL SCREENING OF SENSING MATERIALS

At present, advances in wireless electronic technologies promise to add attractive new capabilities for combinatorial screening of sensing materials. Wireless proximity chemical, biological, and physical sensors provide several key advantages over wired sensors, such as (1) noncontact and noncontamination measurements, (2) operation though packaging, and (3) rapid reading of multiple sensors with a single reader. Several proximity-sensing approaches based on thickness shear-mode (TSM) and surface acoustic-wave (SAW) sensors connected to antennas as well as passive radio-frequency identification (RFID) sensors have been developed (see Figure 3.28) with possible applications for combinatorial screening of sensing materials.

To eliminate the direct wiring of individual TSM sensors and to permit materials evaluation in environments where wiring is not desirable or adds a prohibitively complex design, a wireless TSM sensor array system (Potyrailo and Morris 2007b) was developed in which each sensor resonator was coupled to a receiver antenna coil and an array of these coils was scanned with a transmitter coil (Figure 3.29A). Using this wireless sensor system, sensing materials can be screened for their gas-sorption properties,

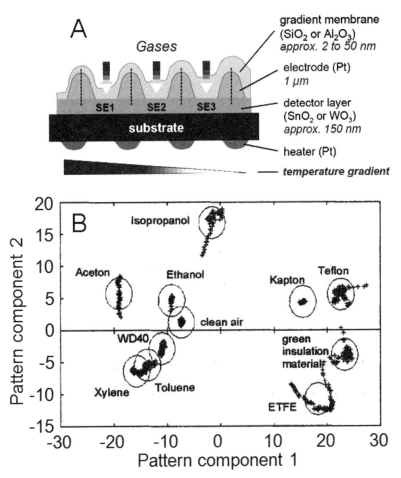

**Figure 3.27.** Double-gradient sensor microarray for selective gas detection. (A) Sensor schematic illustrating a single metal-oxide thin film segmented by electrodes and arranged on a temperature-gradient heater. The sensing film is further covered with a gradient-thickness ceramic membrane. (B) Results of linear discrimination analysis of the signal patterns in practical tests to detect gaseous precursors of smoldering fires induced by overheated cable insulation (ETFE, ethylene tetrafluorine ethylene). [(A) Used with kind permission from Goschnick. (B) Reprinted with permission from Arnold et al. 2002b. Copyright 2002 The Institute of Electrical and Electronics Engineers, Inc.]

analyte binding in liquids, and for changes in chemical and physical properties upon weathering and aging tests. The applicability of the wireless sensor materials screening approach has been demonstrated for the rapid evaluation of the effects of conditioning of polymeric sensing films at different temperatures on the vapor-response patterns. In one set of high-throughput screening experiments, Nafion film-aging effects on the selectivity pattern were studied. Evaluation of this and many other polymeric sensing materials lacks the detailed studies on the change of the chemical selectivity patterns as a function of temperature conditioning and aging. Conditioning of Nafion-coated resonators was performed at 22, 90, and 125°C for 12 h. Temperature-conditioned sensing films were exposed to water ($H_2O$), ethanol (EtOH), and acetonitrile (ACN) vapors, all at concentrations (partial pressures) ranging from 0 to 0.1 of the saturated vapor pressure $P_0$, as shown in Figures 3.29B and 3.29C. It was found that conditioning of sensing films at 125°C compared to room-temperature conditioning provided (1) an improvement in linearity in response to EtOH and ACN vapors, (2) an increase in relative response to ACN, and (3) a 10-fold increase of the contribution to principal component #2. The latter point signifies an improvement in the discrimination ability between different vapors upon conditioning of the sensing material at 125°C. This new knowledge will be critical in designing sensors for practical applications where the need exists to preserve sensor response selectivity over long exploitation time or when there is temperature cycling for accelerated sensor-film recovery after vapor exposure.

Recently, ubiquitous and cost-effective passive RFID tags have been adapted for chemical sensing (Potyrailo and Morris 2007a, Potyrailo et al. 2009b). By applying a sensing material onto the resonant antenna of the RFID tag and measuring the complex impedance of the RFID resonant antenna, it was possible to correlate impedance response to chemical properties of interest. A variety of sensing materials was evaluated, including Nafion polymeric compositions formulated with plasticizers, as shown in Figure 3.30. When a sensing film is deposited onto the resonant antenna (Figure 3.31A), the analyte-induced changes in the dielectric and dimensional properties of this sensing film affect the complex impedance of the antenna circuit through changes in film resistance and capacitance between the antenna turns. Such changes provide selectivity in response of an individual RFID sensor and provide the opportunity to replace a whole array of conventional sensors with a single RFID sensor (Potyrailo and Morris 2007a). For this selective analyte quantitation using individual RFID sensors,

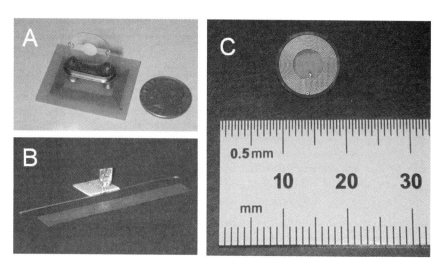

**Figure 3.28.** Examples of proximity wireless sensors applicable for combinatorial screening of sensing materials: (A) TSM sensor resonating at 10 MHz; (B) SAW sensor resonating at 915 MHz; (C) RFID sensor resonating at 13.56 MHz.

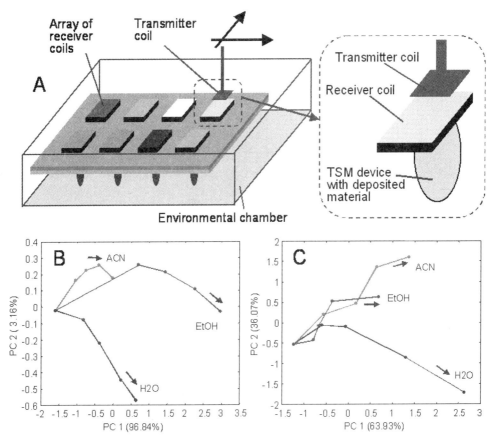

Figure 3.29. Concept for wireless high-throughput screening of materials properties using thickness shear mode resonators. (A) Configuration of a wireless proximity resonant sensor array system for high-throughput screening of sensing materials with a single transmitter coil that scans across an array of receiver coils attached to resonant sensors. (B–D) Evaluation of selectivity of Nafion sensing films to several vapors after conditioning at different temperatures: (B) 22°C and (C) 125°C. Vapors: $H_2O$ (water), EtOH (ethanol), and ACN (acetonitrile). Concentrations of vapors are 0, 0.02, 0.04, 0.07, and 0.10 $P/P_0$. Arrows indicate the increase of concentrations of each vapor. (Reprinted with permission from Potyrailo and Morris 2007b. Copyright 2007 American Institute of Physics.)

complex impedance spectra of the resonant antenna are measured. Several parameters from the measured real and imaginary portions of the complex impedance are further calculated. These parameters include $F_p$ and $Z_p$ (the frequency and magnitude of the maximum of the real part of the complex impedance, respectively) and $F_1$ and $F_2$ (the resonant and antiresonant frequencies of the imaginary part of the complex impedance, respectively). By applying multivariate analysis of the full complex impedance spectra or the calculated parameters, quantitation of analytes and rejection of interferences is performed with individual RFID sensors.

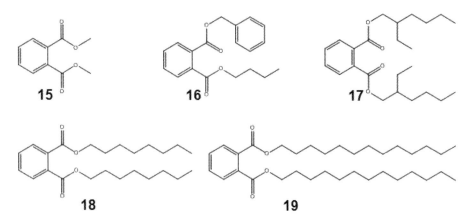

**Figure 3.30.** Structures of plasticizers 15–19 employed for construction of vapor-sensing materials based on Nafion polymer.

Because temperature effects are important at all stages of sensor fabrication, testing, and end use, understanding of temperature effects can provide the ability to build robust temperature-corrected transfer functions of sensor performance in order to preserve response sensitivity, response selectivity, and response baseline stability. RFID sensors were applied for the combinatorial screening of sensing materials to evaluate combined effects of plasticizers in polymeric formulated films and annealing temperature (Potyrailo et al. 2009c). As a model system, a 6 × 8 array of polymer-coated RFID sensors was constructed as shown in Figure 3.31B. A solid polymer electrolyte Nafion was formulated with five types of phthalate plasticizers as shown in Figure 3.30, including dimethyl phthalate (15), butyl benzyl phthalate (16), di-(2-ethylhexyl) phthalate (17), dicapryl phthalate (18), and diisotridecyl phthalate (19). These sensing film formulations and control sensing films without a phthalate plasticizer were deposited onto RFID sensors, exposed to eight temperatures ranging from 40 to 140°C using a gradient-temperature heater, and evaluated for their response stability and gas-selectivity response patterns. The interrogation of RFID sensors in the array was done with a single transmitter (pick-up antenna) coil positioned on an X–Y translation stage and connected to a network analyzer.

To evaluate temperature effects on sensor response selectivity, the 6 × 8 array of temperature-annealed sensing films was exposed to $H_2O$ and ACN vapors. Acetonitrile was selected as a simulant for blood chemical warfare agents (CWAs) (Lee et al. 2005) and water vapor was selected as an interference. In these experiments, the partial pressures of $H_2O$ and ACN vapors were 0.4 of the saturated vapor pressure $P_0$. Nafion sensing films were used previously for detection of humidity (Feng et al. 1997; Tailoka et al. 2003) and organic vapors (Sun and Okada 2001). Conductance and dielectric properties of Nafion have been shown to be vapor-dependent (Cappadonia et al. 1995; Wintersgill and Fontanella 1998). Figures 3.31C and 3.31D show representative $\Delta Z_p$ and $\Delta F_p$ responses to $H_2O$ and ACN upon annealing at two temperatures, 40 and 110°C. The patterns of $\Delta F_1$, $\Delta F_2$, $\Delta F_p$, and $\Delta Z_p$ responses of sensing films to $H_2O$ and ACN vapors upon temperature annealing were further examined using principal components analysis (PCA). From these screening experiments, it was found that different plasticizers affect the response diversity to different extents; however, Nafion sensing films formulated with

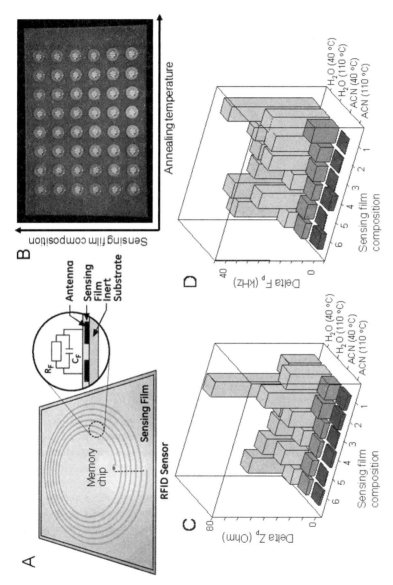

**Figure 3.31.** Combinatorial screening of sensing film compositions using passive RFID sensors. (A) Strategy for adaptation of conventional passive RFID tags for chemical sensing by deposition of a sensing film onto the resonant circuit of the RFID antenna. Inset, analyte-induced changes in the film material affect film resistance ($R_F$) and capacitance ($C_F$) between the antenna turns. (B) Photo of an array of 48 RFID sensors prepared for temperature-gradient evaluations of response of Nafion/phthalates compositions. (C and D) Examples of RFID sensor response to water ($H_2O$) and acetonitrile (ACN) vapors after annealing at different temperatures (40 and 110°C). (C) $\Delta Z_p$ response and (D) $\Delta F_p$ response. Nafion sensing film compositions: 1, control without plasticizer; 2, dimethyl phthalate; 3, butyl benzyl phthalate; 4, di-(2-ethylhexyl) phthalate; 5, dicapryl phthalate; 6, diisotridecyl phthalate. (Color photographs can be seen in the original paper.) (Reprinted with permission from Potyrailo et al. 2009c. Copyright 2009 American Chemical Society.)

the dimethyl phthalate plasticizer improve response diversity of the sensing films to $H_2O$ and ACN. Overall, this study demonstrated that this RFID-based sensing approach permits rapid, cost-effective combinatorial screening of dielectric properties of sensing materials. As was pointed out earlier (Potyrailo et al. 2003d), in general, an increase of the level of environmental stress may be problematic if the correlation with traditional test methods is lost. To avoid this situation, it will be critical to plan detailed accelerated-aging, high-throughput experiments with positive and negative controls.

## 8. SUMMARY AND OUTLOOK

Combinatorial technologies in materials science have been successfully developed by research groups in academia and governmental laboratories that have overcome the entry barrier of dealing with new and emerging aspects of materials research, such as automation and robotics, computer programming, informatics, and materials data mining. The main driving forces for combinatorial materials science in industry include broader and more detailed exploration of materials and process parameters and faster time to market. Industrial research laboratories working on new catalysts and inorganic luminescent materials were among the first adopters of combinatorial methodologies in industry. The classical example, the effort of Mittasch, who spent 10 years (over 1900–1909) conducting 6500 screening experiments with 2500 catalyst candidates to find a catalyst for industrial ammonia synthesis (Ertl, 1990) will never happen again because of the availability and affordability of modern tools for high-throughput synthesis and characterization.

In the area of sensing materials, reported examples of significant screening efforts are less dramatic yet also breathtaking. For example, a decade ago, Cammann, Shulga, and co-workers (Buhlmann et al. 1998) reported an "extensive systematic study" of more than 500 compositions to optimize vapor-sensing polymeric materials. Walt and co-workers (1998) reported screening over 100 polymer candidates for "their ability to serve as sensing matrices" for solvatochromic reagents. Seitz and co-workers (Conway et al. 1997) investigated the influence of multicomponent compositions on the properties of pH-swellable polymers by designing $3 \times 3 \times 3 \times 2$ factorial experiments. Clearly, combinatorial technologies have been introduced at the right time to make the search for new materials more intellectually rewarding. Naturally, numerous academic groups that were involved in the development of new sensing materials turned to combinatorial methodologies to speed up knowledge discovery (Dickinson et al. 1997; Cho et al. 2002; Simon et al. 2002; Apostolidis et al. 2004; Frantzen et al. 2004; Mirsky et al. 2004; Lundström et al. 2007).

Among numerous results achieved using combinatorial and high-throughput methods, the most successful have been in the areas of molecular imprinting, polymeric compositions, catalytic metals for field-effect devices, and metal oxides for conductometric sensors. In those materials, the desired selectivity and sensitivity have been achieved by the exploration of multidimensional chemical composition and process parameters at a previously unavailable level of detail and in a fraction of the time required for conventional one-at-a-time experiments. These new tools have provided the opportunity for more challenging, yet more rewarding, explorations that previously were too time-consuming to pursue.

Future advances in combinatorial development of sensing materials will be related to several key remaining unmet needs that prevent researchers from having a complete combinatorial workflow. At present, data management of the combinatorial workflow is perhaps the weakest link. However, over the last several years, there have been a growing number of reports on data mining in sensing materials (Villoslada and Takeuchi 2005; Frenzer et al. 2006; Mijangos et al. 2006; Potyrailo et al. 2006a). "Searching for a needle in the haystack" was popular in the early days of combinatorial materials science (Jandeleit et al. 1999; Jansen 2002). It has now been realized that screening the entire materials and process parameters space is still too costly and time prohibitive, even with the availability of existing tools. Instead, designing high-throughput experiments to discover relevant descriptors will become more attractive.

A modern combinatorial scientist is acquiring skills as diverse as experimental planning, automated synthesis, basics of high-throughput materials characterization, chemometrics, and data mining. These new skills can be now obtained through the growing network of practitioners and through the new generation of scientists educated across the world in combinatorial methodologies. Combinatorial and high-throughput experimentation have been able to bring together several previously disjoined disciplines and to combine valuable complementary attributes from each of them into a new scientific approach.

## 9. ACKNOWLEDGMENTS

I gratefully acknowledge GE components for support of our combinatorial sensor research.

## REFERENCES

Abraham M.H. (1993) Scales of solute hydrogen bonding: their construction and application to physicochemical and biochemical processes. *Chem. Soc. Rev.* **22**, 73–83.

Akporiaye D.E. (1998) Towards a rational synthesis of large-pore zeolite-type materials? *Angew. Chem. Int. Ed.* **37**, 2456–2457.

Albert K.J., Lewis N.S., Schauer C.L., Sotzing G.A., Stitzel S.E., Vaid T.P., and Walt D.R. (2000) Cross-reactive chemical sensor arrays. *Chem. Rev.* **100**, 2595–2626.

Alivisatos, A.P. (2004) The use of nanocrystals in biological detection. *Nature Biotechnol.* **22**, 47–52.

Amao Y. (2003) Probes and polymers for optical sensing of oxygen. *Microchim. Acta* **143**, 1–12.

Amis E.J. (2004) Combinatorial materials science reaching beyond discovery. *Nature Mater.* **3**, 83–85.

Anderson F.W. and Moser J.H. (1958) Automatic computer program for reduction of routine emission spectrographic data. *Anal. Chem.* **30**, 879–881.

Anker J.N., Hall W.P., Lyandres O., Shah N.C., Zhao J., and Van Duyne R.P. (2008) Biosensing with plasmonic nanosensors. *Nature Mater.* **7**, 442–453.

Apostolidis A., Klimant I., Andrzejewski D., and Wolfbeis O.S. (2004) A combinatorial approach for development of materials for optical sensing of gases. *J. Comb. Chem.* **6**, 325–331.

Arnold C., Andlauer W., Häringer D., Körber R., and Goschnick J. (2002a) Gas analytical gradient microarrays compromising low price and high performance for intelligent consumer products. *Proc. IEEE Sens.* **1**, 426–429.

Arnold C., Harms M., and Goschnick J. (2002b) Air quality monitoring and fire detection with the Karlsruhe electronic micronose KAMINA. *IEEE Sens. J.* **2**, 179–188.

Aronova M.A., Chang K. S., Takeuchi I., Jabs H., Westerheim D., Gonzalez-Martin A., Kim J., and Lewis B. (2003) Combinatorial libraries of semiconductor gas sensors as inorganic electronic noses. *Appl. Phys. Lett.* **83**, 1255–1257.

Baker B.E., Kline N.J., Treado P.J., and Natan M.J. (1996) Solution-based assembly of metal surfaces by combinatorial methods. *J. Am. Chem. Soc.* **118**, 8721–8722.

Barbero C., Acevedo D.F., Salavagione H.J., and Miras M.C. (2003) Synthesis, properties and applications of functionalized conductive polymers. *Jornadas Sam/Conamet/Simposio Materia 2003*, C-12.

Barsan N., Koziej D., and Weimar U. (2007) Metal oxide-based gas sensor research: how to? *Sens. Actuators B* **121**, 18–35.

Barsoukov E. and Macdonald J.R. (Eds.) (2005) *Impedance Spectroscopy: Theory, Experiment, and Applications.* Wiley, Hoboken, NJ.

Basu B.J., Thirumurugan A., Dinesh A.R., Anandan C., and Rajam K.S. (2005) Optical oxygen sensor coating based on the fluorescence quenching of a new pyrene derivative. *Sens. Actuators B* **104**, 15–22.

Bedlek-Anslow J.M., Hubner J.P., Carroll B.F., and Schanze K.S. (2000) Micro-heterogeneous oxygen response in luminescence sensor films. *Langmuir* **16**, 9137–9141.

Beebe K.R., Pell R.J., and Seasholtz M. B. (1998) *Chemometrics: A Practical Guide.* Wiley, New York.

Belmares M., Blanco M., Goddard W.A. III, Ross R.B., Caldwell G., Chou S.-H., Pham J., Olofson P.M., and Thomas C. (2004) Hildebrand and Hansen solubility parameters from molecular dynamics with applications to electronic nose polymer sensors. *J. Comput. Chem.* **25**, 1814–1826.

Bhat R.R., Tomlinson M.R., and Genzer J. (2004) Assembly of nanoparticles using surface-grafted orthogonal polymer gradients. *Macromol. Rapid Commun.* **25**, 270–274.

Bidan G. (1992) Electroconducting conjugated polymers: new sensitive matrices to build up chemical or electrochemical sensors. A review. *Sens. Actuators B* **6**, 45–56.

Birina G.A. and Boitsov K.A. (1974) Experimental use of combinational and factorial plans for optimizing the compositions of electronic materials. *Zavodskaya Laboratoriya (in Russian)* **40**, 855–857.

Bobacka J., Ivaska A., and Lewenstam A. (2003) Potentiometric ion sensors based on conducting polymers. *Electroanalysis* **15**, 366–374.

Briceño G., Chang H., Sun X., Schultz P.G., and Xiang X.-D. (1995) A class of cobalt oxide magnetoresistance materials discovered with combinatorial synthesis. *Science* **270**, 273–275.

Brocchini S., James K., Tangpasuthadol V., and Kohn J. (1997) A combinatorial approach for polymer design. *J. Am. Chem. Soc.* **119**, 4553–4554.

Buhlmann K., Schlatt B., Cammann K., and Shulga A. (1998) Plasticised polymeric electrolytes: new extremely versatile receptor materials for gas sensors (VOC monitoring) and electronic noses (odour identification:discrimination). *Sens. Actuators B* **49**, 156–165.

Bühlmann P., Pretsch E., and Bakker E. (1998) Carrier-based ion-selective electrodes and bulk optodes. 2. Ionophores for potentiometric and optical sensors. *Chem. Rev.* **98**, 1593–1687.

Calvert P. (2001) Inkjet printing for materials and devices. *Chem. Mater.* **13**, 3299–3305.

Cappadonia M., Erning J.W., Niaki S.M.S., and Stimming U. (1995) Conductance of Nafion 117 membranes as a function of temperature and water content. *Solid State Ionics* **77**, 65–69.

Cawse J.N., Olson D., Chisholm B.J., Brennan M., Sun T., Flanagan W., Akhave J., Mehrabi A., and Saunders D. (2003) Combinatorial chemistry methods for coating development V: Generating a combinatorial array of uniform coatings samples. *Prog. Org. Coat.,* **47**, 128–135.

Chang H., Gao C., Takeuchi I., Yoo Y., Wang J., Schultz P.G., Xiang X.-D., Sharma R.P., Downes M., and Venkatesan T. (1998) Combinatorial synthesis and high throughput evaluation of ferroelectric/dielectric thin-film libraries for microwave applications. *Appl. Phys. Lett.* **72**, 2185–2187.

Chen H.M., Pang L., Kher A., and Fainman Y. (2009) Three-dimensional composite metallodielectric nanostructure for enhanced surface plasmon resonance sensing. *Appl. Phys. Lett.* **94**, 073117.

Chisholm B.J., Potyrailo R.A., Cawse J.N., Shaffer R.E., Brennan M.J., Moison C., Whisenhunt D.W., Flanagan W.P.,

Olson D.R., Akhave J.R., Saunders D.L., Mehrabi A., and Licon M. (2002) The development of combinatorial chemistry methods for coating development I. Overview of the experimental factory. *Prog. Org. Coat.* **45**, 313–321.

Cho E.J., Tao Z., Tang Y., Tehan E.C., Bright F.V., Hicks W.L. Jr., Gardella J.A. Jr., and Hard R. (2002) Tools to rapidly produce and screen biodegradable polymer and sol-gel-derived xerogel formulations. *Appl. Spectrosc.* **56**, 1385–1389.

Chojnacki P., Werner T., and Wolfbeis O. S. (2004) Combinatorial approach towards materials for optical ion sensors. *Microchim. Acta* **147**, 87–92.

Christensen D. and Fowers D. (1996) Modeling SPR sensors with the finite-difference time-domain method. *Biosens. Bioelectron.* **11**, 677–684.

Clark K.J.R. and Furey J. (2006) Suitability of selected single-use process monitoring and control technology. *BioProcess Intern.* **4**(6), S16–S20.

Collaudin A.B. and Blum L.J. (1997) Enhanced luminescence response of a fibre-optic sensor for $H_2O_2$ by a high-salt-concentration medium. *Sens. Actuators B* **38–39**, 189–194.

Conway V.L., Hassen K.P., Zhang L., Seitz W.R., and Gross T.S. (1997) The influence of composition on the properties of pH-swellable polymers for chemical sensors. *Sens. Actuators B* **45**, 1–9.

Cooper A.C., McAlexander L.H., Lee D.-H., Torres M.T., and Crabtree R.H. (1998) Reactive dyes as a method for rapid screening of homogeneous catalysts. *J. Am. Chem. Soc.* **120**, 9971–9972.

Cui J., Chu Y. S., Famodu O. O., Furuya Y., Hattrick-Simpers J., James R.D., Ludwig A., Thienhaus S., Wuttig M., Zhang Z., and Takeuchi I. (2006) Combinatorial search of thermoelastic shape-memory alloys with extremely small hysteresis width. *Nature Mater.* **5**, 286–290.

Dai L., Soundarrajan P., and Kim T. (2002) Sensors and sensor arrays based on conjugated polymers and carbon nanotubes. *Pure Appl. Chem.* **74**, 1753–1772.

Danielson E., Devenney M., Giaquinta D.M., Golden J.H., Haushalter R.C., McFarland E.W., Poojary D.M., Reaves C.M., Weinberg W.H., and Wu X.D. (1998) A rare-earth phosphor containing one-dimensional chains identified through combinatorial methods. *Science* **279**, 837–839.

De Gans B.-J. and Schubert U.S. (2004) Inkjet printing of well-defined polymer dots and arrays. *Langmuir* **20**, 7789–7793.

Department of Defense (2005) *Technology Readiness Assessment (TRA) Deskbook, May 2005,* Department of Defense, Prepared by the Deputy Under Secretary of Defense for Science and Technology (DUSD(S&T), https://acc.dau.mil/CommunityBrowser.aspx?id=18545.

Dickinson T.A., Walt D.R., White J., and Kauer J.S. (1997) Generating sensor diversity through combinatorial polymer synthesis. *Anal. Chem.* **69**, 3413–3418.

Draxler S., Lippitsch M.E., Klimant I., Kraus H., and Wolfbeis O.S. (1995) Effects of polymer matrices on the time-resolved luminescence of a ruthenium complex quenched by oxygen. *J. Phys. Chem.* **99**, 3162–3167.

Eash M.A. and Gohlke R.S. (1962) Mass spectrometric analysis. A small computer program for the analysis of mass spectra. *Anal. Chem.* **34**, 713–713.

Eaton K. (2002) A novel colorimetric oxygen sensor: dye redox chemistry in a thin polymer film. *Sens. Actuators B* **85**, 42–51.

Eberhart M.E. and Clougherty D.P. (2004) Looking for design in materials design. *Nature Mater.* **3**, 659–661.

Ekin A. and Webster D.C. (2007) Combinatorial and high-throughput screening of the effect of siloxane composition on the surface properties of crosslinked siloxane-polyurethane coatings. *J. Comb. Chem.* **9**, 178–188.

Eriksson M., Klingvall R., and Lundström I. (2006) A combinatorial method for optimization of materials for gas sensitive field-effect devices. In: *Combinatorial and High-Throughput Discovery and Optimization of Catalysts and Materials,* Potyrailo R.A. and Maier W.F. (eds.). CRC Press, Boca Raton, FL, pp. 85–95.

Eriksson M., Salomonsson A., Lundström I., Briand D., and Åbom A.E. (2005) The influence of the insulator surface properties on the hydrogen response of field-effect gas sensors. *J. Appl. Phys.* **98**, 034903.

Ertl G. (1990) Elementary steps in heterogeneous catalysis. *Angew. Chem. Int. Ed.* **29**, 1219–1227.

Feldmann C. (2001) Polyol mediated synthesis of oxide particle suspensions and their application. *Scripta Mater.* **44**, 2193–2196.

Feng C.-D., Sun S.-L., Wang H., Segre C.U., and Stetter J.R. (1997) Humidity sensing properties of Nafion and sol-gel derived $SiO_2$/Nafion composite thin films. *Sens. Actuators B* **40**, 217–222.

Florescu M. and Katerkamp A. (2004) Optimisation of a polymer membrane used in optical oxygen sensing. *Sens. Actuators B* **97**, 39–44.

Franke M.E., Koplin T.J., and Simon U. (2006) Metal and metal oxide nanoparticles in chemiresistors: does the nanoscale matter? *Small* **2**, 36–50.

Frantzen A., Sanders D., Scheidtmann J., Simon U., and Maier W.F. (2005) A flexible database for combinatorial and high-throughput materials science. *QSAR & Comb. Sci.* **24**, 22–28.

Frantzen A., Scheidtmann J., Frenzer G., Maier W.F., Jockel J., Brinz T., Sanders D., and Simon U. (2004) High-throughput method for the impedance spectroscopic characterization of resistive gas sensors. *Angew. Chem. Int. Ed.* **43**, 752–754.

Frenzer G., Frantzen A., Sanders D., Simon U., and Maier W.F. (2006) Wet chemical synthesis and screening of thick porous oxide films for resistive gas sensing applications. *Sensors* **6**, 1568–1586.

Genet C. and Ebbesen T.W. (2007) Light in tiny holes. *Nature* **445**, 39–46.

Gill I. (2001) Bio-doped nanocomposite polymers: sol-gel bioencapsulates. *Chem. Mater.* **13**, 3404–3421.

Gill I. and Ballesteros A. (1998) Encapsulation of biologicals within silicate, siloxane, and hybrid sol-gel polymers: an efficient and generic approach. *J. Am. Chem. Soc.* **120**, 8587–8598.

Gomez-Romero P. (2001) Hybrid organic-inorganic materials—in search of synergic activity. *Adv. Mater.* **13**, 163–174.

Göpel W. (1998) Chemical imaging: I. Concepts and visions for electronic and bioelectronic noses. *Sens. Actuators B* **52**, 125–142.

Goschnick J. (2001) An electronic nose for intelligent consumer products based on a gas analytical gradient microarray. *Microelectron. Eng.* **57–58**, 693–704.

Goschnick J., Frietsch M., and Schneider T. (1998) Non-uniform $SiO_2$ membranes produced by ion beam-assisted chemical vapor deposition to tune $WO_3$ gas sensor microarrays. *Surf. Coat. Technol.* **108–109**, 292–296.

Goschnick J., Koronczi I., Frietsch M., and Kiselev I. (2005) Water pollution recognition with the electronic nose KAMINA. *Sens. Actuators B* **106**, 182–186.

Grate J.W. (2000) Acoustic wave microsensor arrays for vapor sensing. *Chem. Rev.* **100**, 2627–2648.

Grate J.W. and Abraham M.H. (1991) Solubility interactions and the design of chemically selective sorbent coatings for chemical sensors and arrays. *Sens. Actuators B* **3**, 85–111.

Grate J.W., Abraham H., and McGill R.A. (1997) Sorbent polymer materials for chemical sensors and arrays. In: *Handbook of Biosensors and Electronic Noses. Medicine, Food, and the Environment*, Kress-Rogers E. (ed.). CRC Press, Boca Raton, FL, pp. 593–612.

Greeley J., Jaramillo T.F., Bonde J., Chorkendorff I., and Nørskov J.K. (2006) Computational high-throughput screening of electrocatalytic materials for hydrogen evolution. *Nature Mater.* **5**, 909–913.

Grunlan J.C., Mehrabi A.R., Chavira A.T., Nugent A.B., and Saunders D.L. (2003) Method for combinatorial screening of moisture vapor transmission rate. *J. Comb. Chem.* **5**, 362–368.

Hagleitner C., Hierlemann A., Brand O., and Baltes H. (2002) CMOS single chip gas detection systems—Part I. In: *Sensors Update, Vol. 11,* Baltes H., Göpel W., and Hesse J. (eds.). VCH, Weinheim, pp. 101–155.

Hagleitner C., Hierlemann A., Brand O., and Baltes H. (2003) CMOS single chip gas detection systems—Part II. In: *Sensors Update, Vol. 12,* Baltes H., Göpel W., and Hesse J. (eds.). VCH, Weinheim, pp. 51–120.

Hanak J.J. (1970) The "Multiple-Sample Concept" in materials research: synthesis, compositional analysis and testing of entire multicomponent systems. *J. Mater. Sci.* **5**, 964–971.

Hänsel H., Zettl H., Krausch G., Schmitz C., Kisselev R., Thelakkat M., and Schmidt H.-W. (2002) Combinatorial study of the long-term stability of organic thin-film solar cells. *Appl. Phys. Lett.* **81**, 2106–2108.

Hartmann P. and Trettnak W. (1996) Effects of polymer matrices on calibration functions of luminescent oxygen sensors based on porphyrin ketone complexes. *Anal. Chem.* **68**, 2615–2620.

Hassib L. and Potyrailo R.A. (2004) Combinatorial development of polymer coating formulations for chemical sensor applications. *Polymer Preprints* **45**, 211–212.

Hatchett D.W. and Josowicz M. (2008) Composites of intrinsically conducting polymers as sensing nanomaterials. *Chem. Rev.* **108**, 746–769.

Hierlemann A., Weimar U., Kraus G., Schweizer-Berberich M., and Göpel W. (1995) Polymer-based sensor arrays and multicomponent analysis for the detection of hazardous organic vapours in the environment. *Sens. Actuators B* **26**, 126–134.

Hoffmann R. (2001) Not a library. *Angew. Chem. Int. Ed.* **40**, 3337–3340.

Holzwarth A., Schmidt H.-W., and Maier W. (1998) Detection of catalytic activity in combinatorial libraries of heterogeneous catalysts by IR thermography. *Angew. Chem. Int. Ed.* **37**, 2644–2647.

Honeybourne C.L. (2000) Organic vapor sensors for food quality assessment. *J. Chem. Educ.* **77**, 338–344.

Hoogenboom R., Meier M.A.R., and Schubert U.S. (2003) Combinatorial methods, automated synthesis and high-throughput screening in polymer research: past and present. *Macromol. Rapid Commun.* **24**, 15–32.

Hu Y., Tan O.K., Pan J.S., and Yao X. (2004) A new form of nanosized $SrTiO_3$ material for near-human-body temperature oxygen sensing applications. *J. Phys. Chem. B* **108**, 11214–11218.

Janata J. and Josowicz M. (2002) Conducting polymers in electronic chemical sensors. *Nature Mater.* **2**, 19–24.

Jandeleit B., Schaefer D.J., Powers T.S., Turner H.W., and Weinberg W.H. (1999) Combinatorial materials science and catalysis. *Angew. Chem. Int. Ed.* **38**, 2494–2532.

Jansen M. (2002) A concept for synthesis planning in solid-state chemistry. *Angew. Chem. Int. Ed.* **41**, 3746–3766.

Kahl M., Voges E., Kostrewa S., Viets C., and Hill W. (1998) Periodically structured metallic substrates for SERS. *Sens. Actuators B* **51**, 285–291.

Kennedy K., Stefansky T., Davy G., Zackay V.F., and Parker E.R. (1965) Rapid method for determining ternary-alloy phase diagrams. *J. Appl. Phys.* **36**, 3808–3810.

Klingvall R., Lundstrom I., Lofdahl M., and Eriksson M. (2003) 2D-evaluation of the gas response of a RhPd-MIS device. *Proc. IEEE Sensors* **2**, 1114–1115.

Klingvall R., Lundström I., Löfdahl M., and Eriksson M. (2005) A combinatorial approach for field-effect gas sensor research and development. *IEEE Sensors J.* **5**, 995–1003.

Ko M.C. and Meyer G.J. (1999) Photoluminescence of inorganic semiconductors for chemical sensor applications. In: *Optoelectronic Properties of Inorganic Compounds*, Roundhill D.M. and Fackler J.P. Jr. (eds.). Plenum Press, New York, pp. 269–315. Koinuma H. and Takeuchi I. (2004) Combinatorial solid state chemistry of inorganic materials. *Nature Mater.* **3**, 429–438.

Kolytcheva N.V., Müller H., and Marstalerz J. (1999) Influence of the organic matrix on the properties of membrane coated ion sensor field-effect transistors. *Sens. Actuators B* **58**, 456–463.

Koronczi I., Ziegler K., Kruger U., and Goschnick J. (2002) Medical diagnosis with the gradient microarray of the KAMINA. *IEEE Sensors J.* **2**, 254–259.

Korotcenkov G. (2005) Gas response control through structural and chemical modification of metal oxide films: state of the art and approaches. *Sens. Actuators B* **107**, 209–232.

Krishnan A., Thio T., Kim T.J., Lezec H.J., Ebbesen T.W., Wolff P.A., Pendry J., Martín-Moreno L., and Garcia-Vidal F.J. (2001) Evanescently coupled resonance in surface plasmon enhanced transmission. *Opt. Commun.* **200**, 1–7.

Kulikov V. and Mirsky V.M. (2004) Equipment for combinatorial electrochemical polymerization and high-throughput investigation of electrical properties of the synthesized polymers. *Meas. Sci. Technol.* **15**, 49–54.

Lai R., Kang B.S., and Gavalas G.R. (2001) Parallel synthesis of ZSM-5 zeolite films from clear organic-free solutions. *Angew. Chem., Int. Ed.* **40**, 408–411.

Lavigne J.J. and Anslyn E.V. (2001) Sensing a paradigm shift in the field of molecular recognition: from selective to differential receptors. *Angew. Chem. Int. Ed.* **40**, 3119–3130.

Leach A.M. and Potyrailo R.A. (2006) Gas sensor materials based on semiconductor nanocrystal/polymer composite films. In: *Combinatorial Methods and Informatics in Materials Science, MRS Symposium Proceedings*, Vol. 894, Wang Q., Potyrailo R.A., Fasolka M., Chikyow T., Schubert U.S., and Korkin A. (eds.). Materials Research Society, Warrendale, PA, pp. 237–243.

Leclerc M. (1999) Optical and electrochemical transducers based on functionalized conjugated polymers. *Adv. Mater.* **11**, 1491–1498.

Lee W.S., Lee S.C., Lee S.J., Lee D.D., Huh J.S., Jun H.K., and Kim J.C. (2005) The sensing behavior of $SnO_2$-based thick-film gas sensors at a low concentration of chemical agent simulants. *Sens. Actuators B* **108**, 148–153.

Lee K.-L., Lee C.-W., Wang W.-S., and Wei P.-K. (2007) Sensitive biosensor array using surface plasmon resonance on metallic nanoslits. *J. Biomed. Opt.* **12**, art. no. 044023.

Lee K.-L., Wang W.-S., and Wei P.-K. (2008a) Comparisons of surface plasmon sensitivities in periodic gold nanostructures. *Plasmonics* **3**, 119–125.

Lee K.-L., Wang W.-S., and Wei P.-K. (2008b) Sensitive label-free biosensors by using gap plasmons in gold nanoslits. *Biosens. Bioelectron.* **24**, 210–215.

Legin A., Makarychev-Mikhailov S., Goryacheva O., Kirsanov D., and Vlasov Y. (2002) Cross-sensitive chemical sensors based on tetraphenylporphyrin and phthalocyanine. *Anal. Chim. Acta* **457**, 297–303.

Lemmo A.V., Fisher J.T., Geysen H.M., and Rose D.J. (1997) Characterization of an inkjet chemical microdispenser for combinatorial library synthesis. *Anal. Chem.* **69**, 543–551.

Lemmon J.P., Wroczynski R.J., Whisenhunt D.W. Jr., and Flanagan W.P. (2001) High throughput strategies for monomer and polymer synthesis and characterization. *Polymer Preprints* **42**, 630–631.

Levitsky I., Krivoshlykov S.G., and Grate J.W. (2001) Rational design of a nile red/polymer composite film for fluorescence sensing of organophosphonate vapors using hydrogen bond acidic polymers. *Anal. Chem.* **73**, 3441–3448.

Lim S.-H., Raorane D., Satyanarayana S., and Majumdar A. (2006) Nano-chemo-mechanical sensor array platform for high-throughput chemical analysis. *Sens. Actuators B* **119**, 466–474.

Lisensky G.C., Meyer G.J., and Ellis A.B. (1988) Selective detector for gas chromatography based on adduct-modulated semiconductor photoluminescence. *Anal. Chem.*, **60**, 2531–2534.

Löfdahl M., Eriksson M., and Lundström I. (2000) Chemical images. *Sens. Actuators B* **70**, 77–82.

Lundström I., Shivaraman S., Svensson C., and Lundkvist L. (1975) A hydrogen-sensitive MOS field-effect transistor. *Appl. Phys. Lett.* **26**, 55–57.

Lundström I., Erlandsson R., Frykman U., Hedborg E., Spetz A., Sundgren H., Welin S., and Winquist F. (1991) Artificial 'olfactory' images from a chemical sensor using a light pulse technique. *Nature* **352**, 47–50.

Lundström I., Sundgren H., Winquist F., Eriksson M., Krantz-Rülcker C., and Lloyd-Spetz A. (2007) Twenty-five years of field effect gas sensor research in Linköping. *Sens. Actuators B* **121**, 247–262.

Lundström I., Svensson C., Spetz A., Sundgren H., and Winquist F. (1993) From hydrogen sensors to olfactory images—twenty yerars with catalytic field-effect devices. *Sens. Actuators B* **13–14**, 16–23.

MacLean D., Baldwin J.J., Ivanov V.T., Kato Y., Shaw A., Schneider P., and Gordon E.M. (2000) Glossary of terms used in combinatorial chemistry. *J. Comb. Chem.* **2**, 562–578.

Maranas C.D. (1996) Optimal computer-aided molecular design: a polymer design case study. *Ind. Eng. Chem. Res.* **35**, 3403–3414.

Martin P.D., Wilson T.D., Wilson I.D., and Jones G.R. (2001) An unexpected selectivity of a propranolol-derived molecular imprint for tamoxifen. *Analyst* **126**, 757–759.

Martín-Moreno L., García-Vidal F.J., Lezec H.J., Pellerin K.M., Thio T., Pendry J.B., and Ebbesen T.W. (2001) Theory of extraordinary optical transmission through subwavelength hole arrays. *Phys. Rev. Lett.* **86**, 1114–1117.

Matteo J.A., Fromm D.P., Yuen Y., Schuck P.J., Moerner W.E., and Hesselink L. (2004) Spectral analysis of strongly enhanced visible light transmission through single C-shaped nanoapertures. *Appl. Phys. Lett.* **85**, 648–650.

McKusick B.C., Heckert R.E., Cairns T.L., Coffman D.D., and Mower H.F. (1958) Cyanocarbon chemistry. VI. Tricyanovinylamines. *J. Am. Chem. Soc.* **80**, 2806–2815.

McMahon J.M., Henzie J., Odom T.W., Schatz G.C., and Gray S.K. (2007) Tailoring the sensing capabilities of nanohole arrays in gold films with Rayleigh anomaly-surface plasmon polaritons. *Opt. Express* **15**, 18119–18129.

McQuade D.T., Pullen A.E., and Swager T.M. (2000) Conjugated polymer-based chemical sensors. *Chem. Rev.* **100**, 2537–2574.

Medintz I.L., Uyeda H.T., Goldman E.R., and Mattoussi H. (2005) Quantum dot bioconjugates for imaging, labelling and sensing. *Nature Mater.* **4**, 435–446.

Meyerhoff M.E. (1993) In vivo blood-gas and electrolyte sensors: progress and challenges. *Trends Anal. Chem.* **12**, 257–266.

Mijangos I., Navarro-Villoslada F., Guerreiro A., Piletska E., Chianella I., Karim K., Turner A., and Piletsky S. (2006) Influence of initiator and different polymerisation conditions on performance of molecularly imprinted polymers. *Biosens. Bioelectron.* **22**, 381–387.

Mills A. (1998) Controlling the sensitivity of optical oxygen sensors. *Sens. Actuators B* **51**, 60–68.

Mills A., Lepre A., and Wild L. (1998) Effect of plasticizer-polymer compatibility on the response characteristics of optical thin $CO_2$ and $O_2$ sensing films. *Anal. Chim. Acta* **362**, 193–202.

Mirsky V.M. and Kulikov V. (2003) Combinatorial electropolymerization: concept, equipment and applications. In: *High Throughput Analysis: A Tool for Combinatorial Materials Science,* Potyrailo R.A. and Amis E.J. (eds.). Kluwer Academic/Plenum, New York, pp.431–446.

Mirsky V.M., Kulikov V., Hao Q., and Wolfbeis O.S. (2004) Multiparameter high throughput characterization of combinatorial chemical microarrays of chemosensitive polymers. *Macromol. Rapid Commun.* **25**, 253–258.

Mohr G.J. and Wolfbeis O.S. (1996) Effects of the polymer matrix on an optical nitrate sensor based on a polarity-sensitive dye. *Sens. Actuators B* **37**, 103–109.

Montgomery J.M., Lee T.-W., and Gray S.K. (2008) Theory and modeling of light interactions with metallic nanostructures. *J. Phys.: Condens. Matter.* **20**, art. no. 323201.

Nazzal A.Y., Qu L., Peng X., and Xiao M. (2003) Photoactivated CdSe nanocrystals as nanosensors for gases. *Nano Lett.* **3**, 819–822.

Newman J.D. and Turner A.P.F. (2005) Home blood glucose biosensors: a commercial perspective. *Biosens. Bioelectron.* **20**, 2435–2453.

Newnham R.E. (1988) Structure-property relationships in sensors. *Cryst. Rev.,* **1**, 253–280.

Njagi J., Warner J., and Andreescu S. (2007) A bioanalytical chemistry experiment for undergraduate students: biosensors based on metal nanoparticles. *J. Chem. Educ.* **84**, 1180–1182.

Olk C.H. (2005) Combinatorial approach to material synthesis and screening of hydrogen storage alloys. *Meas. Sci. Technol.* **16**, 14–20.

Otto M. (1999) *Chemometrics: Statistics and Computer Application in Analytical Chemistry.* Wiley-VCH, Weinheim, Germany.

Papkovsky D.B., Ponomarev G.V., Trettnak W., and O'Leary P. (1995) Phosphorescent complexes of porphyrin ketones: optical properties and applications to oxygen sensing. *Anal. Chem.* **67**, 4112–4117.

Pedersen C.J. (1967) Cyclic polyethers and their complexes with metal salts. *J. Am. Chem. Soc.* **89**, 7017–7036.

Penco M., Sartore L., Bignotti F., Sciucca S.D., Ferrari V., Crescini P., and D'Antone S. (2004) Use of compatible blends to fabricate carbon black composite vapor detectors. *J. Appl. Polymer Sci.* **91**, 1816–1821.

Peper S., Ceresa A., Qin Y., and Bakker E. (2003) Plasticizer-free microspheres for ionophore-based sensing and extraction based on a methyl methacrylate-decyl methacrylate copolymer matrix. *Anal. Chim. Acta* **500**, 127–136.

Pickup J.C. and Alcock S. (1991) Clinicians' requirements for chemical sensors for *in vivo* monitoring: a multinational survey. *Biosens. Bioelectron.* **6**, 639–646.

Potyrailo R.A. (2001) Combinatorial screening. In: *Encyclopedia of Materials: Science and Technology*, Vol. 2, Buschow K.H.J., Cahn R.W., Flemings M.C., Ilschner B., Kramer E.J., and Mahajan S. (eds.). Elsevier, Amsterdam, pp. 1329–1343.

Potyrailo R.A. (2004c) Expanding combinatorial methods from automotive to sensor coatings. *Polymeric Mater Sci. Eng. Polymer Preprints* **90**, 797–798.

Potyrailo R.A. and Amis E.J. (eds.) (2003b) *High Throughput Analysis: A Tool for Combinatorial Materials Science.* Kluwer Academic/Plenum, New York.

Potyrailo R.A. and Brennan M.J. (2004b) Method and apparatus for characterizing the barrier properties of members of combinatorial libraries, U.S. Patent 6,684,683(B2).

Potyrailo R.A. and Hassib L. (2005) Analytical instrumentation infrastructure for combinatorial and high-throughput development of formulated discrete and gradient polymeric sensor materials arrays. *Rev. Sci. Instrum.* **76**, 062225.

Potyrailo R.A. and Leach A.M. (2006) Selective gas nanosensors with multisize CdSe nanocrystal/polymer composite films and dynamic pattern recognition. *Appl. Phys. Lett.*, **88**, 134110.

Potyrailo R.A. and Maier W.F. (eds.) (2006) *Combinatorial and High-Throughput Discovery and Optimization of Catalysts and Materials.* CRC Press, Boca Raton, FL.

Potyrailo R.A. and Mirsky V.M. (2008) Combinatorial and high-throughput development of sensing materials: the first ten years. *Chem. Rev.* **108**, 770–813.

Potyrailo R.A. and Mirsky V.M. (2009) Introduction to combinatorial methods for chemical and biological sensors. In: *Combinatorial Methods for Chemical and Biological Sensors,* Potyrailo R.A. and Mirsky V.M. (eds.). Springer-Verlag, New York, pp. 3–24.

Potyrailo R.A. and Morris W.G. (2007a) Multianalyte chemical identification and quantitation using a single radio frequency identification sensor. *Anal. Chem.* **79**, 45–51.

Potyrailo R.A. and Morris W.G. (2007b) Wireless resonant sensor array for high-throughput screening of materials. *Rev. Sci. Instrum.* **78**, 072214.

Potyrailo R.A. and Sivavec T.M. (2004) Boosting sensitivity of organic vapor detection with silicone block polyimide polymers. *Anal. Chem.* **76**, 7023–7027.

Potyrailo R.A. and Takeuchi I. (2005a) Role of high-throughput characterization tools in combinatorial materials science. *Meas. Sci. Technol.* **16**, 1–4.

Potyrailo R.A. and Takeuchi I. (eds.) (2005b) Special feature on combinatorial and high-throughput materials research. *Meas. Sci. Technol.* **16**, 1–316.

Potyrailo, R.A. and Wroczynski, R.J. (2005) Spectroscopic and imaging approaches for evaluation of properties of one-dimensional arrays of formulated polymeric materials fabricated in a combinatorial microextruder system. *Rev. Sci. Instrum.* **76**, 062222.

Potyrailo R.A., Sivavec T.M., and Bracco A.A. (1999) Field evaluation of acoustic wave chemical sensors for monitoring of organic solvents in groundwater. *Proc. SPIE*, **3856**, 140–147.

Potyrailo R.A., Chisholm B.J., Olson D.R., Brennan M.J., and Molaison C.A. (2002) Development of combinatorial chemistry methods for coatings: high-throughput screening of abrasion resistance of coatings libraries. *Anal. Chem.* **74**, 5105–5111.

Potyrailo R.A., Chisholm B.J., Morris W.G., Cawse J.N., Flanagan W.P., Hassib L., Molaison C.A., Ezbiansky K.,

Medford G., and Reitz H. (2003a) Development of combinatorial chemistry methods for coatings: high-throughput adhesion evaluation and scale-up of combinatorial leads. *J. Comb. Chem.* **5**, 472–478.

Potyrailo R.A., Morris W.G., and Wroczynski R.J. (2003b) Acoustic-wave sensors for high-throughput screening of materials. In: *High Throughput Analysis: A Tool for Combinatorial Materials Science,* Potyrailo R.A. and Amis E.J. (eds.). Kluwer Academic/Plenum, New York, pp. 219–246.

Potyrailo R.A., Olson D.R., Brennan M.J., Akhave J.R., Licon M.A., Mehrabi A.R., Saunders D.L., and Chisholm B.J. (2003c) Systems and methods for the deposition and curing of coating compositions. U.S. Patent 6,544,334 B1.

Potyrailo R.A., Wroczynski R.J., Pickett J.E., and Rubinsztajn M. (2003d) High-throughput fabrication, performance testing, and characterization of one-dimensional libraries of polymeric compositions. *Macromol. Rapid Commun.* **24**, 123–130.

Potyrailo R.A., Karim A., Wang Q., and Chikyow T. (eds.) (2004a) *Combinatorial and Artificial Intelligence Methods in Materials Science II.* Materials Research Society, Warrendale, PA.

Potyrailo R.A., May R.J., and Sivavec T.M. (2004b) Recognition and quantification of perchloroethylene, trichloroethylene, vinyl chloride, and three isomers of dichloroethylene using acoustic-wave sensor array. *Sensor Lett.* **2**, 31–36.

Potyrailo R.A., Morris W.G., and Wroczynski R.J. (2004c) Multifunctional sensor system for high-throughput primary, secondary, and tertiary screening of combinatorially developed materials. *Rev. Sci. Instrum.* **75**, 2177–2186.

Potyrailo R.A., Morris W.G., Wroczynski R.J., and McCloskey P.J. (2004d) Resonant multisensor system for high-throughput determinations of solvent-polymer interactions. *J. Comb. Chem.* **6**, 869–873.

Potyrailo R.A., McCloskey P.J., Ramesh N., and Surman C.M. (2005a) Sensor devices containing co-polymer substrates for analysis of chemical and biological species in water and air. U.S. Patent Application 2005133697.

Potyrailo R.A., Szumlas A.W., Danielson T.L., Johnson M., and Hieftje G.M. (2005b) A dual-parameter optical sensor fabricated by gradient axial doping of an optical fibre. *Meas. Sci. Technol.* **16**, 235–241.

Potyrailo R.A., McCloskey P.J., Wroczynski R.J., and Morris W.G. (2006a) High-throughput determination of quantitative structure-property relationships using resonant multisensor system: solvent-resistance of bisphenol A polycarbonate copolymers. *Anal. Chem.,* **78**, 3090-3096.

Potyrailo R.A., Morris W.G., Leach A.M., Sivavec T.M., Wisnudel M.B., and Boyette S. (2006b) Analog signal acquisition from computer optical drives for quantitative chemical sensing. *Anal. Chem.* **78**, 5893–5899.

Potyrailo R.A., Morris W.G., Leach A.M., Sivavec T.M., Wisnudel M.B., Krishnan K., Surman C., Hassib L., Wroczynski R., Boyette S., Xiao C., Agree A., and Cecconie T. (2007a) A multiplexed, ubiquitous chem/bio detection platform on DVD. *Am. Lab.* 32–35.

Potyrailo R.A., Morris W.G., Leach A.M., Hassib L., Krishnan K., Surman C., Wroczynski R., Boyette S., Xiao C., Shrikhande P., Agree A., and Cecconie T. (2007b) Theory and practice of ubiquitous quantitative chemical analysis using conventional computer optical disk drives. *Appl. Opt.* **46**, 7007–7017.

Potyrailo R.A., Barash E., Dovidenko K., and Lorraine P.W. (2008a) Methods and systems for detecting biological and chemical materials on a submicron structured substrate. U.S. Patent Application 20080280374.

Potyrailo R.A., Ding Z., Butts M.D., Genovese S.E., and Deng T. (2008b) Selective chemical sensing using structurally colored core-shell colloidal crystal films. *IEEE Sensors J.* **8**, 815–822.

Potyrailo R.A., Dovidenko K., Le Tarte L.A., Surman C., and Pris A. (2009a) Theoretical and experimental development of label-free biosensors based on localized plasmon resonances on nanohole and nanopillar arrays. *Proc. SPIE* **7322**, 73220M1-11.

Potyrailo R.A., Morris W.G., Sivavec T., Tomlinson H.W., Klensmeden S., and Lindh K. (2009b) RFID sensors based on ubiquitous passive 13.56-MHz RFID tags and complex impedance detection. *Wireless Commun. Mobile Comput.* **2009**, DOI: 10.1002/wcm.711.

Potyrailo R.A., Surman C., and Morris W.G. (2009c) Combinatorial screening of polymeric sensing materials using RFID sensors: combined effects of plasticizers and temperature. *J. Comb. Chem.* **11**, 598–603.

Preininger C., Mohr G., Klimant I., and Wolfbeis O.S. (1996) Ammonia fluorosensors based on reversible lactonization of polymer-entrapped rhodamine dyes, and the effects of plasticizers. *Anal. Chim. Acta* **334**, 113–123.

Ramirez A.G. and Saha R. (2004) Combinatorial studies for determining properties of thin-film gold–cobalt alloys. *Appl. Phys. Lett.* **85**, 5215–5217.

Raorane D., Lim S.-H., and Majumdar A. (2008) Nanomechanical assay to investigate the selectivity of binding interactions between volatile benzene derivatives. *Nano Lett.* **8**, 2229–2235.

Reddington E., Sapienza A., Gurau B., Viswanathan R., Sarangapani S., Smotkin E.S., and Mallouk T.E. (1998) Combinatorial electrochemistry: a highly parallel, optical screening method for discovery of better electrocatalysts. *Science* **280**, 1735–1737.

Sanders D. and Simon U. (2007) High-throughput gas sensing screening of surface-doped $In_2O_3$. *J. Comb. Chem.* **9**, 53–61.

Schatz G.C. (2007) Using theory and computation to model nanoscale properties. *Proc. Natl. Acad. Sci. USA* **104**, 6885–6892.

Scheidtmann J., Frantzen A., Frenzer G., and Maier W.F. (2005) A combinatorial technique for the search of solid state gas sensor materials. *Meas. Sci. Technol.* **16**, 119–127.

Schena M. (2003) *Microarray Analysis*. Wiley, Hoboken, NJ.

Schneider. T., Betsarkis K., Trouillet V., and Goschnick J. (2004) Platinum-doped nanogranular tin dioxide layers prepared by spin-coating from colloidal dispersions as basis for gradient gas sensor micro arrays. *Proc. IEEE Sensors* **1**, 196–197.

Schultz P.G. (2003) Commentary on combinatorial chemistry. *Appl. Catal. A* **254**, 3–4.

Seker F., Meeker K., Kuech T.F., and Ellis A.B. (2000) Surface chemistry of prototypical bulk II-VI and III-V semiconductors and implications for chemical sensing. *Chem. Rev.,* **100**, 2505–2536.

Semancik S. (2002) Correlation of Chemisorption and Electronic Effects for Metal Oxide Interfaces: Transducing Principles for Temperature Programmed Gas Microsensors. Final Technical Report Project Number: EMSP 65421, Grant Number: 07-98ER62709, U.S. Department of Energy Information Bridge.

Semancik S. (2003) Temperature-dependent materials research with micromachined array platforms. In: *Combinatorial Materials Synthesis,* Xiang X.-D. and Takeuchi I. (eds.). Marcel Dekker, New York, pp. 263–295.

Setasuwon P., Menbangpung L., and Sahasithiwat S. (2008) Eight combinatorial stacks of three layers of carbon black/PVA-carbon black/EVA composite as a vapor detector array. *J. Comb. Chem.* **10**, 959–965.

Shtoyko T., Zudans I., Seliskar C.J., Heineman W.R., and Richardson J.N. (2004) An attenuated total reflectance sensor for copper: an experiment for analytical or physical chemistry. *J. Chem. Educ.* **81**, 1617–1619.

Sieg S., Stutz B., Schmidt T., Hamprecht F., and Maier W.F. (2006) A QCAR approach to materials modeling. *J. Molec. Mod.* **12**, 611–619.

Sieg S.C., Suh C., Schmidt T., Stukowski M., Rajan K., and Maier W.F. (2007) Principal component analysis of catalytic functions in the composition space of heterogeneous catalysts. *QSAR & Comb. Sci.* **26**, 528–535.

Siemons M., Koplin T.J., and Simon U. (2007) Advances in high throughput screening of gas sensing materials. *Appl. Surf. Sci.* **254**, 669–676.

Siemons M. and Simon U. (2006) Preparation and gas sensing properties of nanocrystalline La-doped $CoTiO_3$. *Sens. Actuators B* **120**, 110–118.

Siemons M. and Simon U. (2007) Gas sensing properties of volume-doped $CoTiO_3$ synthesized via polyol method. *Sens. Actuators B* **126**, 595–603.

Simon U., Sanders D., Jockel J., and Brinz T. (2005) Setup for high-throughput impedance screening of gas-sensing materials. *J. Comb. Chem.* **7**, 682–687.

Simon U., Sanders D., Jockel J., Heppel C., and Brinz T. (2002) Design strategies for multielectrode arrays applicable for high-throughput impedance spectroscopy on novel gas sensor materials. *J. Comb. Chem.,* **4**, 511–515.

Sivavec T.M. and Potyrailo R.A. (2002) *Polymer Coatings for Chemical Sensors,* U.S. Patent 6,357,278 B1.

Stafslien S.J., Bahr J.A., Feser J.M., Weisz J.C., Chisholm B.J., Ready T.E., and Boudjouk P. (2006) Combinatorial materials research applied to the development of new surface coatings I: a multiwell plate screening method for the high-throughput assessment of bacterial biofilm retention on surfaces. *J. Comb. Chem.* **8**, 156–162.

Steinle E.D., Amemiya S., Bühlmann P., and Meyerhoff M.E. (2000) Origin of non-Nernstian anion response slopes of metalloporphyrin-based liquid/polymer membrane electrodes. *Anal. Chem.* **72**, 5766–5773.

Stewart M.E., Anderton C.R., Thompson L.B., Maria J., Gray S.K., Rogers J.A., and Nuzzo R.G. (2008) Nanostructured plasmonic sensors. *Chem. Rev.* **108**, 494–521.

Stewart M.E., Mack N.H., Malyarchuk V., Soares J.A.N.T., Lee T.-W., Gray S.K., Nuzzo R.G., and Rogers J.A. (2006) Quantitative multispectral biosensing and 1D imaging using quasi-3D plasmonic crystals. *Proc. Natl. Acad. Sci. USA* **103**, 17143–17148.

Suman M., Freddi M., Massera C., Ugozzoli F., and Dalcanale E. (2003) Rational design of cavitand receptors for mass sensors. *J. Am. Chem. Soc.* **125**, 12068–12069.

Sun L.-X. and Okada T. (2001) Studies on interactions between Nafion and organic vapours by quartz crystal microbalance. *J. Memb. Sci.* **183**, 213–221.

Suzuki H. (2000) Advances in the microfabrication of electrochemical sensors and systems. *Electroanalysis*, **12**, 703–715.

Svetlicic V., Schmidt A.J., and Miller L.L. (1998) Conductometric sensors based on the hypersensitive response of plasticized polyaniline films to organic vapors. *Chem. Mater.* **10**, 3305–3307.

Sysoev V.V., Kiselev I., Frietsch M., and Goschnick J. (2004) Temperature gradient effect on gas discrimination power of a metal-oxide thin-film sensor microarray. *Sensors* **4**, 37–46.

Tailoka F., Fray D.J., and Kumar R. V. (2003) Application of Nafion electrolytes for the detection of humidity in a corrosive atmosphere. *Solid State Ionics* **161**, 267–277.

Takeuchi I., Famodu O.O., Read J.C., Aronova M.A., Chang K.-S., Craciunescu C., Lofland S.E., Wuttig M., Wellstood F.C., Knauss L., and Orozco A. (2003) Identification of novel compositions of ferromagnetic shape-memory alloys using composition spreads. *Nature Mater.* **2**, 180–184.

Takeuchi I., Newsam J.M., Wille L.T., Koinuma H., and Amis E.J. (eds.) (2002) *Combinatorial and Artificial Intelligence Methods in Materials Science*. Materials Research Society, Warrendale, PA.

Taylor C.J. and Semancik,S. (2002) Use of microhotplate arrays as microdeposition substrates for materials exploration. *Chem. Mater.* **14**, 1671–1677.

Turcu F., Hartwich G., Schäfer D., and Schuhmann W. (2005) Ink-jet microdispensing for the formation of gradients of immobilised enzyme activity. *Macromol. Rapid Commun.* **26**, 325–330.

Ulmer C.W. II, Smith D.A., Sumpter B.G., and Noid D.I. (1998) Computational neural networks and the rational design of polymeric materials: the next generation polycarbonates. *Comput. Theor. Polym. Sci.* **8**, 311–321.

Vassiltsova O.V., Zhao Z., Petrukhina M.A., and Carpenter M.A. (2007) Surface-functionalized CdSe quantum dots for the detection of hydrocarbons. *Sens. Actuators B* **123**, 522–529.

Villoslada F.N. and Takeuchi T. (2005) Multivariate analysis and experimental design in the screening of combinatorial libraries of molecular imprinted polymers. *Bull. Chem. Soc. Jpn.* **78**, 1354–1361.

Walt D. R., Dickinson T., White J., Kauer J., Johnson S., Engelhardt H., Sutter J., and Jurs P. (1998) Optical sensor arrays for odor recognition. *Biosens. Bioelectron.*, **13**, 697–699.

Wang J. (2002) Electrochemical detection for microscale analytical systems: a review. *Talanta* **56**, 223–231.

Wang J., Musameh M., and Lin Y. (2003) Solubilization of carbon nanotubes by Nafion toward the preparation of amperometric biosensors. *J. Am. Chem. Soc.* **125**, 2408–2409.

Wicks D.A. and Bach H. (2002) The coming revolution for coatings science: high throughput screening for formulations. In: *Proceedings of the 29th International Waterborne, High-Solids, and Powder Coatings Symposium*, **29**, p. 1–24.

Wintersgill M.C. and Fontanella J.J. (1998) Complex impedance measurements on Nafion. *Electrochim. Acta* **43**, 1533–1538.

Wise B.M., Gallagher N.B., and Grate J.W. (2003) Analysis of combined mass- and volume-transducing sensor arrays. *J. Chemometrics* **17**, 463–469.

Wong D.W. and Robertson G.H. (1999) Combinatorial chemistry and its applications in agriculture and food. *Adv. Exp. Med. Biol.* **464**, 91–105.

Wu X., Zhang J., Chen J., Zhao C., and Gong Q. (2009) Refractive index sensor based on surface-plasmon interference. *Opt. Lett.* **34**, 392–394.

Xiang X.-D., Sun X., Briceño G., Lou Y., Wang K.-A., Chang H., Wallace-Freedman W.G., Chen S.-W., and Schultz P.G. (1995) A combinatorial approach to materials discovery. *Science* **268**, 1738–1740.

Xiang X.-D. and Takeuchi I. (eds.) (2003) *Combinatorial Materials Synthesis*. Marcel Dekker, New York.

Yang J.-C., Ji J., Hogle J.M., and Larson D.N. (2008) Metallic nanohole arrays on fluoropolymer substrates as small label-free real-time bioprobes. *Nano Lett.* **8**, 2718–2724.

Yoon J., Jung Y.-S., and Kim J.-M. (2009) A combinatorial approach for colorimetric differentiation of organic solvents based on conjugated polymer-embedded electrospun fibers. *Adv. Funct. Mater.* **19**, 209–214.

Zemel J.N. (1990) Microfabricated nonoptical chemical sensors. *Rev. Sci. Instrum.* **61**, 1579–1606.

Zhao J., Pinchuk A.O., McMahon J.M., Li S., Ausman L.K., Atkinson A.L., and Schatz G.C. (2008) Methods for describing the electromagnetic properties of silver and gold nanoparticles. *Acc. Chem. Res.* **41**, 1710–1720.

Zhao J.-C. (2001) A combinatorial approach for structural materials. *Adv. Eng. Mater.* **3**, 143–147.

Zheng Y.B., Juluri B.K., Mao X., Walker T.R., and Huang T.J. (2008) Systematic investigation of localized surface plasmon resonance of long-range ordered Au nanodisk arrays. *J. Appl. Phys.,* **103**, 014308.

Zou L., Savvate'ev V., Booher J., Kim C.-H., and Shinar J. (2001) Combinatorial fabrication and studies of intense efficient ultraviolet–violet organic light-emitting device arrays. *Appl. Phys. Lett.* **79**, 2282–2284.

# INDEX

2-D composition, 194
absorption-based optical sensors, 39
absorption-based sensors, 39
absorption fiber optic chemical sensors, 39
acetonitrile (ACN), 198–202
ACN. *See* acetonitrile
acoustic wave propagation, 27, 29
acoustic wave sensor, 24, 27, 29, 30
active and passive filters, 355
active surface area, 131, 134, 139, 140, 142
ADFETs. *See* adsorption-based FETs
adsorption, 9, 17, 20, 24, 32, 70, 71, 75–77, 79–85, 88, 90, 94, 96, 98, 105, 112, 119, 120, 121, 124, 129, 132, 135, 136, 186, 259, 260, 262, 263, 315, 320, 322, 326, 327, 332, 344, 348, 350, 356, 361, 363, 364, 365, 367
adsorption ability, 64, 129
adsorption-based FETs (ADFETs), 12
adsorption/desorption parameters, 79, 80–82, 315
adsorption processes, 64, 76, 81
adsorption properties, 70, 124, 315
aerosol phase, 245, 246, 248, 249, 252
AFM. *See* atomic force microscopy
Ag, 90, 171, 172, 174, 218, 260, 265, 332, 345, 346, 348–350
agglomerate, 131, 140–142, 144, 219, 227, 229, 257, 318, 319, 324, 326
agglomerated, 140–142, 257, 258, 278, 285, 318, 319

agglomeration, 23, 121, 131, 140, 142, 144, 257, 269, 271, 282, 283, 285, 288, 318, 319, 321, 324, 347, 365
$Al_2O_3$, 66–68, 71–73, 76, 79, 85, 89, 90, 93, 109, 111, 112, 143, 188, 196, 218, 229, 233, 238, 239, 242, 243, 254, 260, 271, 282, 283, 325, 338, 345, 346, 356, 357, 364
alcohol condensation, 273, 274
alloy, 83, 131, 161, 217, 220, 231, 241, 242, 351, 352
amperometric sensor, 2–4
analyte, 2, 3, 8, 9, 17, 21–26, 33, 34, 36–39, 42, 44, 64, 74, 84, 111, 127, 177, 180, 182, 184, 186, 192, 194, 197–199, 201
aqueous solutions, 259, 260, 262, 272, 290, 351
atomic force microscopy (AFM), 32, 174, 327, 360
atomization techniques, 248
Au nanoparticles, 127, 171, 174

band gap, 9, 69–71, 74, 106–109, 169, 342
baseline sensor drift, 24
biological sensors, 32, 51, 169
bridging oxygen, 103, 317
bulk stoichiometry, 321
calcination, 135, 188, 277, 278, 284, 333, 336
calorimetric sensors, 48
cantilever, 9, 32, 33, 83, 177, 179, 180
capacitance sensors, 8, 9
carbon-containing species, 320

215

# INDEX

carbon nanotubes, 21, 330
catalyst, 19, 20, 23, 45, 46, 47, 82, 83, 105, 124, 125, 127, 128, 131, 161, 186, 188, 202, 253, 260, 267, 274, 276, 291, 314, 320, 332, 337, 347, 351, 352, 353, 357, 358, 368
catalytic activity, 64, 70, 82, 84, 87, 88, 101, 105, 121, 127, 186, 215, 317, 331, 334, 339, 340, 349, 350, 356, 358
catalytic activity of sensing materials, 84
catalytic bead, 19
catalytic sensors, 19
cationic site, 75
CBD. *See* chemical bath deposition
CdS, 78, 271
CdSe, 165–168
CFS. *See* combustion flame synthesis
$CH_4$, 40, 96, 121, 339, 350, 355, 356, 362, 368
charging agent, 270
chemFET, 1, 12, 15
chemFET-based sensors, 15
chemical activity, 92, 93, 235, 346
 of metal oxides, 93
chemical bath deposition (CBD), 259–262, 264, 265, 269, 271, 272
chemical polymerization, 289, 292, 293
chemical precursor, 237, 241, 244, 284, 299
chemical sensitization, 349
chemical sensor, 1, 17, 19, 25, 31, 32, 33, 34, 36, 37, 38, 39, 45, 52, 63, 64, 65, 69, 70, 78, 80, 84, 88, 90, 92, 94, 95, 96, 97, 103, 106, 107, 109, 110, 112, 113, 120, 144, 215, 216, 224, 229, 231, 245, 246, 267, 272, 275, 276, 279, 288, 289, 292, 294, 298, 301, 327, 328, 329, 330, 332, 336
chemical stability, 92, 229, 243
chemical vapor condensation (CVC), 280, 282–284
chemical vapor deposition (CVD), 189, 235, 236, 237, 238, 239, 241, 242, 243, 244, 245, 247, 253, 282, 289, 292, 294, 295, 298, 299, 300, 314, 322, 326, 346
chemical warfare agents (CWAs), 200

chemiluminescence, 33, 34, 35
chemisorption, 21, 22, 69, 76, 77, 79, 80, 88, 96–98, 101–103, 121, 124, 238, 315, 317, 318, 344, 361, 364
cladding-based chemical fiber-optic sensors, 37
Clark oxygen sensor, 3
CO, 3, 9, 40, 70, 79–84, 87, 96, 98, 103, 111, 127, 128, 133, 135, 136, 139, 142, 188, 190, 191, 238, 323, 327, 339, 341, 349–353, 355, 356, 362, 368
coating design and tooling, 295
colorimetric sensors, 39
combinatorial experimentation, 165
combinatorial library, 163, 195
combinatorial materials screening, 160, 184
combinatorial screening, 160, 172, 177, 183, 186, 188, 196, 198, 200, 202
combinatorial technologies, 160, 165, 202
combustion flame synthesis (CFS), 280, 281, 283–285
complexant, 260, 261
composite, 101, 105, 186, 187, 215, 216, 232, 235, 288, 289, 291, 292, 297
conducting polymer, 110, 111, 169, 184, 245, 290, 292
conductivity, 1, 4, 5, 20, 21, 23, 24, 27, 28, 46, 47, 48, 64, 70, 74, 79, 80, 82, 92, 95, 96, 97, 98, 99, 100, 101, 102, 103, 104, 105, 106, 107, 108, 109, 110, 111, 114, 115, 128, 131, 169, 188, 226, 250, 252, 265, 320, 334, 339, 341, 343, 344, 356, 362
conductivity type, 101, 102
conductometric sensor, 4, 20–23, 107, 109, 196, 202
conjugated organic monomers, 169
corona spray pyrolysis, 250
$CoTiO_3$, 191
counter electrode, 3, 4, 250, 265
$Cr_{2-x}Ti_xO_3$, 103
cracking effect, 276
crystal faceting, 318
crystallite, 88, 95, 98, 113, 115, 116, 118, 119, 120, 121, 128, 132, 135, 136, 139, 141, 142,

186, 219, 226, 227, 229, 254, 256, 257, 269, 314, 315, 318, 319, 321, 324, 325, 326, 332, 333, 336, 338, 341, 354, 358, 364, 366
crystal shape, 120
crystal structure, 65, 67, 103, 104, 120, 226, 234, 241, 283, 313, 338, 341
CVC. *See* chemical vapor condensation
CVD. *See* chemical vapor deposition
CWAs. *See* chemical warfare agents

DC and RF sputtering, 222
Debye length ($L_D$), 114
dendrogram, 167
deposition from aerosol phase, 245
deposition parameters, 226, 232, 239, 240, 254, 257, 315, 323
deposition process, 16, 124, 217, 218, 223, 231, 235, 239, 242, 245, 246, 249, 253, 265, 269, 270, 293, 295
deposition rate, 220–225, 232, 234, 237, 239, 241, 242, 252, 253, 264, 267, 270, 284, 294, 298, 300
deposition technology, 121, 215, 293, 297
desorption, 77, 79–82, 98, 119–121, 186, 238, 315, 322, 332, 334, 351, 352, 364, 365
detection mechanism, 38, 116, 361
device application, 358
dielectric constant, 8, 9, 43, 106, 112, 114, 186
diffusion, 18, 20, 26, 67, 96–98, 100, 103, 120, 124, 131, 132, 138–142, 192, 196, 225, 227, 233, 236, 237, 239, 245, 252, 261, 313, 335, 353, 356, 359, 365
diffusion-layer thickness, 196
dimension effect, 114, 115
discrete arrays, 165, 166
dissociation, 18, 44, 75, 89, 105, 124, 138, 221, 236, 260, 345, 357, 360, 361
dissociative chemisorption, 124, 317
doping, 13, 23, 82, 105, 116, 126, 127, 130, 138, 169, 188, 191, 291, 314, 331, 333–342, 361, 364
doping process, 342
droplet size, 246–248

ED. *See* electroless deposition
EDOC. *See* electrochemical deposition under oxidizing conditions
electrical energy transduction, 183, 184
electrical energy transduction sensors, 183
electrochemical cell, 2–5, 98
electrochemical deposition, 259, 264, 267, 270
electrochemical deposition under oxidizing conditions (EDOC), 259, 267, 269
electrochemical polymerization, 288–291, 297
electrochemical sensors, 1–3, 38, 64, 94, 105, 110
electroconductivity, 69, 70, 74, 95, 107, 109, 110, 141, 342, 344, 351
electroless deposition (ED), 259, 260, 265, 269, 272
electrolyte, 2–5, 16, 95, 98, 169, 200, 266–268, 290, 297, 298, 300
electron-beam evaporation, 216, 234
electron configuration, 68, 69, 74, 87
electron exchange, 186, 332, 334, 340
electronic "nose," 368
electronic properties, 71–73, 77, 78, 96, 186
   of metal oxide surfaces, 77
electronic sensitization, 349
electronic structure, 68, 70, 74, 106, 129, 351
   of metal oxides, 68, 70, 106
electrophoretic deposition (EPD), 270–272, 300
electrophysical properties, 68, 97, 113, 215
   of sensing materials, 97
electrostatic spray deposition (ESD), 250–252
electrostatic spray pyrolysis. *See* electrostatic spray deposition
environmental conditions, 358, 361
EPD. *See* electrophoretic deposition
ESD. *See* electrostatic spray deposition
ethanol (EtOH), 198, 199
EtOH. *See* ethanol
evaporation rate, 216, 281

$Fe_2O_3$, 66, 68–70, 72, 73, 78, 85, 87–91, 93, 105, 108, 111, 131, 238, 254, 260, 277, 331, 345, 346, 350

217

# INDEX

Fermi level, 9, 77, 78, 109, 130, 315
Ferrite plating, 269, 272
FET. *See* field-effect transistor
$F$ (flat) faces, 317
fiber, 34–39, 41, 109, 110, 112, 169–171, 173, 272, 273, 292, 294–297
fiber Bragg gratings, 39
fiber coating, 294, 296
fiber optic chemical sensor, 36–39
fiber optic sensor, 36, 38
field-effect transistor (FET), 11–15, 51, 94
field-effect transistor sensors, 12
filament thermistor, 48
film agglomeration, 324
film deposition, 128, 216, 218–220, 224, 225, 232, 233, 239, 245, 247, 248, 250, 253, 260, 264, 269, 271, 290, 291, 294, 321, 324, 325, 340
  by thermal evaporation, 218
film microstructure, 233
film texture, 23, 127–129, 256
film thickness influence, 128, 337
finely dispersed fraction, 118, 358
flame ionization detector, 49
flame ionization sensor, 49, 50
flammable gas, 19, 140
flexural plate wave (FPW), 25, 31
fluorescence-based optical chemical sensor, 38
fluorescence fiber optic chemical sensors, 38
fluorescent chemical sensors, 34
FPW. *See* flexural plate wave

$Ga_2O_3$, 19, 66, 71–73, 85, 87, 89, 93, 98, 100, 102, 104, 109, 229, 243, 328, 334, 345, 356, 364
GaAs, 19, 78, 94, 95, 108, 111, 112
GaN, 19, 95, 108
gap plasmons, 177
gas condensation, 280
gas processing condensation (GPC), 280, 283
gas-sensing effect, 63, 79, 80, 115, 121, 123, 131, 140, 142, 318, 344, 347, 357, 358, 365
gas-sensing properties, 114, 115, 121, 129, 137, 188, 314–316, 319, 322, 334, 337, 339, 344, 354, 367
  of metal oxides, 314, 315, 316, 348
  of multicomponent metal oxides, 337
gas sensor, 2, 9, 11, 13, 14, 18, 21–24, 27, 31, 45, 63, 64, 69, 70, 74, 77–79, 81, 82, 84, 87, 90, 92, 94–101, 103, 106, 107, 109–113, 116, 119–121, 123, 124, 127, 130, 132, 134, 135, 138, 139, 165, 167, 172, 225, 227, 243, 250, 270, 272, 276, 285, 313, 314, 315, 318, 322, 323, 325, 326, 329, 334, 337–339, 341, 343, 348, 351–353, 356, 358–361, 364, 365, 367–370
GCS. *See* grating coupler sensor
gelation, 276
GPC. *See* gas processing condensation
gradient arrays, 166, 192
gradient sensor material, 192
grain, 21–23, 88, 97, 98, 100, 107, 113–121, 127–135, 141–144, 188, 219, 225–227, 234, 238, 255, 256, 264, 271, 277, 278, 286, 287, 314–316, 318, 319, 321, 323–328, 333–338, 341, 343, 345, 347, 354–356, 358–367
grain boundary, 22, 98, 100, 115, 120, 121, 128, 188, 334–336, 338
grain faceting, 121
grain-growth inhibitors, 367
grain size, 23, 113–121, 129, 131–135, 142–144, 225–227, 234, 238, 255, 256, 271, 277, 278, 286, 287, 315, 316, 318, 319, 321, 324, 326–328, 333, 334, 336–338, 345, 358–365, 367
grain-size growth, 256, 277
"grains" model, 113, 115
grating coupler sensor, (GCS), 35
growth directions, 328

$H_2$, 5, 70, 76, 81, 83, 87, 88, 90, 111, 117, 119, 124, 133, 134, 137–139, 142, 188, 190, 191, 238, 327, 338–340, 348, 350, 352, 353, 355–357, 359, 360–363, 368
$H_2S$, 3, 92, 105, 238, 341, 345, 350, 357, 358, 366, 368

# INDEX

heat of formation of the most stable oxide, 88
humidity, 9, 24, 37, 39, 41, 48, 49, 64, 94, 96, 97, 99, 101–103, 106, 109, 111, 112, 119, 121, 131, 139, 180, 200, 313, 343, 355, 361, 363, 364
hybrid materials, 279, 280
hydrolysis, 97, 117, 135, 237, 242, 260, 261, 264, 273, 274, 276, 277, 279, 280, 292
hydroxyl groups, 77, 119, 269, 364

IBAD. *See* ion-beam–assisted deposition
IDTs. *See* interdigital transducers
$In_2O_3$, 66, 72, 73, 74, 85, 89, 93, 96, 98, 101, 102, 109, 115–117, 119, 128, 129, 134, 136, 137, 139, 188, 190, 225, 226, 229, 238, 239, 243, 254, 255, 260, 265, 271, 276, 277, 325–328, 331, 333, 334, 339–345, 351–353, 357, 359–365, 367
indium tin oxide (ITO), 373
initial reduction temperature, 88, 90
InP, 19, 94, 95, 108, 111, 112
interagglomerate contact, 141, 142, 144, 318, 319
interdigital transducers (IDTs), 27, 29–31
interdigitated electrode, 8, 23, 186, 187
intergrain contact, 115, 315, 343
ion-beam–assisted deposition (IBAD), 233–235
ionic drift, 95
ionosorption, 74
ion-selective electrode (ISE), 5–7
ion-sensitive field-effect transistor (ISFET), 12, 15, 16
IR spectroscopy, 39–41, 76, 111
ISE. *See* ion-selective electrode
ISFET. *See* ion-sensitive field-effect transistor
ITO. *See* indium tin oxide

Kelvin probe, 10, 11
kinetics of sensor response, 324, 364
K (kinked) face, 334
K-nearest-neighbor (KNN) cluster analysis, 167, 168
KNN. See *K*-nearest-neighbor cluster analysis
Knudsen cell, 217

Lamb wave sensors, 30
Langmuir-Blodgett (LB) films, 51, 52
Langmuir-Blodgett (LB) sensors, 1, 50
Langmuir-Blodgett (LB) technique, 292
laser ablation, 21, 229, 230, 232, 299, 300
lattice oxygen, 88, 103, 105, 106, 338, 344
LB. *See* Langmuir-Blodgett
Lewis acidity, 70
Lewis basicity, 71, 76
LFD. *See* liquid flow deposition
liquid flow deposition (LFD), 269, 272
liquid-phase deposition (LPD), 188, 259, 260, 264, 272, 297
long-term stability, 25, 64, 91, 94, 95, 97, 99, 142, 160, 180, 188, 275, 358
Love-mode acoustic wave sensors, 30
lower explosive limit (% LEL), 20
LPD. *See* liquid-phase deposition

magnetron sputtering, 196, 222–224, 245, 252, 296, 298, 324, 326
MAPLE. *See* matrix-assisted pulsed laser evaporation
mass-sensitive chemical sensors, 31
material selection, 182
materials engineering, 23, 313, 368
matrix-assisted pulsed laser evaporation (MAPLE), 293, 294
mechanical energy transduction, 177
mechanical energy transduction sensors, 177
mechanical milling, 124, 142–144, 280, 285–287
melting temperature, 67, 88, 91, 94, 216, 217, 219, 237, 286, 326, 336
membrane, 2, 3, 5–7, 9, 15, 16, 30, 31, 36, 47, 88, 101, 110, 196, 197, 355–357, 368
MEMS. *See* microelectromechanical systems
mesoporous materials, 216, 330
metal-ion coordination numbers, 65
metal-organic chemical vapor deposition (MOCVD), 242, 243, 245, 253, 299
metal-organic precursor, 237, 243, 275, 282, 283

metal oxide, 9, 14, 21, 23, 24, 46, 63–72, 74–79, 81–90, 93, 95–98, 100–112, 114–116, 118, 119, 121, 124–127, 129–131, 134, 135, 138–140, 142, 144, 186, 188, 190, 196, 202, 216, 221, 225–227, 229, 232–235, 237, 238, 242, 243, 245, 246, 250, 253, 255, 258–260, 262–265, 267, 269, 277, 278, 313–316, 318–325, 328–339, 343–348, 350, 351, 354, 356–361, 363–370
  composition, 331
  growth, 255
  surface, 65, 75, 77, 79, 84, 127
metal-oxide semiconductor field-effect transistor (MOSFET), 12–16
metathesis polymerization, 288, 291
microchemical vapor deposition, 243
microelectromechanical systems (MEMS), 32
microsensors, 9, 16
microwave plasma processing (MPP), 280, 283
MOCVD. *See* metal-organic chemical vapor deposition
MOSFET. *See* metal-oxide semiconductor field-effect transistor
MPP. *See* microwave plasma processing
multicomponent materials, 186, 224, 298, 331
multicomponent metal oxides, 333, 336, 337
multielectrode array, 188

Nafion, 94, 197–201
nanobelt, 123, 328–330
nanocomposite, 70, 144, 216, 279, 334, 336, 338, 339
nanocrystal shape, 315
nanopowder collection, 285
nanorod, 328
nanoslit, 177, 178
nanotube, 21, 144, 216, 328–330
nanowire, 21, 123, 144, 216, 328, 330
"necks" model, 113, 115
Nernst equation, 5, 7
NIR spectroscopy, 39–41
NO, 3, 45, 76, 103, 124, 188, 190, 191, 350, 357

$NO_2$, 3, 11, 21, 40, 43, 77, 106, 116, 128, 129, 135, 142, 188, 190, 266, 268, 269, 327, 357
noble metal, 2, 20, 76, 82, 83, 125–127, 131, 246, 261, 263–268, 323, 331, 332, 345–348, 351, 353–358, 365, 366, 368
noble-metal surface clustering, 125
non–transition-metal oxides, 68–70
nucleophilic attack, 274

OIHM. *See* organic–inorganic hybrid material
one-dimensional structures, 113, 121, 123, 216, 328–330
operating characteristics, 113, 120
operating temperature, 19, 25, 80, 81, 84, 91, 95, 97–100, 106, 107, 118, 132, 196, 315, 323, 329, 342, 345, 348, 353, 357, 358, 365, 368
optical sensor, 33–35, 38, 39, 44, 64
optode, 37, 38
organic fluorophore, 165
organic–inorganic hybrid material (OIHM), 279
oxidation, 3, 19, 20, 21, 46, 47, 68, 69, 74, 75, 81, 84, 87, 90, 94, 96, 102, 103, 105, 106, 109, 121, 124, 126–128, 131, 167, 169, 188, 225, 227–229, 237, 242, 262, 265, 267, 269, 279, 289–291, 334, 338, 339, 344, 345, 349, 352, 357, 358, 366, 368
oxidation potential, 169, 290
oxidizing gases, 94, 101, 102, 106, 116, 144
oxygen chemisorption, 21, 102, 317, 318
oxygen diffusion, 97, 98, 365
oxygen states, 344
oxygen vacancies, 75, 97, 100, 124, 129, 131, 317, 320

PACVD. *See* photo-assisted chemical vapor deposition
passive membrane, 356, 357
PCA. *See* principal-components analysis
PCD. *See* photochemical deposition
Pd, 13, 14, 18, 66, 82, 90, 125–127, 130, 189, 194, 195, 218, 238, 253, 254, 260, 265–268, 271, 332, 345–358, 365, 367, 368

# INDEX

PECVD. *See* plasma-enhanced chemical vapor deposition
pellistor, 19, 20, 48, 140
peripheral measuring device, 103, 368
phase modification, 332, 334, 343
  of metal oxides, 331, 332, 343
photoacoustic sensors, 45
photo-assisted chemical vapor deposition (PACVD), 244, 245, 299
photochemical deposition (PCD), 271
photochemical polymerization, 288, 289, 291
photolytic mechanism, 244
photothermal mechanism, 244
pH sensor, 2, 6, 272
physical sputtering, 220
physical vapor deposition (PVD), 232–235, 239, 294, 296, 298–300
piezoelectric material, 24, 28
plasma-enhanced chemical vapor deposition (PECVD), 241, 242, 299
plasma polymerization, 288, 291
plasmonic nanostructure, 171
plasticizer, 169, 171, 172, 198, 200–202, 276
PLD. *See* pulsed laser deposition
point defect, 65–68, 97, 98, 107, 315, 326
polycrystalline film, 129, 141, 243, 318, 319, 363
polymer, 8, 9, 21, 24, 30, 34, 36, 37, 47, 91, 92, 94, 97, 108, 110–112, 161, 166–173, 176, 180–183, 185–187, 192, 193, 200, 202, 224, 245, 259, 261, 262, 276, 288–294, 296, 297
polymer film, 9, 94, 166, 186, 187, 276, 288, 292, 293
polymerization, 94, 107, 161, 166, 169, 184, 245, 252, 273, 275, 280, 288–293, 297
polymer sensor, 94, 97, 110
polymer synthesis, 186, 288
polymer technology, 288
polypyrrole (PPY), 94, 108, 112, 245, 289, 291
porosity, 23, 121, 128, 131–135, 137–139, 142, 225, 227, 229, 272, 274, 276, 280, 318, 319, 324, 334, 338, 341, 345, 356, 360, 365
porosity and active surface area, 131, 134

postdeposition annealing, 326
post–transition-metal oxides, 68, 69, 74
potentiometric sensor, 3, 5, 6, 95
powder, 40, 116–118, 132, 133, 135, 136, 139, 140, 143, 215, 236, 250, 262, 264, 270–273, 278, 280–289, 292, 298, 338, 346, 351, 359, 364
powder technology, 280, 287
PPY. *See* polypyrrole
pressure of the saturated vapor, 216
pre–transition-metal oxides, 68
principal-components analysis (PCA), 182, 200
Pt, 13, 18–20, 23, 47, 82, 90, 91, 96, 109, 117, 125–127, 133, 138, 139, 189, 192–194, 196, 218, 238, 244, 253, 265–268, 271, 290, 332, 345–348, 350, 352, 355–358, 365, 368
pulsed laser deposition (PLD), 229–232, 293
pulsed magnetron sputtering, 223, 224
PVD. *See* physical vapor deposition
pyroelectric effect, 46
pyroelectric sensor, 47
pyrolysis, 21, 116, 119, 121, 129, 137, 140, 141, 237, 242, 243, 246–251, 253–259, 273, 280, 283, 288, 289, 292, 293, 298, 300, 315, 316, 321–325, 339–341, 347, 349, 353, 360–362
pyrolysis temperature, 119, 243, 256, 259, 321–324, 361, 362

QCM. *See* quartz crystal microbalance
QCM crystal, 25, 26
quartz crystal microbalance (QCM), 25, 26

radiant energy transduction, 165
radiant energy transduction sensor, 165
radio-frequency identification (RFID) sensor, 196, 198–202
rate of sensor response, 365
reaction chemistry, 237
reactivity of elements toward oxygen, 90
reagent, 37, 38, 41, 82, 92–94, 165, 192–194, 202, 259, 263, 264, 274, 276, 289, 290, 346
recovery time, 1, 80, 81, 99, 117, 128, 129, 139, 140, 142, 165, 180, 329, 365, 368

221

# INDEX

"redox" mechanism, 101, 361
redox reaction, 10, 74, 368
reduction, 3, 28, 69, 74, 88–90, 95, 98, 102, 103, 120, 121, 125, 127, 129, 135, 186, 224, 237, 263, 267, 269, 270, 276, 281, 317, 332, 333, 339, 344, 358
reference electrode, 2, 6, 7, 15, 16, 37, 95, 290
refractive index, 34–37, 39, 43, 44, 71, 111, 134, 172, 176, 296
refractometric fiber optic chemical sensors, 39
refractometric optical sensors, 39
reothaxial growth and thermal oxidation (RGTO), 126, 131, 227, 229
response time, 44, 63, 98–100, 116, 128, 135, 137, 139, 140, 144, 186, 276, 324, 329, 345, 357, 361–363, 365
Rayleigh sensor. *See* surface acoustic wave (SAW) or Rayleigh sensor
*RFID*. *See* radio-frequency identification sensor
RF sputtering, 222, 231
RGTO. *See* reothaxial growth and thermal oxidation

Sauerbrey equation, 25, 27
SAW. *See* surface acoustic wave sensor
scanning light pulse technique (SLPT), 194, 195
Schottky barrier, 17, 18, 78
Schottky diode, 1, 16–18, 19
Schottky diode–based sensor, 16
screen printing, 21, 166, 292, 300
Seebeck effect, 45
selective ion-layer adsorption and reaction (SILAR), 259, 260, 262
self-lithography, 244
semiconducting metal oxides, 96, 100, 186, 196
sensing film, 11, 141, 167, 177, 188, 192–194, 197–202, 244
sensing layer, 21, 22, 24, 31, 132, 135, 225, 276, 318, 325
sensing material, 38, 46, 63, 65, 77, 79, 84, 87, 88, 92, 94, 97, 101, 102, 107–112, 115, 130–134, 138, 140, 144, 159–166, 169–172, 177, 180, 181, 183, 185, 186, 188, 191–193, 196–200, 202, 203, 231, 244, 254, 274, 276, 295–297, 318, 330, 361, 367
sensing properties, 21, 70, 114, 115, 121, 129, 137, 188, 276, 314–316, 319, 322, 334, 339, 344, 354, 367
sensitivity, 1, 12–14, 16, 18, 21–25, 27, 28, 30, 32, 34, 37–40, 44, 46, 63, 64, 78, 79, 81, 84, 86–88, 91, 92, 94–96, 98, 99–102, 104–106, 109–112, 115–117, 119, 124, 125, 127, 129, 131–134, 137–139, 142, 144, 166, 169, 172, 177, 183, 185, 186, 188–190, 194, 200, 202, 313–315, 319, 320, 324, 327, 329, 330, 338–342, 344–352, 355–359, 361, 362, 364–366, 368–370
to air humidity, 102, 119, 131
sensor array, 99, 177, 180–184, 188, 190, 192, 196, 199, 244
sensor parameters, 95, 96, 119, 129, 140, 279, 280, 328, 366, 369
sensor response, 17, 18, 24, 28, 80–84, 86, 88, 114–117, 121, 123, 126, 128, 129, 131, 134–137, 140–144, 192, 193, 196, 198, 200, 201, 314, 315, 318, 322–324, 327, 337, 340, 342, 345, 347, 348, 350, 353, 356, 364, 365, 367
control of, 331, 344
selectivity of, 198, 200, 367
sensor stability, 95, 195
serial scanning analysis, 165
SERS. *See* surface-enhanced Raman spectroscopic sensor
SH-APM. *See* shear-horizontal acoustic plate mode (SH-APM) sensor
shear-horizontal acoustic plate mode (SH-APM) sensor, 29
shear-horizontal surface acoustic wave (SH-SAW) sensor, 29, 30
SH-SAW. *See* shear-horizontal surface acoustic wave (SH-SAW) sensor
Si, 19, 46, 72, 78, 90, 94, 95, 108, 111, 112, 196, 242, 243, 274, 279, 325, 332, 346, 347, 367
SiC, 14, 15, 19, 89, 95, 108, 112

# INDEX

SILAR. *See* selective ion-layer adsorption and reaction
SILD. *See* successive ionic-layer deposition
sintering, 21, 66, 115, 128, 134–136, 274, 275, 278, 282, 285, 298, 300, 334, 338
SLPT. *See* scanning light pulse technique
$SnO_2$, 11, 23, 66, 68, 69, 72–74, 77–82, 85–87, 89, 91–93, 95, 96, 98, 101–105, 109, 111, 112, 115–119, 121–124, 126, 127, 129–144, 188, 189, 196, 225–227, 229, 233, 234, 238, 239, 242–244, 254, 256–262, 270, 271, 277, 278, 282, 315–329, 331, 333–345, 347–355, 357, 358, 360–366
    crystallographic planes, 121
$SnO_2$-CuO, 92, 341, 342
sol-gel method, 116, 140, 245, 246, 275–278, 333, 339
sol-gel polymerization, 275
solid electrolytes, 95, 298, 300
solid-state gas sensor, 63, 64, 78, 81, 87, 90, 95–98, 106, 113, 119, 121, 124, 130, 285, 313, 314, 322, 323, 334, 358, 359, 364, 365, 367–370
space-charge region, 115
specific-ion electrodes, 6
SPR. *See* surface plasmon resonance
spray nozzle, 246, 248
spray pyrolysis, 298, 300
spray pyrolysis deposition, 255, 341
sputtering, 21, 196, 217, 219–226, 231, 233, 234, 245, 252, 291, 293, 296, 298, 300, 314, 324, 326, 347, 355
    rate, 220, 223, 224
    techniques/technology, 219, 221, 224, 324
$SrF_eO_3$, 100
stability, 21, 23–25, 44, 64, 69, 76, 78, 88–92, 94–97, 99–102, 109, 111, 113, 118, 119, 121, 123, 129, 131, 134, 140, 142, 160, 162, 167, 169, 180, 186, 188, 192–195, 200, 229, 243, 270, 271, 275, 293, 313, 314, 323, 326, 328, 329, 336, 338, 345, 346, 348, 353, 356, 358, 359, 361, 364, 366, 367, 369

structural engineering, 119, 314, 319, 321, 322, 327, 334, 337, 359, 369
structural properties, 21, 100, 134, 333
successive ionic-layer deposition (SILD), 142, 144, 259, 260, 262–264, 272, 319, 349
surface acoustic wave (SAW) or Rayleigh sensor, 25, 27–30, 83, 84, 110, 196, 198, 294
surface additive, 355, 357
surface catalysis, 22, 96, 340, 348
surface charge, 15, 46, 79, 80, 129, 271
surface clustering, 125
surface clusters, 125, 349, 350, 353
surface diffusion, 124, 131, 353
surface-enhanced Raman spectroscopic (SERS) sensor, 171, 172, 174
surface geometry, 23, 123–125
surface modification, 23, 103, 125, 231, 245, 264, 323, 326, 345–352, 354–357, 367
    of metal oxides, 264, 344, 346, 350, 356
surface morphology, 124, 172, 234, 241, 354
surface plasmon, 1, 35, 43, 44, 111, 171, 178
surface plasmon resonance (SPR), 1, 43–45, 111, 171, 177
surface plasmon resonance sensor, 43
surface plasmon resonance technique, 43
surface poisoning, 82, 103
surface processes, 70, 369, 370
    of sensing materials, 77
surface properties, 70, 77, 79, 94, 95, 118, 130, 139, 195, 296, 337, 338, 343, 367
surface reaction, 23, 83, 88, 98, 103, 121, 124, 270, 317, 339, 348, 364, 366
surface segregation, 334–336
surface solubility, 335
surface state, 63, 74, 77–80, 121, 130, 186, 315, 332, 342, 345, 346
surface stoichiometry, 130, 131, 342, 345, 346
    disordering, 129
swelling, 8, 24, 184

technology readiness level (TRL), 162, 163
temporal drift, 95, 98

223

# INDEX

thermal conductivity, 1, 47, 48
thermal conductivity sensor, 47, 48
thermal evaporation, 216–219, 221, 224, 227, 233, 280, 282, 294
thermalization, 224, 225
thermally activated CVD, 241, 242, 245
thermal program reduction (TPR), 88–90
thermal stability, 78, 89, 91, 94, 102, 131, 134, 293, 326, 336, 345, 348, 359, 366
thermodynamic stability, 64, 88–90
thermoelectric effect, 45
thermoelectric sensor, 45, 46
thick film, 118, 119, 138, 139, 188, 290, 359, 365
thick-film sensor, 21, 134, 140
thickness shear mode (TSM) sensor, 25, 27, 180–183, 196, 198
thin film, 21, 23, 28, 47, 50, 109, 116, 118, 119, 123, 141, 166, 196, 197, 219, 221, 229, 231–233, 242, 245, 246, 251, 259, 262, 264, 271–273, 275, 282, 290, 293, 300, 314, 324, 347, 359, 365
thin-film sensor, 21, 129, 144, 368
threshold voltage, 12–16
$TiO_2$, 66, 68, 69, 72–76, 78, 85, 87, 89, 90, 93, 95, 98, 102, 104, 108, 111, 112, 125–128, 131, 226, 238, 242, 243, 254, 260, 265, 271, 277, 281, 297, 328, 331, 333, 334, 336
TPR. *See* thermal program reduction
transducer, 25, 27, 29, 45, 162, 177, 184, 341, 350, 358
transition-metal oxide, 66, 68–70, 74, 87, 106, 127, 265, 331, 334, 345, 357
transmittance, 40–42
TRL. *See* technology readiness level

TSM. *See* thickness shear-mode sensor

ultrasonic agitation, 271
ultrasonic atomizer, 246
UV absorption, 42
UV irradiation, 94, 173, 291, 293, 346
UV spectroscopy, 42

vacuum deposition, 216, 293
vacuum evaporation, 216–218, 245, 293, 346
vacuum polymerization, 289, 293
van der Waals interaction (force), 24, 177, 180
vapor pressure, 198, 200, 216, 217, 221, 236, 237, 281, 291, 298

water adsorption, 119, 124, 363, 364
water condensation, 139, 273, 274
WE. *See* working electrode
Wheatstone bridge, 5, 19
wireless technologies, 196
$WO_3$, 66, 69, 70, 72, 84, 85, 89, 93, 98, 100–102, 108, 111, 112, 115, 124, 260, 277, 331, 361
work function, 1, 9, 10, 11, 13, 15, 17, 18, 77, 78, 106, 194, 346
work-function sensor, 9, 84, 88, 109
working electrode (WE), 2–4

ZnO, 30, 66, 68–70, 72–74, 78, 85, 87, 89, 90, 93, 102, 109, 111, 112, 127, 128, 130, 131, 225, 226, 229, 242, 243, 254, 260, 265, 269, 271, 328, 331, 345, 364
ZnS, 78
$ZrO_2$, 66, 72, 73, 76, 78, 85, 89, 90, 93, 98, 108, 111, 112, 126, 127, 218, 238, 242, 243, 260, 277, 281, 283, 356

# 丛书书目
## Series Catalog

## 化学传感器：传感材料基础
## Chemical Sensors: Fundamentals of Sensing Materials

### 第1册 化学传感器基本原理及其材料
### Basic Principles and Materials of Chemical Sensors

*SENSORS TECHNOLOGY SERIES* VOLUME 1:
1. 化学传感器的基本工作原理
2. 传感材料应具备的特性
3. 传感材料的组合概念

### 第2册 传感材料的合成及改性
### Synthesis and Modification of the Sensing Material

*SENSORS TECHNOLOGY SERIES* VOLUME 1:
4. 传感器材料的合成和沉积
5. 传感材料的改性：金属氧化物材料工程

### 第3册 准一维材料的合成、特性及应用
### Quasi-one-dimensional Materials: Synthesis, Properties and Applications

*SENSORS TECHNOLOGY SERIES* VOLUME 2:
1. 纳米材料与纳米技术简介
2. 准一维金属氧化物结构：合成、表征及应用
3. 化学传感器中的碳纳米管和富勒烯
4. 基于单层包覆金属纳米颗粒的传感器

### 第4册 多孔纳米材料的特性及应用
#### The Characteristics and Applications of Porous Nanomaterials

*SENSORS TECHNOLOGY SERIES* VOLUME 2:

5 多孔半导体材料及用于气体传感器的优缺点
6 有序介孔膜的合成、性质及其在气体传感器中的应用
7 以沸石为基本材料的化学传感器
8 纳米复合材料的制备及其在化学传感器中的应用

### 第5册 聚合物与其他材料的特性及应用
#### The Characteristics and Applications of Polymers and Other Materials

*SENSORS TECHNOLOGY SERIES* VOLUME 3:

1 聚合物化学传感器
2 分子印迹（模板）——一种有前景的聚合物基化学传感器的设计方法
3 杯芳烃基材料化学传感器
4 化学传感器中的生物与仿生系统
5 新型半导体材料对化学传感器的促进
6 离子导体及其在化学传感器中的应用
7 传感器材料：选择指南

# 化学传感器：传感器技术
# Chemical Sensors: Comprehensive Sensor Technologies

### 第6册 固态器件 I
#### Solid State Devices I

*SENSORS TECHNOLOGY SERIES* VOLUME 4:

1 化学传感器技术简介
2 传感与取样方法

3 电导型金属氧化物气体传感器：工作原理与制备方法

4 基于逸出功的气体传感器：肖特基和场效应晶体管器件

5 电容型化学传感器

## 第7册 固态器件 II
**Solid State Devices II**

*SENSORS TECHNOLOGY SERIES* VOLUME 4:

6 运用热释电和热电效应的气体传感器

7 量热式传感器

8 基于微悬臂梁的化学传感器

9 石英晶体微天平

10 声表面波传感器的化学应用

11 集成化学传感器

## 第8册 电化学传感器
**Electrochemical Sensors**

*SENSORS TECHNOLOGY SERIES* VOLUME 5:

1 电化学气体传感器：基本原理、制备及参数

2 稳定的氧化锆基气体传感器

3 用于液态环境的电化学气体传感器

4 离子敏感场效应晶体管(ISFET)型化学传感器

5 微流控芯片作为电化学传感器的新平台

## 第9册 光学传感器
**Optical Sensors**

*SENSORS TECHNOLOGY SERIES* VOLUME 5:

6 光学与光纤化学传感器

7 化学发光传感器：基本工作原理和水污染物控制的应用

第10册 化学传感器的应用

**The Application of Chemical Sensors**

*SENSORS TECHNOLOGY SERIES* VOLUME 6:

1 化学气体混合物分析与电子鼻：现状与未来发展趋势
2 电子舌：方法与成果
3 无线化学传感器
4 远程化学传感：应用于大气监测
5 化学传感器在我们生活中的应用
6 化学传感器在工业、农业、交通运输中的应用
7 化学传感器的选择与操作指南